For Patricia

Chapters

Contents

Preface

I am fortunate to teach a number of different subjects at my college. Courses in management, ethics, information technology or health care almost always have a receptive audience. I found that there is one subject that doesn't fit the usual student optimism - students dread a course in statistics. In terms of negative experiences, it falls somewhere between having a root canal and trying to understand the theory of relativity. Few if any would take a statistics course if it were not required.

The antipathy toward statistics is troublesome since the subject is an important one. Most statistical texts begin by pointing out all the places we encounter statistics. The authors make cogent arguments on how valuable statistics are in making decisions or analyzing a problem. In my own experience, statistics has been a fascinating discipline. As a student, practicing statistician and teacher of statistics I have loved the discipline. One only hopes that such enthusiasm can be transferred to the student.

This text is not written for the professional statistician. The pages are not dense with formulae, Greek letters and complex mathematical symbols. Such texts have their place, but not when the goal is to generate statistical appreciation and understanding rather than produce a practicing statistician's manual. Few students who take a required introductory course in statistics intend to become active researchers let alone practicing statisticians. For most, knowledge of how statistical methods can be useful and how statistical results should be interpreted is a sufficient objective.

Consequently, the goals for this text are the following:

1. Present statistical principles in a simple yet informative fashion so that student fear and distaste of statistics is overcome.
2. Develop an awareness of the abuses and misuses of statistics. Hopefully students will know when they're being "tricked" by impressive numbers that lack substance. Too often people are overwhelmed by numbers that may distort the truth.

3. Combine research and statistical practices because the two disciplines are so closely linked that a student can obtain a more comprehensive understanding of both fields by integrating them.
4. Finally, as noted above, there is no intent to train statisticians, but students need enough awareness to know when they have a statistical application in their working worlds. If a person is not familiar with a subject area there is no way to know when the techniques of that field can be useful.

This text uses the following instructional techniques:

1. A liberal number of examples and illustrations amplify a statistical principle or idea. Graphs and descriptive material reinforce ideas and permit students to see how statistical practices apply to their individual situations.
2. A set of problems at the end of each chapter provides a review of what was discussed in the text. More importantly, the problem sets are comprehensive - i.e. they include problems from earlier chapters as well as the current chapter. Repetition is important in learning and a quick review of past material enhances a student's mastery of the material. Consequently, it is recommended that the student do all the problems at the end of each chapter.
3. Equations are introduced only if they communicate a useful concept or idea. Equations with many mathematical symbols are not just intimidating to the non-mathematician; they cause confusion rather than clarity. They are not included in the text. However, an equation is presented when it helps to explain and increases understanding of a concept or technique.
4. A reduction in formulae means a reduction in computational emphasis. Time spend manipulating numbers is better spent learning how to interpret statistical results. In the age of computers, number crunching can be left to machines. Yet the question always is - what computer software to use? There are a number of excellent statistical programs available. As a professional statistician, I have my own favorite - other statisticians have their preferences as well. Learning a powerful statistical program would make sense for people going into a full-time research career or advanced statistical training. I fear that for most students, the time and effort to learn the statistical program would not be a reasonable trade off.

    A more practical choice is a spreadsheet program. Recent versions of Microsoft Excel™ (versions 5.0 and higher) include routines for most of the statistical features that will be presented in this text. Furthermore there is a very good chance a student already possesses the Excel program and has some expertise in its use. There is also a much greater chance that students can also use spreadsheets for other applications in their scholarly or professional career.

    It is important to add that the Excel instruction is presented only after the statistical material is presented. Therefore, if instructors prefer to bypass the Excel instructions they may do so.

The text consists of 15 chapters, in addition to the Preface which includes an Introduction to Excel section and is at the end of the Preface. There are two types of chapters - those that deal primarily with statistical methods (chapters 1, 2, 3, 4, 5, 7, 8, 9. 10, 11, 13 and 14) and those that focus on research principles (chapters 2, 6, 8, 12 and 15). Some chapters are included in both lists since they contain topics where there is not a sharp line between statistics and research. An instructor may therefore choose to use the text only for a statistics course or a research course by assigning only one set of chapters. The early chapters centered on statistics (chapters 1-5) are aimed at establishing a foundation for the later chapters (chapters 7-10) which explore inferential statistical procedures. The later chapters (11-15) deal with non-parametric, regression and qualitative methods as well as research design and ethical issues. In the statistical set of chapters instructors may prefer to omit, certain chapters that they do not feel are necessary to properly understand and interpret statistical findings. In particular chapter 3 (on probabilities) and the first portion of chapter 4 (on permutations and combinations) may be omitted if an instructor feels their value is not worth the effort. However, before rejecting these chapters the instructor is urged to read them over since the presentation approach may convince the instructor that the material can be given in an interesting and informative manner.

The chapters on statistical methods and those for research methods are intentionally interspersed. The research chapters are introduced when it is felt they will be most useful to appreciating a statistical topic. However, week after week being immersed in mathematical thought and numerical manipulation can be exhausting and the research chapters offer a respite from a strictly quantitative environment.

As any text, an instructor will have to add material that he or she feels has been overlooked. In some cases a topic was not included, although it might be found in other texts, simply because from personal experience the knowledge never seemed to be particularly useful. For example, the hypergeometric distribution or measures of skewness/kurtosis are interesting, but their utilization level is low and so these elements were intentionally omitted. Other procedures were left out because the mathematics involved was difficult and there were no routines in Excel to get around doing the tedious mathematics. An example of an omission for this reason is discriminant analysis and the analysis of covariance. Perhaps in a later edition the breadth of the material will be widened to include these methods.

At the end of each chapter, answers to some of the problems are provided. A complete set of answer is available for instructors by sending a request on their institution's stationary for the Instructor's Answer Booklet to the address below. There is no charge for the booklet.

Ronald Gauch, Ph.D.
Dyson Center
Marist College
Poughkeepsie, NY 12601

An Introduction to Excel

It is assumed you know how to work in a Windows environment; can save, retrieve and name a file; are familiar with how a mouse works; know what the characters on the keyboard do and understand the features of a typical Windows screen. If you are new to Excel and intend to use the program for other spreadsheet applications, you should invest in one of the many manuals available in bookstores. The instruction will show how to use Excel, version 97, to perform basic statistical procedures discussed in the text. Excel provides alternative ways to do many of the statistical operations. The instructions will describe only the simplest and most direct methods.

An Excel Workbook

STRUCTURE OF AN EXCEL WORKBOOK. A workbook is made up of one or more worksheets. A worksheet consists of rows (the horizontal lines) and columns (the vertical lines). The rows are numbered (1, 2, 3 etc.) and the columns are lettered (A, B, C, etc.). The place where a row and a column intersect is called a cell. Cells are named for the column and row where they are located. Cell A1 means it is at column A and row 1. You move around (i.e., navigate) a worksheet using the mouse, scroll bars or keys (e.g., the arrow keys). A single worksheet can contain up to 65,536 rows and 256 columns, but you will probably prefer to use multiple worksheets for each assignment rather than placing all material onto one worksheet.

The essential features of an Excel screen include the elements listed below. Also, refer to the display on the next page.

Menu Bar - a list of commands that can be performed (e.g., File, Edit, etc.).

Toolbar(s) - a series of icons which perform functions such as opening or saving a file, copying an entry in the worksheet, etc. By moving the mouse across the icons on a toolbar, the name or function the icon performs is given. The number of Toolbars will vary depending on the operator's wishes. The type of Toolbar that appears is set by selecting "View" from the Menu Bar, followed by "Toolbars" from the drop down menu. Use the mouse to check or uncheck the toolbar you want on your screen.

The most useful Toolbars are:

Standard Toolbar- This toolbar contains the most basic functions for file handling and data manipulation such as copying or moving data. There are also useful statistical tools that can add up a column of numbers or create charts on this toolbar.

Format Toolbar- as the name indicates there are many aids to design the appearance of your worksheet on this toolbar. For example, you can select different font types, font sizes or different data alignments with this toolbar.

Title Bar - the name of the workbook that you are working on is shown in this area.

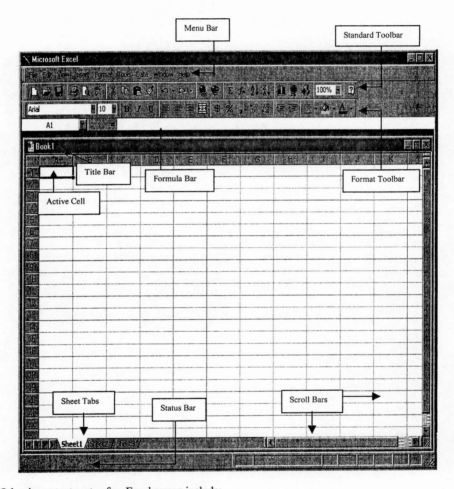

Other important parts of an Excel screen include:
Formula Bar – displays the data or equation you are entering or have entered in a cell. The formula area is located above the Column headings with a box on the left showing the cell that is active (e.g. A1 in the display).

Status Bar - located at the bottom of the screen, which shows the condition Excel is in. For instance in the display, the entry "Ready" is shown and means that Excel is ready for your next instruction.

Scroll Bars - on the far right side of the screen is the vertical scroll bar and just above the Status Bar is the horizontal scroll bar. You use these bars to navigate around the screen.

Your location on a worksheet is identified by the active cell which has a double border and a white background (Cell A1 is the active cell in the display). In addition, on the left-hand side of the horizontal scroll bar there are sheet tabs for you to select different sheets in your workbook. The sheet you are working on is highlighted. In the display, the active sheet is called Sheet1. Excel has extensive help menus, which can be accessed by clicking the Help button on the far right hand side of the Menu bar. The basic operations of Excel are described below.

OPENING, SAVING AND PRINTING A FILE. To open a file, move the mouse pointer to the Standard Toolbar and click on the new file icon, ▢. Excel creates a workbook and you now can begin to make entries in the first sheet of your workbook. Excel automatically gives your workbook a name (e.g. Book1), but you should assign the workbook a name of your choice. To save a file and give it a name that will help you identify it, click the "File" command from the Menu Bar. A drop down menu appears (see below).

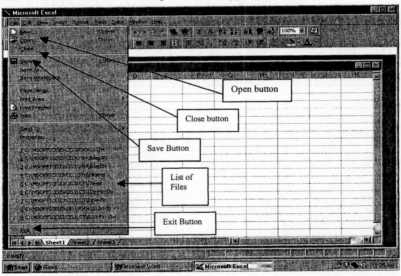

Select the "Save" option and a new dialog box called "Save As" will appear. Click on the arrow in the "Save in" box to locate the drive you want to use to save your file. A selection of drives will appear and assuming you are not working on your own computer, you will need to save your work on a floppy disk which should be placed in the A drive. Therefore, select the "3 ½ floppy (A)" drive option. In the "File name" box, type the name you wish to use (e.g. "Statistics"). When you are finished, the "Save As" dialog box should look like this:

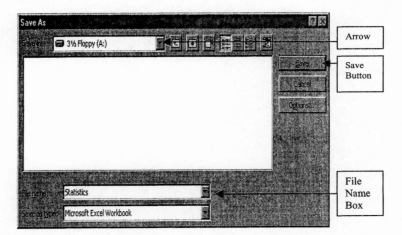

Click on the Save button and you should now be back at the worksheet and ready to enter data into the workbook titled Statistics. As you work on your file, you are well advised to save your work periodically. This can be done easily by selecting the Save icon on the Standard toolbar - it's the third icon from the left and looks like a floppy disk. (▦)

Once you have completed your work be sure you save it, before exiting from EXCEL. To leave the Statistics Workbook, select the "File" command from the Menu Bar and chose the "Close" option from the drop down menu (see the display on the preceding page). To exit from the entire Excel program, select "File" once more and this time choose the "Exit" button at the bottom of the drop down menu.

To re-open a file, select the File command and near the bottom of the drop down menu there is a listing of files — the name of your file should be there - select it and your workbook will then appear on the screen. If you do not find your file listed, you will need to click the "Open" button on the File menu (see the display on the preceding page). In the Open dialog box go to the "Look in" section, find your file, click on it and then hit the "Open" button. Your file should then appear on your screen and you may resume your work.

To print a file, select the Print icon from the Standard toolbar - it's the one that looks like a printer (⧉). Simply click the icon and your workbook should be printed. You may resume working in Excel when the Status Bar indicates "Ready".

Be aware that when an hourglass figure appears on the screen, it indicates that the system is working on your document - be patient and do not hit other keys as you wait. When the cross or arrow re-appears on the worksheet, you may resume your work.

DATA ENTRY. To enter data, move to the cell, which will contain your first entry and click the mouse. This is now considered the active cell. Type the information you want and it will appear in the cell, as well as in the formula box near the top of the worksheet. When you have finished, hit the Enter key to confirm that your entry is what you want. If you have a very long entry, Excel may not be able to display all the characters in the cell, but don't worry, Excel will retain all the information. For example, if there is nothing in the adjacent cell, Excel allows information that exceeds the cell length to spill over into the adjacent cell's space.

Numbers (which are referred to as "values" by Excel) are treated differently than text. Excel classifies an entry as a value or text on the basis of the first character entered. If you enter a digit (e.g., 1, etc.) Excel will classify the entry as a value, otherwise, it will assume it's a text entry with a few exceptions. For instance, Excel interprets the equal sign (=) as the beginning of a calculation. (Remember to begin all calculations with the = sign). You can enter a number, as text rather than a value, by preceding the number with an apostrophe (').

You may wish to use upper and lower case when typing text into a worksheet. However, when you are interacting with Excel in terms of providing names for functions or cell locations, lower or upper case will invoke the same response.

You'll also notice that values are right justified in a cell and text is left justified. However, you can change that arrangement by using the alignment icons on the Formatting Toolbar. If your screen is not displaying the Formatting Toolbar select View from the Menu Bar. Select Toolbars from the drop down menu and then use the mouse to check the Formatting Toolbar. On the Formatting Toolbar there are icons for left justified (⧉), right justified (⧉) and centered (⧉). To change the alignment of a cell, highlight it and then point and click on the alignment icon you need.

EDITING. To edit a cell, go to the cell. The entry will appear in the formula box. Move your pointer to the entry and double click at the point you want to make a change. Delete, add or change characters as needed and when you are through, hit Enter. After hitting the Enter key, you will be free to move around the worksheet. To delete the entire contents of a cell simply highlight the cell and hit the Delete key.

Creating a Database

A small database will be used to illustrate how Excel can assist you in performing statistical operations. The database represents information you might find in a typical personnel file. The database (see page xxvii) contains information on 39 hypothetical people for whom 14 different characteristics are tabulated. An explanation of the data items included in the data base is also provided (see page xxviii). The database is intentionally small so, if you wish, you can check your work by doing an exercise manually.

CREATING THE WORKING FILE. The best way to begin building the Excel workbook is to enter the titles of the personal file onto a worksheet. Start with the "DATABASE" title and type that title in cell A1 - we'll worry about centering the title later. As noted previously, remember to always hit Enter after you finished typing to confirm that the correct information has been entered. Excel will reposition the cursor at the cell directly below the one you just used. However, if you prefer to end up in a different cell use the arrow keys rather than the Enter key (e.g. the right arrow key will reposition you in the cell to the right of the one you just used). Move to cell A2 so you can start entering the abbreviated titles for the 14 characteristics that were collected on the staff (see page xxvii). In cell A2 type ID and in cell A3 type No. Leave cell B2 blank and enter SEX in cell B3. Continue to enter the headings (e.g. AGE, SPC AWDS etc.) in rows 2 and 3. When you get to cell L2 type the general heading: PC PROFICIENCY into that cell. The use of a common heading for columns K, L and M is reasonable because the data in these columns all refer to the proficiency scores. In cell K3 type JUNE, in L3 type JULY and finish with cell M3 in which you enter AUG. Your last heading, EDUC, goes in cell N3.

It's generally easier to enter data from top to bottom so for the rest of the database, enter all the data for each column. In Cell A4 through the last cell we will need for Column A, cell A42, we want to enter the numbers 1 through 39. It is not necessary to enter each number one at a time since they follow an order that Excel can recognize. Type "1" in cell A4. Then select the range of cells A4 to A42. You select a range by highlighting the first cell and, while holding down the left mouse button, you move the pointer to the last cell of your range (A42). You can also highlight cells using the F8 key. Highlight the first cell in a range, hit the F8 key and use the PgDn and arrow keys to reach the last cell of the desired range. In either case the selected range of cells will be identified by a border around the set of cells and all but the first cell will appear black.

From the Menu Bar choose "Edit", from the Edit drop down menu choose "Fill" and from the next menu choose "Series". See the diagram at the top of the next page which shows you what your screen will look like at this point. Right below that diagram there is another screen, which displays the Series dialog box which will be produced after you have selected Series. In the Series dialog screen, check to make sure that the "Series in" box has a dot in the "Columns" option and that the "Type" box has a dot in the "Linear" option. You enter or delete a dot by clicking on the circle. Also be sure the step value box has a "1" in it. Change any entry that needs to be changed by clicking on the appropriate circle to satisfy these requirements. Then click OK. The numbers 1 through 39 will appear in cells A4 to A42 respectfully. The cells will still be highlighted in case you want to perform other operations with the range. We do not wish to do more so we finish our operation by hitting the left mouse button. At this point we will not place any entries in row 43. Instead we will continue entering the raw data.

Statistical Methods for Researchers

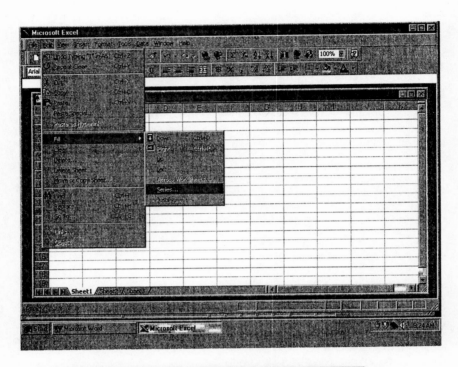

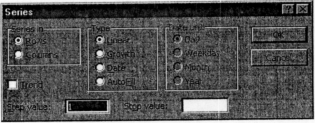

The entries that are to go into the B column refer to sex where F represents female and M represents male. These entries will have to be entered one at a time, beginning with an F in cell B4 and ending with an F in cell B42. Column C should contain data on the AGE variable so just type in the numbers shown in the database for age. The rest of the entries are straight forward, but two comments are still necessary. First, be sure to leave cells J11 and J24 blank since there is no data for these two entries. You may also find that the salary information (Column E with the heading "PAY K$") does not always contain a decimal value. To have a consistent format select all the numeric cells in this column (cell E4 through cell E42) and then select "Format" from the Menu Bar. Click on "Cells" in the drop down menu that follows and you will then have the Format Cells dialogue box (see below)

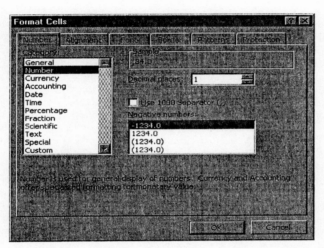

You should be at the "Numbers" tab, but if you are not, click on that tab. In the category window that appears select "Number". If necessary, adjust the Decimal place box so it reads 1, by clicking on the up or down arrow buttons. Finish by selecting OK. All your entries in Column E will now show values to one decimal place.

CHECKING YOUR FILE. As a way to check on whether you've entered all the numeric data correctly the last row of the database on page xxvii contains column totals. Do not copy the totals shown on the database into row 43. We will have Excel calculate those totals for us. If the calculated totals do not equal the totals shown in the database sheet, you have one or more data entry errors. Type the heading, "TOTAL", in cell A43. We will place the sum of values for column C (the age variable) in cell C43. Go to cell C43 and select the AutoSum button (⬚) from the Standard Toolbar. Hit Enter and the value 1501 will appear in Cell C43. If you did not get 1501 it means you have one or more data entry errors. Check the individual entries against the original database to identify and correct any errors.

As you may already know, there is more than one way to carry out a task with Excel. For example, if you activate Cell C43 you will see that the entry in that cell is the expression:

=SUM(C4:C42). In fact you could have simply typed that expression in C43 and the result 1501 would also have appeared in Cell C43. Excel refers to the entry in Cell C43 as a function. All functions begin with an equal sign (=), require a command (e.g., SUM), and are completed by entering, in parentheses, specific information required by the function. For the function Sum, it is necessary to identify the range of cells you want added. You list the first and last cell of the range, separated by a colon. In our example our complete entry is: =SUM(C4:C42). We will use other functions later.

To obtain totals for the other numeric columns we could repeat the steps we performed for AGE. However, since we want the same function performed for all columns it's easier just to copy the function used for AGE to the other columns. Activate cell C43 (the cell that contains the AGE total), click the Copy button on the Standard Toolbar (the icon is two sheets of paper overlapping slightly i.e.,). Place the pointer in cell C43 and drag the arrow across row 43 from Column C through Column N. Drag is a common operation in windows. You activate the first cell of the range you're interested in, hold down the left mouse key and move to the last cell of the range. Then release the mouse button and hit Enter. (Remember that you may also use the F8 key option to highlight cells). The correct sums appear in Row 43 for all Columns. Note that Columns H and N have no numeric values; they contain only alphabetical characters and their totals are therefore 0. The 0 results are unnecessary and best eliminated. This can be done by activating cell H43 and hitting the Delete key. Repeat the steps for cell N43.

FORMATTING THE WORKSHEET. To complete the data input tasks, there are a few cosmetic activities we can perform. We can center the title, DATABASE, by activating cells A1 through N1 and then clicking the Center Across Columns button on the Format Toolbar. The icon has an a in the center i.e. . We can also activate cells K2 to M2 and again select the Center Across Columns button to center the PC PROFICIENCY column heading.

The spacing of columns widths can be optimized by using an automatic fitting procedure built into Excel. Highlight the cells A1 through N43. Click on Format from the Menu Bar. Select Column from the submenu and then in the next menu click on the AutoFit Selection. The columns should now be at their optimal width.

Now that we have all the data entered we can proceed with using the data in subsequent exercises. Remember before leaving EXCEL to save your file and it's a good idea to make a copy of it since you will not want to re-enter all your data if the original file is lost or corrupted in some manner.

DATABASE

ID NO.	SEX	AGE	SPC AWDS	PAY K$	SERV YRS	TEST SCR	ADV Y/N	CRTY CODE	SNSV LEVEL	PC PROFICIENCY JUNE	JULY	AUG	EDUC
1	F	45	3	29.0	10	55.0	N	3	4	12	16	15	BS
2	M	45	4	38.6	3	69.0	Y	4	2	9	14	22	MS
3	M	52	6	45.2	18	78.4	N	1	2	13	18	22	BS
4	F	60	4	43.4	38	58.8	Y	5	3	9	15	18	BS
5	F	17	0	18.2	0	56.8	N	1	4	11	11	13	HS
6	M	36	5	33.8	13	75.0	N	2	2	10	12	16	HS
7	F	46	3	35.0	6	71.4	Y	4	5	10	13	14	MS
8	M	20	2	19.4	2	50.2	Y	3		7	13	18	HS
9	M	34	2	44.6	6	74.1	N	2	1	11	17	20	MS
10	F	31	6	51.2	9	80.4	Y	1	5	11	10	14	MS
11	F	23	1	22.4	4	67.4	N	1	4	18	18	23	HS
12	M	31	4	33.2	10	87.8	Y	3	1	13	11	15	BS
13	F	62	5	44.6	18	66.0	Y	3	4	11	14	19	HS
14	M	64	5	48.2	23	67.5	Y	3	3	7	12	16	MS
15	F	21	1	23.6	1	71.2	N	1	2	10	16	21	BS
16	M	29	0	20.0	0	63.4	N	3	4	14	20	25	HS
17	F	39	3	30.8	11	59.2	N	5	4	13	22	25	BS
18	F	42	4	53.0	8	98.2	Y	2	4	12	22	25	HS
19	F	59	5	44.0	41	82.0	Y	3	3	9	13	18	HS
20	F	22	1	25.4	0	91.7	N	5	3	10	13	16	BS
21	F	46	4	38.9	14	92.1	N	5		8	12	12	BS
22	F	28	3	29.0	3	79.5	Y	3	4	8	14	17	MS
23	M	27	5	42.8	3	93.8	Y	2	3	4	9	15	MS
24	F	32	3	31.4	2	73.6	N	2	2	6	11	13	MS
25	F	25	3	27.2	1	77.5	N	5	4	11	20	26	HS
26	M	41	4	36.8	6	80.2	Y	1	1	7	12	18	BS
27	M	28	3	24.2	2	64.7	Y	1	3	9	16	17	BS
28	F	25	3	27.2	2	81.5	Y	3	4	5	10	22	BS
29	M	33	4	29.5	5	75.8	Y	2	4	9	17	16	BS
30	M	49	7	59.4	20	96.9	Y	4	5	8	10	24	MS
31	F	53	5	48.8	14	89.5	N	5	2	11	11	13	MS
32	M	35	5	41.6	7	94.8	Y	2	4	9	9	15	MS
33	M	19	2	18.8	1	76.2	Y	4	2	7	12	14	HS
34	F	66	5	47.0	28	73.6	Y	2	3	8	15	17	HS
35	M	26	2	21.8	5	54.5	Y	2	3	7	20	22	HS
36	F	62	6	51.8	25	87.4	N	4	3	14	17	19	BS
37	M	38	5	50.1	12	86.6	N	3	5	9	10	25	MS
38	F	39	2	37.0	12	79.5	N	4	5	8	14	19	MS
39	F	51	1	43.1	21	91.1	N	1	2	4	15	18	HS
TOTAL		1501	136	1410.0	404	2972.3		110	119	372	554	717	

DATABASE - EXPLANATION of FIELDS

Your database consists of records, which corresponds to the information on different people in a hypothetical organization. For each record the following information is provided:

ID NO.   Identification number
SEX   M = male and F = female
AGE   In years
SPC AWDS   Special awards received while a member of the organization.
PAY K$ Annual pay expressed in thousands (K) of dollars ($). A salary of $39,800 is expressed as 39.8.
SERV YRS   Service years with the organization. A 0 indicates less than one year of service.
TEST SCR   Test score for a recent aptitude test given to all members of the organization.
ADV Y/N   Advancement recommendation made by a person's superior where Y = yes (person has potential for advancement) and N = no (person does not have potential for advancement).
CRTY CODE   The result of a creativity test given to all people at the time they are hired which uses the following coding scheme:
                1 = very low level
                2 = low level
                3 = moderate level
                4 = high level
                5 = very high level
SNSV LEVEL   The evaluation by a psychologist concerning the person's ability to be sensitive (SNSV) to the needs of others. The evaluations are coded as follows:
                1 = Insensitive and indifferent to the needs of others
                2 = Has concern for others, but does not respond to their needs
                3 = Has some empathy for others and can be responsive to their needs
                4 = Has high concern for others and is usually approachable
                5 = An extremely sensitive person.
PC PROFICIENCY Computer test scores for June, July, and August
EDUC   Highest educational degree where:
                HS = High School
                BS = Bachelor's degree
                MS = Master's degree

Chapter 1

Descriptive Statistics

Introduction

We can define statistics as a field of study that involves planning a study and obtaining data which is then organized, displayed, analyzed and interpreted according to a set of rules, principles and guidelines. Why study statistics? It's a good question to ask. For many, the answer is that statistics is a required course. If you surveyed a typical group of students perhaps no more than ten percent would voluntarily take a statistics course. That's unfortunate because statistics are everywhere - in sports, in economics, in education, in the stock market, in health care decision making and so on.

A **statistic** is a number that provides information about some characteristics in a data set. A **data set** is simply a collection of observations about a phenomenon in which we have an interest. A statistic, when viewed as a simple count, is easily understood. There are 12 females in the class. The average GPA for those who graduate is 2.85. Four out of five students who begin a master's degree program graduate from it. It's when we begin talking about probability theory or the t test that the concerns about statistics usually surface.

We also have a pretty good idea of the strengths and weaknesses of statistics. Most people know how to interpret statisitcs when they are derived from familiar processes. For example, surveys are a frequent source of statistical information and we know we must be skeptical about how we interpret the results. A poll taken when Ronald Reagan was President included the question: Do you think Ronald Reagan is trustworthy? The result was that 38% answered "yes". In order to interpret this response, there are some questions we would like answered. Who was polled, when was the survey conducted, were the "No opinion" responses included in the calculation of 38%? It would also be helpful to know if any other questions were asked about trustworthiness so that the 38% response might be put in perspective. The 38% seems low, but if we found out that in the same poll only

25% feel religious leaders are trustworthy we might rethink our interpretation. The point is that without knowing how a statistic is produced, it is difficult to know what it means.

We have become very dependent on research in our advanced society and statistical methods are a major component of the way research is conducted. The statistical approach, when included as a part of a research methodology, is based on appealing logic and not just esoteric mathematical manipulations. For example, if someone came up with a new product that appeared to stimulate hair growth and wanted to test the product in human beings we would begin by asking some questions. Who would we want to test - clearly the subjects chosen should have hair loss. What will we measure - hair growth is a reasonable choice but even this obvious selection will need a more specific definition. How will we know if any change is due to the new product - perhaps we want to add a control agent such as a placebo in the experiment? Note that none of these questions involve statistics in the sense of mathematical numbers. They do involve logic and much of statistics is nothing more than the presentation and interpretation of information in a rational way.

There's an appropriate story to illustrate this last point. In a certain area of Europe rabbit and horse pie is considered a delicacy as long as the rabbit is fresh. During a rabbit shortage only one restaurant continued to serve fresh rabbit and horse pie. One day a customer, unsophisticated in the ways of statistics, asked what ratio of rabbit and horse was being used. The owner answered 50/50 and the customer gladly ordered the rabbit pie. A more statistically oriented customer overheard the conversation and when the first customer left, he asked the really important question - What does 50/50 mean? Not unexpectedly the answer was one rabbit and one horse.

It's also valuable to appreciate the fact that statistics is a data reduction technique. Humans can not extract much meaning out of a large set of numbers that is unorganized. We frequently refer to unorganized data as **raw data**, which is a good description since the data have not been processed, arranged or ordered. They are still in their "natural" or raw state. In that condition our minds simply can not make sense out of the array we see before us. Summarization and making new arrangements of the data gives us a way to understand the information we have. Nevertheless, in the process information is lost. As an example, it is more efficient to summarize the age distribution for a class of 40 students by reporting the mean rather than citing each value. However, even though we learn a lot from the mean, we have no idea how many people are above the mean or below the mean, we don't know the age of the youngest or oldest person, and we are unaware of whether the age distribution is uniform or uneven. We may even find that there is not a single student in the class whose age matches the mean.

Terminology

All disciplines have a specialized vocabulary. In this text, common statistical terms will be introduced and an explanation provided. For instance we will soon be talking about variables. In statistical parlance a **variable** is an observable attribute that when measured/viewed can have more than one value/assessment. The characteristic can be almost anything we observe - a person's gender, the rating a manager gives to a subordinate, the cost of a gallon of gasoline, etc. As shown below, for each attribute there can be more than one value or state:

Attribute	Values or States
Gender	Male, Female
Performance Rating	Excellent, Good, Fair, Poor
Gasoline Cost	A range from $1.00 to $2.00

Since there is more than one value that an attribute can assume we could say it varies - hence the term variable.

In a research study variables can be classified in different ways. One of the most prevalent classification schemes is to refer to **independent variables** and **dependent variables**. In a typical study, a researcher introduces an agent and sees what effect it has on some object. For example, a new drug is given (the agent) and the effect on a person's blood pressure (the object) is observed. The term independent variable is used to refer to the agent being introduced. The term dependent variable is used to refer to the object being affected. Since the researcher is free to use different doses of a drug, he can be said to be acting in an independent fashion. On the other hand, the person's blood pressure response is changed depending on how much drug is given. Thus an independent variable is the one that the researcher controls and the dependent variable is the one in which a change in response occurs. The relationship can often be seen in a cause and effect relationship. The independent variable is the cause and the dependent variable is the effect.

If we wanted to study the effect of high and low concentrations of a fertilizer on crop growth, the amount of fertilizer is the independent variable and crop growth is the dependent variable. It should be obvious that there must be a time relationship between the two variables. The independent variable must precede the dependent variable.

Descriptive Statistics

The branch of statistics that organizes data, primarily for illustrative purposes, is called **descriptive statistics**. The types of functions involved include arranging, summarizing, tabulating and graphing data as opposed to analyzing the data. A popular way to arrange data is to create a distribution.

Distribution

A **distribution** is a listing of all the elements in a data set that are classified/assigned to specific categories. Perhaps the best way to explain a distribution is to look at some data.

The set of data shown below represents the sex of 39 employees in a hypothetical organization. The F stands for female and the M for male.

F	M	F	M	F	M	F	M
M	F	M	F	F	F	M	M
M	M	F	F	M	F	M	F
F	M	M	M	F	M	F	F
F	M	F	F	F	M	F	

In their current state we can't do much with the data. At the very least we'd like to know how many are female and how many are male. We can obtain this information by counting the number of females and males and presenting the result in the following tabular format.

Female	Male
21	18

The table we created, in statistical terminology, is called a **frequency table** because the data are organized in terms of how frequent a characteristic occurs. The first row of the table contains the headings for the categories (Female and Male). The second row contains the frequencies associated with each heading category.

The data set on gender involves a **discrete variable**. Male and female are distinct, there are no gradations, and hence sex is considered a discrete variable. A discrete variable such as gender can be used to categorize people. When we work with what are called continuous variables we have to do a little more organizing.

For example, assume the numbers given in the following table are weights in kilograms.

55	75	68	63	92	80	89	83
69	71	87	59	79	65	95	79
78	68	66	98	93	81	67	91
59	74	67	82	74	76	74	88
57	80	72	92	87	86	55	

Weight is a **continuous variable** since it can take on almost any value depending on the precision of the measurement instrument we used. If the scale is very sensitive we can obtain very precise weights - and we would need a number of decimal places to record that precision. In that sense the value that represents weight is continuous, the number we present can go on and on depending on the precision of the scale. We frequently limit the number we use to represent weight to the nearest pound or kilogram, but that's an arbitrary decision on our part - the actual weight could be expressed with greater precision. If all we can do is count an item then it's usually safe to conclude it is a discrete variable. On the other hand, if we can measure an item, then it is usually a continuous variable.

In our example of sex, we came up with a way to condense the data. What can we do to condense or summarize our weight data? We could phrase the question somewhat differently - How do we interpret the numbers associated with a continuous variable? What can we say about the set of data in the previous table? As we glance through the set we may feel that the data look like "typical weights". But we probably should say something that sounds a little more informative and to do that we need to provide the data in a more organized format. One thing we can do is to rearrange the data so they are in order - the smallest value to the largest one.

If we did this we would end up with the arrangement shown in the following table.

55	63	68	74	78	81	87	92
55	65	68	74	79	82	88	93
57	66	69	74	79	83	89	95
59	67	71	75	80	86	91	98
59	67	72	76	80	87	92	

Grouping and Classifying Data

Unfortunately both of the weight tables are rather large and even though the last table is an improvement over the earlier table, it still is not easy to interpret. Imagine what it would be like if we had to deal with two or three hundred values. What we frequently do in this situation is to begin to summarize our data. The summary usually groups the data in classes. For weights the classes could be all those who weigh less than 60 Kg in one group, those who weigh 60 to 69 Kg in another group and so on until all the weights have a category into which they can be placed. The groupings are then presented in a table format. The table below gives the results from the grouping described above.

Category	Frequency
Less than 60	5
60 - 69	8
70 – 79	10
80 – 89	10
More than 89	6
Total	39

This table allows us to see the data in a more efficient way. While using a lot less space and mental effort, we can get a sense of how heavy this group of people is. However, the more compact representation comes at a price - we lose detail. We know 10 people weigh in the 70s, but we don't know if they were all in the low 70s, high 70s or distributed evenly. This exercise illustrates one of the advantages of statistics and one of its disadvantages. Grouping the data makes it more manageable for us to interpret, but in the process we lose detail. For most people it's a reasonable trade-off, but there are occasions where the loss of detail is important and can lead to misinterpretation.

Essentially, the table above is also a frequency distribution. We have a count of the number of people in each of the categories we've set up. Although our data set is small we might want to add another piece of information - the percent of people in each weight category. If we had even more numbers, the value of including the percents would have even greater value. The following table shows the results of adding percentages.

Category	Frequency	Percent
Less than 60	5	12.8
60 – 69	8	20.5
70 – 79	10	25.6
80 – 89	10	25.6
More than 89	6	15.4
Total	39	99.9

Due to rounding the total percent is 99.9 rather than 100. In addition to percentages some people find it is helpful to have **cumulative frequencies** included in a table. Cumulative frequencies simply begin with the first category and then add the result from the next category to it. This sum is then added to the next category and the process continues until all categories have been included. A table with cumulative totals is shown below.

Category	Frequency	Cumulative Percent	Frequency	Cumulative Percent
Less than 60	5	12.8	5	12.8
60 – 69	8	20.5	13	33.3
70 – 79	10	25.6	23	59.0
80 – 89	10	25.6	33	84.6
More than 89	6	15.4	39	100.0
Total	39	99.9		

The advantage of the **cumulative frequency table** is that it can provide quick answers to question such as how many people weigh 79 kilograms or less. From this table we see the answer is 23 or 59 percent. If we wanted to know the percentage of people who weigh more than 79 kilograms we can find that answer by adding the percents for the "80-89" category and the "More than 89" percent category together. Our answer is 41 percent (25.6+15.4 = 41.0).

Summarizing data with tables is an improvement over the raw number display. However, we can even make a further enhancement that many find easier to interpret. We can use a graph to display our data. We will devote more attention to that topic in chapter 2.

Central Tendency Measures

In addition to organizing the data we can also get a quick idea of the kind of distribution we have by finding the center or central value of the set. By finding a single value that is in the middle of the data set we learn quite a bit about the entire set.

As it turns out there is more than one way to determine the middle of a data set. In statistical terms we can calculate or "measure" the center by different methods. Hence in statistical texts we usually find a chapter on "Measures of Central Tendency". There are three competing ways to determine the center. The terms associated with the different approaches are the mean, median and mode.

Mean

Perhaps the most common way to define the center is to calculate a mean. By definition the **mean** is the sum of the values in the data set divided by the number of values you have. The formula for the mean is:

$$\overline{X} = \frac{\Sigma X_i}{N} = \frac{X_1 + X_2 + X_3 + \ldots X_n}{N}$$

The X with a bar over it is pronounced "X bar" and is the symbol for a sample mean. The Greek symbol μ (pronounced "mu") is used for a mean called the population mean. The same formula is used for μ and $\overline{X}$ and in that sense they are equivalent, but remember that μ represents a population mean and $\overline{X}$ a sample mean. The difference between a sample and population is explained in a later chapter. Following the = sign, the expression ΣX_i appears in the numerator (i.e. the upper portion of the fraction). The X_i in that expression refers to the elements in the data set. If we were interested in a mean age of students, the elements would be student ages. If we wanted the mean for salaries of police officers, the elements would be salaries. The little i after the X is a way to number each element in the data set. The first element in the data set is X_1, the second X_2, etc. The last item is identified as X_n. The Σ sign in front of the X_i is a summation sign - it means sum or add up the X elements. The N in the denominator stands for the number of elements in the data set - e.g., number of students for whom an age was included or the number of police salaries included. The last part of the formula for the mean is just a restatement of the formula and uses the notation "$X_1 + X_2 + X_3 + \ldots X_n$" to represent the process of adding the elements together and then dividing the sum by the total number of elements or N.

If, in an age data set, there are 39 entries and the sum of the ages is 1501 then by dividing the sum (1501) by the number of entries (39) gives us the mean (38.5).

Median

The median is another way to define the center of a data set. In this case the **median** is the middle value, after the set is rearranged in sequential order. In the next table, we arranged a data set of ages in sequential order - the smallest to the largest value.

17	23	28	32	38	45	51	62
19	25	28	33	39	45	52	62
20	25	29	34	39	45	53	64
21	26	31	35	41	46	59	66
22	27	31	36	42	49	60	

There are 39 values in the set so we want the 20th value in the ordered sequence. The 20th value has 19 values below it and 19 values above it. From the table we find that the 20th value is 36. The median is pretty straight forward when we have an odd number of values in a set. What to do if there are an even number of values? Let's drop off the last value in our set of ages. Now we have 38 values and there is no single value sitting right in the middle. We have though two values that are as close as we can get - the 19th and the 20th value. In this example they correspond to the ages 35 and 36. The logical thing to do is find the midpoint between those two values, which we can obtain by calculating their mean. The sum of the 19th and 20th value is 71 (35+36) and we divide that sum by 2 since there are two values involved in the calculation. The result (71/2) is 35.5 - the median for the set with only 38 items.

Mode

The third alternative is to determine the mode. The **mode** is simply the most frequently occurring value. By using the arrangement in the above table, we can find that the value that occurs most often is 45 - it occurs 3 times. Hence, 45 is the mode. If we have a tie for the most frequent occurring value, we simply report that state by saying the distribution is bimodal - i.e., there are two modes. Of course, if three values tied for most frequent then we would say we have a trimodal distribution and so on.

The mean, median and mode, as well as to other numbers that can be derived from a data set, can be considered **statistics** since they represent a useful feature or characteristic of a data set.

Although we are after the central point of a distribution, it is also true that we want that point to be consistent with the most "typical" or representative value in a data set. None of the measures for central tendency is difficult to understand, nor calculate, but which one is the "best" measure is not obvious. We are left with a possible dilemma - what value should we use when? As a rule of thumb, the decision should be made on the basis of which of the three statistics provides the best representation of the whole data set. The goal here is to avoid using a statistic that causes distortion. In general, it may be safest to report all three values. However, that may not be the most efficient thing to do - why cite three values when one may be sufficient?

To appreciate the difference between the competing statistics let us look at three different situations and make an arbitrary rule that we can only cite one of them - under what circumstances would we chose the mean, median or mode? Be warned that our examples represent some unusual distributions and, in those cases, we would want to describe the distribution as accurately as possible by citing two or even three of the measures of central tendency. However, for our immediate purposes, we will have to select only one statistic.

If we were dealing with a distribution (see Figure 1-A) that rose quickly, peaked and then had a long tail we might feel that the mode would be our best choice. Distributions that have this appearance are often called skewed distributions. Note that in this circumstance, the median, and especially the mean, are out on the long tail and are undesirable choices for representing the central location.

Figure 1-A

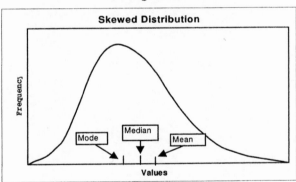

In Figure 1-B on the next page, the shape of the distribution is symmetrical (the left and right halves are mirror images of each other) and there is a single peak in the middle. Under this arrangement the mean, median and mode will all be at the same place and so it may not matter what value you select - they all communicate a similar situation. If you only can report one value, most people would be content with the mean.

Figure 1-B

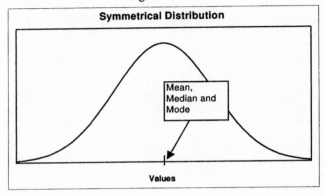

What if the shape of distribution is similar to that shown in Figure 1C? Notice that this distribution has a small blip representing one value way out on the right hand side.

Figure 1-C

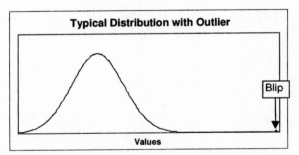

The data set could be 100 household incomes with all families in the $30,000 to $70,000 range except for seven families that have an income of $150,000. In this situation, the mean would be around $60,000, but the median would only be about $50,000. Under the circumstances, the median would probably be the single best choice - it comes closest to being the most "representative" statistic and is preferred when there are a few unusually extreme values in a data set.

The point of this small diversion is not to argue for using one statistic or another, it is to warn the reader that using a single statistic to summarize a data set can distort the presentation of the set. Therefore, be alert to the need to present statistics that represent the data set honestly and accurately without overwhelming the reader with unnecessary numbers.

On page 7 we introduced our first formula, the formula for the mean. Many people have an aversion to formulae. However, formulae serve a useful purpose. They tell us, in an efficient way, the steps that are performed to arrive at an answer to a mathematical question. The question may be: What is the mean of the distribution? The formula for the mean is repeated below:

$$\bar{X} = \frac{\Sigma \, X_i}{N}$$

The equation tells us that the mean ($\bar{X}$) is equal to the sum (Σ) of the individual elements (X_i) divided by the number of elements in the data set (N). The formula is a kind of short hand that is much more efficient than spelling out the calculation step by step in English. In this day and age, computer programs can do almost all the work so there is no need to memorize a formula. Consequently, some may feel it isn't necessary even to present formulae. Wrong - there is still a value in presenting a formula because a formula is a powerful communication tool; it tells us the operations that are occurring in a calculation. We can have a greater appreciation for a process if we understand what is taking place and that's what a formula can tell us.

Variation

In some respects we could say that statistics is the process of dealing with variation. We recognize variation in our everyday life. Two people see the same thing but report it differently. I measure a board and my son then measures the same board - we very likely will have two different lengths. Even when I measure the same board twice I too frequently end up with lengths that are a little different. In statistics we measure these inconsistencies and refer to them as "error". When statisticians talk about **error** however they're not using error as meaning a mistake - they use the term to refer to the random inconsistent result observed with many phenomenon. Almost all data sets have an element of error - test grades, weight, blood pH levels, etc. The results are not consistent time after time - they vary. Variation is clearly related to the concept we call consistency. If a person scores 78, 75, 72, and 75 on a series of tests we'd be inclined to say the student's grades were pretty consistent. On the other hand if a person scored 50, 95, 65, and 90 on the set of tests, we'd say that that person was inconsistent. Note that in both cases the students scored the same mean - 75.

To understand a data set well, we need more than just a measure of central tendency to describe it. Even when the three measures of central tendency are all at the same point, we can have very differently looking distributions. The distribution shown in figures 1-D and 1-E have the same mean, median and mode yet they don't look alike. Why? Because they are spread out differently - in figure 1-D the values are close together. but they are more dispersed in figure 1-E. The challenge becomes how can we express this property of spread, or using the statistical term, how can we measure variation?

Figure 1-D

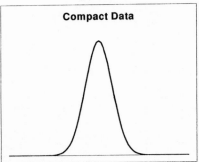

Figure 1-E

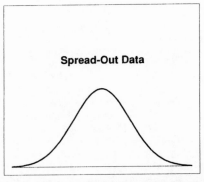

When we sought a measure of the central position in a distribution, we ended up with a choice of ways to do this, e.g., the mean and median. Each of these statistics was represented by a single value. It would be nice to be as efficient and come up with a single statistic to measure variation. But before we search for a statistic to measure variation, let's point out that there are some data sets in which the variation may be the most important property of the set. An example is a surgical instrument in which a uniform product is essential. Let us assume we want an instrument that makes a small opening - we'd ideally like that opening to be 0.15 mm in diameter, but can tolerate an opening as small as 0.12 mm or as wide as 0.18 mm. A hole too small will require a new procedure and a hole too large will cause surgical problems. Imagine that we have two models of the instrument and that after repeated testing we obtain the following results:

Model	Mean	Lowest Value	Highest Value
1	0.15	0.10	0.20
2	0.16	0.14	0.18

Based on our needs, model 2 is the preferred product - not because the mean is closer to our ideal requirement (0.15 mm), but because it has less variation. Note that model 1 exceeds the upper (0.18 mm) and lower (0.12 mm) tolerance limits of the ideal instrument, but this never occurs with model 2.

Range

How can we express the fluctuation in results exhibited by the two models? Usually the first thing that comes to mind is the range. The **range** is simply the interval between the lowest and highest value in a data set. For example, in the table above, the range for model 1 is 0.10 (0.10 - .20). For model 2 the range is 0.04 (0.14 - 0.18).

We see that the range served us well in differentiating between the two surgical instruments. Yet, by and large, statisticians tend to be dissatisfied with the range. In spite of its virtues, it has two drawbacks. First of all, it tells us only about extremes. Remember the example of when to choose the median or mean - in that case the range could have an upper limit of about $70,000 or one of $150,000 by including or excluding only a few values.

The second concern with using the range as the primary measure for variation is that it varies with the sample size. If you selected 10 items from a data set you'd get one range, but if you'd selected 100 items you'd get a different and, almost assuredly, a larger range. It would be better to find a statistic that wasn't so vulnerable to change due to the number of elements chosen or what statisticians call the **sample size** (i.e., the number of elements selected in a data set).

To continue our search, let's take the three small data sets below and use different strategies to develop a variation measure.

Statistical Methods for Researchers

	Set 1	Set 2	Set 3
Raw data:	5,5,5,5,5	6,4,5,5,5	6,4,5,10,0
Mean:	5	5	5
Variation:	None	Small	Large

In each set we can calculate the mean and median. When we do this, we obtain a value of 5 for each set. We can safely say that the three sets have the same center. Our intuition also allows us to comment on the variation - we can say there is no variation in set 1, a small amount of variation in set 2 and a large amount of variation in set 3.

Average Deviation

What options do we have to convert our sense of variation into a mathematical form? One thing we could do is to find, on average, out how far each value is from the center of its distribution. In other words lets subtract each value in a set from the mean of the set. We can then calculate the mean of the difference scores which we will call deviations. If we do this we get the results shown below.

	Set 1	Set 2	Set 3
Deviations	0,0,0,0,0	1,-1,0,0,0	1,-1,0,5,-5
Mean of Deviations	0	0	0

Unfortunately when we do this all means turn out to be 0 and this obviously is not a satisfactory way to differentiate the variation displayed in the three sets. What is happening is that the values above the mean are offset by the values below the mean - the plus values are canceled out by the minus values. We can overcome this problem by counting a deviation above the mean the same as a deviation below the mean. In mathematical terms we use what is known as the absolute value. We obtain an absolute value by ignoring the minus sign so all values are positive numbers (i.e., a -1 becomes 1). When we use the absolute value of our deviations our results become:

	Set 1	Set 2	Set 3
Absolute Values	0,0,0,0,0	1,1,0,0,0	1,1,0,5,5
Mean of Absolute Values:	0	0.4	2.4

This strategy is appealing. Set 1 has 0 as the mean of the absolute values and fits our notion of no variation. Set 2 with a mean value of .4 (2/5) and set 3 with a mean value of 2.4 (12/5) also have results that would seem consistent with our intuitive concepts of little variation and a lot of variation respectively. The calculation we just performed has a name - it's called the **average deviation**.

Standard Deviation

As appealing as the average deviation is, statisticians have taken the strategy one step further. We eliminated the minus signs by using absolute values, but there is another way to get rid of the minus sign. We can take each of the deviations and square them. In this case -1 times -1 becomes 1 and -5 times -5 becomes 25. Squaring numbers gives only positive values. The squared deviation results are shown below:

	Set 1	Set 2	Set 3
Squared Deviations	0,0,0,0,0	1,1,0,0,0	1,1,0,25,25

We can now calculate the mean of the squared deviations and get:

	Set 1	Set 2	Set 3
Mean of Squared Deviations	0	0.4	10.4

The approach we've used results in a statistic called the **variance**. Note that the variance is simply the mean value for a data set, which is created by squaring the difference between each original value and the original mean value of the data set. In our example the variance of the data sets is 0, 0.4 and 10.4 for set 1, 2 and 3 respectively. Note that in the process of doing this calculation, we squared the deviations to eliminate the negative sign. In that sense the original values have been transformed. We can undo that manipulation by now taking the square root of the variance. When we do that, we get a statistic called the **standard deviation** (e.g., for set 2, the square root of 0.4 is .63 and for set 3 the square root of 10.4 is 3.23).

	Set 1	Set 2	Set 3
Standard Deviation	0	0.63	3.23

Finally we are at the end of our quest - the standard deviation is the statistic that is used most often to show variation. We will see that it has many uses.

We use the standard deviation to tell us how much spread we have in a data set. The spread gives us an idea about how stable the values in the data set are. A large standard deviation means the data fluctuate a great deal. A large standard deviation indicates that the measurements we're making are not very consistent. It's very critical to have a good understanding of the standard deviation. A small standard deviation isn't necessarily better or worse than a large one since the standard deviation just represents a basic characteristic of a data set. Some phenomenon are simply more constant than others are.

The standard deviation is clearly related to the variance. If you have a standard deviation all you need to do to obtain the variance is to square the standard deviation value. In set 2, when we square the standard deviation of .63 we produce a variance of .40. In set 3, 3.23 times 3.23 gives us a variance of 10.4.

Technically the term **population standard deviation** is reserved for the standard deviation we just calculated. The mathematical symbol for the population standard deviation is σ, the Greek letter corresponding to our s and called sigma. We'll see later that there is also a **sample standard deviation**, which is calculated a little differently than the population standard deviation. The sample standard deviation uses the symbol s. However, for now σ will be adequate for our purposes.

The **population variance** is the squared value of the population standard deviation, σ. The symbol for the population variance is σ^2 and pronounced "sigma squared". There is also a **sample variance**, which uses the s^2 symbol.

We worked through the calculation of the variance and standard deviation step by step. We can condense all that description into a single formula. The formula for the variance is:

$$\sigma^2 = \frac{\Sigma(X_i - \mu)^2}{N}$$

Note the μ symbol in the formula which, as noted earlier, is used to represent the mean. The numerator of the formula says sum (Σ) the squared differences between the individual data items and the mean $(X_i - \mu)^2$. This sum is then divided by N - the number of items in the data set. The formula says the same thing as our step by step explanation, but it does it quicker. We sometimes can be intimidated by formulae, but just think of them as an efficient way to give us a lot of instruction.

The standard deviation is simply the square root of the variance so, as a formula, all we need to do is enclose the expression for the variance under a square root sign.

$$\sigma = \sqrt{\frac{\Sigma(X_i - \mu)^2}{N}}$$

Since the standard deviation and the variance are so closely related, we might wonder why we need both. The value of the standard deviation is that it is measured in the same units as the original data. If the unit of measurement for the raw data is minutes, minutes are also the unit of measurement for the standard deviation. However the units of measurement for the variance would be minutes squared, which is a rather unusual form. Nevertheless, the variance turns out to be useful for doing certain statistical calculations, as we will see later. In summary, the standard deviation is more useful for interpretation and the variance is more useful for calculation.

Summary

In this chapter we learned that statistics involves more than number crunching – it also entails logic and rationality. We were introduced to some of the common terms used in statistics (e.g., data set, distribution, variable, frequency table, etc.). We also examined measures of central tendency (e.g., mean, median and mode) as well as measures of variation (e.g., range, average deviation and standard deviation/variation).

PROBLEMS

1. For the weight data set on page 4, calculate the mean, median, mode, range, average deviation, population standard deviation and the population variance.

2. Without doing the calculations determine what happens to the mean, median, mode and standard deviation if we add the value 89 to the data set on weights referred to in problem 1 (note that the new data set will have 40 rather than 39 observations). Your options, for the type of change that occurs, are whether the statistic will become: larger, smaller, stay the same or you can't tell.

3. To see if you obtained the correct answers to question 2, calculate the mean, median, mode, range and standard deviation.

4. Without doing the calculations determine what happens to the mean, median, mode and standard deviation if we add the values 70 and 95 to the original data set on weights referred to in problem 1 (note that the new data set will have 41 rather than 39 observations). Your options, for the type of change that occurs, are whether the statistic will become: larger, smaller, stay the same or you can't tell.

5. To see if you obtained the correct answers to question 4, calculate the mean, median, mode, range and standard deviation.

6. The following data set represent the number of days it takes the town police department to process a complaint against an officer in the department. Prepare a table showing frequencies, cumulative frequencies, percents and cumulative percents.

12	5	3	20	12	30	8	10	12	28
7	21	12	16	26	10	2	19	20	23
22	2	9	29	21	18	13	27	5	21
26	18	2	8	6	21	7	21	3	23
12	9	11	9	1	12	13	13	30	15

7. The following data report the number of service calls the local public utility organization received during the first 10 days of February and the first 10 days of June. For each set calculate the mean, median and population standard deviation.

February	226	256	278	187	203
Calls	135	266	282	244	231
June	158	153	202	166	215
Calls	178	211	205	168	176

What month has the greatest number of calls and in which month is the number of calls more uniform?

Excel Instruction

In this section we will work with the 39 employee database created in the Preface. Start by making a copy of the original data base file. Working with a copy of the original database protects the original version from unintentional errors. To make a copy, open the file of the original database. Then select "File" from the Menu Bar followed by the "Save As" command. Give the file a new name (e.g., STAT2) - a name different then the one you used for the original file. Your original database file will now be secure under the original name and all your work will be done on the backup file named STAT2.

We begin by learning how to prepare a table for a discrete variable such as sex. This task is done in Excel by constructing what Excel calls a PivotTable. As our first exercise we will create a table of the sex distribution for the people in our database. If in the process of preparing a table you find you made a mistake, you can eliminate an erroneous PivotTable by highlighting the whole pivot table, clicking "Edit" on the Menu Bar and then selecting "Clear". A new box with a list of options will be presented. Select the option "All" and the table will be deleted.

To construct a pivot table begin by selecting "Data" from the Menu Bar and then selecting "PivotTable Report" from the options that occur on the next menu. The "Pivot Table Wizard - Step 1 of 4" dialog box opens.

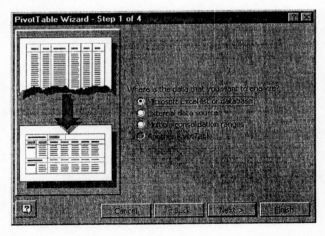

Be sure the circle for "Microsoft Excel list or database" has a dot in it - if not, click on the circle so a dot appears. Click "Next >" at the bottom of the box to reach the "Pivot Table Wizard - Step 2 of 4" dialog box (see next page).

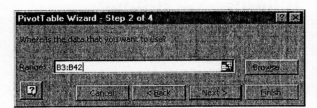

The question in the dialog box requests the range of cells that contain the data you want in a table. This box may be empty, but Excel may anticipate the answer and provide a cell range. In either case, we want the cell range for the SEX variable, which is B3:B42. If that range is not specified, enter B3:B42 in the box. Cell B3 is included since it contains the name of the variable and will be useful in subsequent steps. Click on the "Next >" button and you are at a screen called "Pivot Table Wizard - Step 3 of 4".

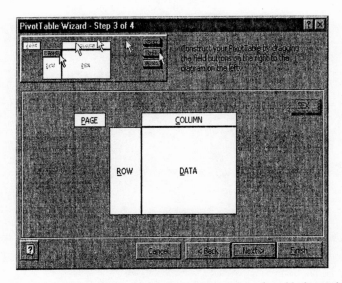

This screen contains instructions at the top, a representation of a table layout below the instructions and a button on the right hand side that is named SEX. Note that this is the label entered in cell B3. In this case there is only one button since we included only one variable (i.e. column) in our input statement. We now place the pointer on the SEX button and drag that button to the ROW area of the table. This is a one variable table so we will not have an entry for COLUMN. But we still need to provide information for the DATA portion of the table. Again click and drag the SEX button from the listing on the right side to the DATA portion of the table. It is important that after you complete this operation that the DATA section contain a box that reads "Count of SEX". If that section contains some other wording, double click on the button you just moved into the DATA area. You will now have on you screen a dialog box called "PivotTable Field" (see below). This dialog box gives you a choice of statistics that can be calculated.

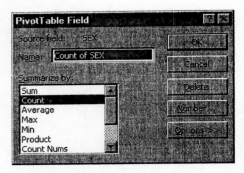

Since we want frequency counts, select Count from the "Summarize by:" box. Click OK and your screen should look as follows:

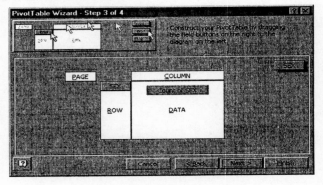

When your step 3 screen has the above appearance click "Next >". You will now be at the last window which reads "PivotTable Wizard - Step 4 of 4" (see below).

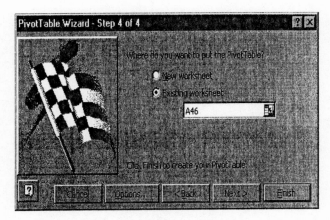

You need to identify a place on your worksheet for your table. The table we want to produce is small so select the "Existing worksheet" option. Next identify a cell on the worksheet that is free of data since your table will overwrite anything that exists. In our case we could pick cell A46 since our database ends at Row 43. Click Finish and starting at cell A46 you should have the sex frequency table shown below.

Count of SEX	
SEX	Total
F	22
M	17
Grand Total	39

For our next exercise we will rearrange the tests scores from high to low score. We do not want to introduce errors onto our original database so we will make a copy of the test scores. Let's copy test scores to column P of the worksheet. This can be done by highlighting cells G2 to G42. Since we want to copy these cells, click on the copy icon, 📋 which is on the Standard Toolbar. Move the mouse pointer to cell P2 and click on the paste button, 📋. Your test scores should now all be replicated in Column P. Alternatively you could simply hit the Enter key after you position the cursor in the P cell. The data in column P are highlighted so click on any cell to complete the copying operation.

Since we set out to sort the test scores, highlight them (i.e., highlight cells P4:P42). Select the "Data" option from the Main Menu. From the dropdown menu select "Sort". The screen that appears is the SORT dialog box (see below).

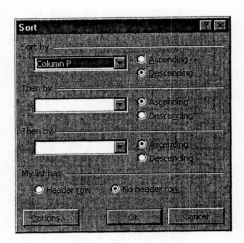

The first box labeled "Sort by" will be filled in with the words SCR if the box at the bottom of the screen labeled "My list has" has a dot in the "Header row "circle. It may contain the entry "Column P" if the "No header row" circle has a dot in it. Either arrangement is satisfactory.

We want scores listed in descending order (from highest to lowest) so click on the descending circle, if that circle does not have a dot in it. It does not matter what the next two boxes have in them since we are only sorting one variable. Finish the operation by clicking on OK. Column P will now present the scores from highest (98.2) to lowest (50.2).

There is another way to do a sort - by using the sort icons on the toolbar. There are two icons - one with the letters A at the top and Z at the bottom, [▨], and another with Z at the top and A at the bottom, [▨]. As you might guess the A to Z arrangement means sort the cells in ascending order and the Z to A pattern means the sort should be done in descending order. Highlight the test scores in column P (P4:P42) and see how you can change the order by clicking the sort icons.

We can also arrange our test scores in a frequency table. To prepare a frequency table for test scores (a continuous variable) you need to create categories. It's necessary to predetermine what seem like reasonable intervals for your categories. For test scores it's best to organize the categories in units of 10. We thus want as categories: Under 60, 60-69, 70-79, 80-89, 90 and over. Excel refers to the categories as bins and only requires the largest number in a category specified. When Excel does a count it includes all entries up to and including the number listed as a bin. For example, if we enter 59 as a bin, Excel will include all entries less than and equal to 59 in that bin. If we wanted all values in the 60s in our next category, we would enter 69 as our bin. Our second bin will therefore contain all entries starting with 60 and ending with 69. We could then add bins of 79 and 89, but we also want a bin to record values over 89. Excel automatically adds an upper bin, which is titled "More", when the frequency table is produced. The values that exceed 89 will be automatically counted in this bin and, therefore, we do not need to specify an upper range category.

Find a blank portion of the worksheet to list the bins we have chosen. Cells G48 through G51 should be empty and, if they are, enter our bin numbers: 59, 69, 79, and 89 in those cells. Select "Tools" from the Main Menu followed by "Data Analysis" from the drop down menu. If "Data Analysis" is not listed under "Tools" you need to install it. To do this, select "Tools" and then click on the "Add-ins" option. From the next menu click on the box next to "Analysis ToolPak" and finish with OK. "Data Analysis" should now be an option under the "Tools" command. Return to the Main Menu Bar, click on "Tools" and then select "Data Analysis". In the "Analysis Tools" listing, you will have a choice of a large number of statistical procedures to select from - the one we will use for this exercise is called "Histogram". After highlighting "Histogram" and clicking "OK" you will have the Histogram dialog box (see below — your box will not have any entries).

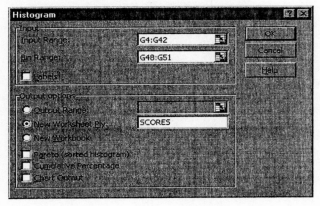

You need to identify, in the "Input Range" box, what data you want summarized. In our case it is test scores, which appear in cells G4:G42. After entering the input range (i.e., "G4:G42"), click the "Bin Range" box. Since we have established bins for the test score data, enter "G48:G51". Be sure there is no check mark, ✓, in the "Labels" box since we did not include a heading label in our Input Range. If there is a "✓" in the "Labels" box click on the "✓" and it will change to a blank entry. For the lower portion of the dialog box, let us assume that you want to have the table appear on a separate sheet from the database. Go to the "Output options" section and make sure the circle next to "New Worksheet Ply" has a dot in it. In the adjacent box give the sheet a name - "SCORES" would be an appropriate choice. If you leave the box blank Excel will automatically assign it to a sheet, but it's best to give the sheet a name so we can recognize and retrieve the worksheet easily. Leave the other boxes blank and your screen should match the previous display. Click "OK". Excel creates a table on the new worksheet called SCORES. The Excel table produced is shown below:

Bin	Frequency
59.0	5
69.0	7
79.0	10
89.0	9
More	8

We can add percents to our test score data. In cell A1 of the worksheet named SCORES the heading "*Bin*" appears. It is followed by the five bin categories in cells A2 to A6. Column B contains the frequencies for each category. To show the matching percentages, we need to perform a calculation. That calculation requires that we first determine the total count for our table. The value we want is the sum of the entries in cells B2 through B6. To obtain the sum, go to cell B7. Select the AutoSum button (⊞) on the Standard Toolbar and then hit Enter. The value 39 will appear in cell B7, which is the sum of the frequencies in our table.

We will show the percentages in the column next to the frequency counts, i.e. in Column C. It is probably wise to give the percent column a title so in cell C1 type "Percent". In cell C2 you need to write a formula for the percent calculation. What we first want to find is the percent of the employees who scored less than 60. That answer can be obtained by the formula for a percentage calculation. We will write the formula in cell C2. In Excel all formulae begin with an equal sign (=). They then can be followed by a cell reference, command (e.g. SUM) or a number. In addition you can incorporate mathematical operations in a formula. The symbols for the most common operations are:

+   for addition
-    for subtraction
*   for multiplication
/   for division

There are other symbols, which are discussed in an Excel manual, but the set just described will serve us well for the time being. In way of illustration, if we want to multiply a value in cell A50 by 3 we can write our formula as "=A50*3". If the value of A50 was 7 and we wrote our formula in cell B50 we would get a result of 21 in cell B50 after we hit Enter.

To obtain our percent we will first calculate a proportion and multiply that result by 100 to get the percent. Click cell C2 and type "=B2/$B$7". We want to divide all our individual counts by the same grand total number, which is in cell B7. The $ signs in cell C2 are used to keep the B7 entry (39) fixed for all the calculations. The dollar signs before the letter B and the number 7 tell Excel that cell B7 shouldn't change as we copy its contents (i.e. formula) to other cells. You will find that when Excel automatically lists cell ranges it frequently uses $ notations.

After writing "=B2/$B$7" in cell C2 and hitting Enter, the value .128205 will appear in cell C2. You may have more or fewer decimals showing depending on the width and format of cell C2 on your worksheet. You can always modify the number of decimal places by selecting "Format" from the Menu Bar. You then need to follow the instructions given in the preceding chapter for setting the number of decimal places you want to appear in a cell. Alternatively, you can increase or decrease the number of decimals by using icons in the Format Toolbar. By clicking on the 🔢 icon you increase the number of decimal places by one decimal place for each click. When you click on the 🔢 icon, you reduce the number of decimal places.

The value .128205 is a proportion and needs to be multiplied by 100 to become a percent. You can do this by returning to cell C2 and inserting 100* between the = sign and the C2 cell reference so that the formula reads "=100*B2/B$7". After hitting Enter, cell C2 will contain the value 12.82051. Again the number of decimals appearing will vary based on the width and format of column C. If an answer won't fit, Excel displays a string of #s. To remedy this condition you can decrease the number of decimal places or increase the width of the cell.

Thus far we have only calculated a percent for the first entry. We can obtain percents for the other categories by simply copying our formula in cell C2 to cells C3 through cell C6. Click on cell C2 and then click on the Copy icon. Highlight cell C3 to cell C6 and hit Enter. The percents corresponding to the other categories appear and we have produced the all the information we need for our table.

You can check to see if you have the correct result by summing the percents - they should total 100. Use the AutoSum icon in cell C7 to do that.

Finally, we need to provide cumulative values for our frequency and percent data. Begin by typing a heading in cells D1 and E1. In cell D1 type "Cum Freq." and in cell E1 type "Cum Percent". The headings are not essential, but they help remind us of the type of information contained in a column.

The first cumulative frequency entry is identical to the first frequency entry (i.e. 5) so we can simply type "=B2" in the D2 cell and hit Enter. In cell D3 we want to combine the values in cell D2 and B3 so in cell D3 type "=D2+B3". The value 12 appears in cell D3. You can continue using this same strategy to complete the remaining cells or simply copy what you wrote in cell D3 to the other cells - in either case you would get the correct values. The operation for cumulative percents is a repeat of the steps we performed for cumulative frequencies except we use columns E and C as our data source. You can again save yourself some work by copying the formulae in cells D2 to D6 into cells E2 through E6. The display on the next page shows what your final table should look like.

Bin	Frequency	Per Cent	Cum Freq	Cum Per Cent	
59.0	5	12.82051	5	12.82051	
69.0	7	17.94872	12	30.76923	
79.0	10	25.64103	22	56.41026	
89.0	9	23.07692	31	79.48718	
More	8	20.51282	39	100	
	39	100.0000			

Our last exercise will be to calculate some descriptive statistics. The statistics we want to calculate are shown in the following table.

Median	Mode	Mean
Range [i.e., the minimum and maximum values in the data set]	Average Deviation	Standard Deviation

Let's obtain these statistics for the AGE variable which is on sheet 1. Click on sheet 1 and select a place to record the statistics: The five cells C52 to C58 are a good location. It would help if we identify the statistics so in column A52 type "Median", in A53 "Mode", in A54 "Mean", in A55 "Range-Min", in A56 "Range-Max", in A57 "Avg. Dev." and in A58 "Std. Dev." We'll begin by providing the median value. We will want the median to appear in cell C52 so click on that cell.

All of the descriptive statistics we want are automatically determined in Excel by using the Function icon on the Standard Toolbar. The Function icon has an *f* followed by a small x on it [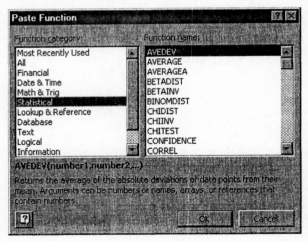]. Click on that icon and the screen "Paste Function" will appear (see below).

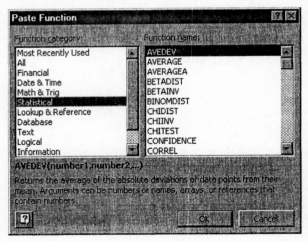

In the area called "Function category" find "Statistical". Click on "Statistical" and on the right hand side a series of terms will appear in the "Function name" portion. Find "MEDIAN" in the "Function name" listing and click on it. Hit the "OK" button at the bottom of the box. The screen "MEDIAN" will appear (see below).

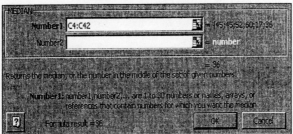

For the "Number1" box give the range for the age data (i.e., "C4:C42"). This is the only variable we are using so click on the "OK" button. In cell C52 the median age will be shown. If you click on cell C52 you will see that the command for it appears in the row right below the icon toolbar. The command reads "=MEDIAN(C4:C42)". In fact once you know the names of the statistics you want you can type them directly into the destination cell (in this case C52) instead of using the Function Wizard. In way of illustration, to find the mode click on cell C53, and type "=MODE(C4:C42)". When you hit enter the value for the mode will appear in cell C53. The Excel terms for the other statistics we want are as follows:

AVERAGE for the mean	MIN for the minimal value in a data set
MAX for the maximum value in a data set	AVEDEV for average deviation
STDEVP for the standard deviation	

The P at the end of "STDEVP" stands for Population, which is one of the two ways a standard deviation can be calculated - we'll study the other choice at a later time.

For the remaining statistics determine the appropriate value for the age variable using the Function icon. When you are done your results should appear as follows:

Median	36
Mode	45
Mean	38.49
Range – Min	17
Range – Max	66
Average Deviation	11.70
Standard Deviation	13.83

You may have a different number of decimals for some of the values which, if you wish, you can change by using the methods previously described.

Here are a few exercises for you to practice on using your Excel database.

1. Use the Pivot Table command to prepare a table for Advancement Potential.
2. Use the Sort command to arrange ages in descending order.
3. Prepare a table for the July PC Proficiency scores showing Frequency, Percent, Cumulative Frequency and Cumulative Percent.

Chapter 2

Data Presentation

Introduction

In studying statistics there is a tendency to devote too little time on ways to present data. There are probably several reasons for this habit. One - there are no generally accepted rules on what to do. Secondly, the subject of data presentation is not based on mathematics and many texts written by mathematical statisticians would be disinclined to include non-quantitative material in their work. However, good data presentation is critically important and requires attention in spite of the lack of standards.

A good data display enhances understanding - a poor display diminishes comprehension. The primary purpose of a data display is to communicate information. Misleading or confusing tables and figures retard communication. Well-constructed tables and figures inform the reader.

In this discussion, the term "table" will refer to a display that contains only text and numbers whereas the term "figure" will refer to a display that incorporates a graphic - i.e., a picture or some object. Graph and chart are synonyms for the term figure. The expression "exhibit" will be used when the reference applies to both tables and figures.

Without firm rules, the best one can do is discuss this subject in terms of goals, guidelines or possibly "rules of thumb". In large part the writer will end up using common sense since data display is mostly art, with little science involved. It's also true that data display incorporates what we often call taste - what appeals to one person may be rather unimpressive to another. Yet there are some sound guidelines and they should help restrain writers in letting their taste overwhelm good sense. In the discussion that follows, I will express my personal views on what to do when decisions have to be made - you may agree or you may not. That's all right since whether you agree or not is a matter of preference. What is important, however, is to realize that there are choices to be made - decisions such as how much information should be included in a title, how many decimal places should be included when presenting a number, etc.

As noted above, perhaps the most important goal or guideline when presenting data is to design your exhibit so it informs the reader. A corollary of this guideline would be - make sure your exhibit can stand on its own. In other words, the reader should be able to understand your exhibit without having to rely on explanatory information in the text that accompanies the exhibit.

Preparing Tables

There are three main parts to a table.
1. The title - a description of the subject matter for the table.
2. Headings for the columns and rows which identify the variables being displayed. Each heading is then broken down into a set of categories.
3. The cells of the table - technically a cell is the intersection of a column and row and usually contains a number that was collected or calculated.

An example of a table in which the main parts are identified is given below.

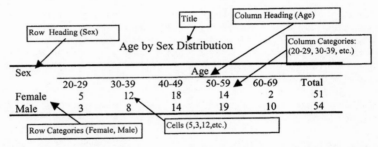

	Title				Column Heading (Age)
Row Heading (Sex)					
	Age by Sex Distribution				Column Categories: (20-29, 30-39, etc.)

Sex	Age					
	20-29	30-39	40-49	50-59	60-69	Total
Female	5	12	18	14	2	51
Male	3	8	14	19	10	54

Row Categories (Female, Male)   Cells (5,3,12,etc.)

Clarity in exhibits is often achieved by only including essential material. This guideline especially applies to formatting - do not include unnecessary lines, symbols or other notation that is not essential. The construction of a good table can generally be created using only three horizontal lines - a line between the title and the column headings, a line after the column headings and a line at the end of the last row of data. Additional horizontal or vertical lines should be avoided although, there may be occasion to include a vertical line between the row headings and the first column of cells. Some people like to box in their exhibits with vertical lines at the beginning of the row titles and at the end of the last set of column entries - this is not wrong but it seems an unnecessary addition.

In keeping with the desire to simplify a table to increase understanding, avoid long wordy titles. In an effort to be comprehensive, there is a temptation to include too much information in the title. Too frequently this means the reader scans the title and misses the point of the exhibit or ignores the title due to its length. For example refer to the following table title:

NUMBER AND PERCENTAGE OF TIMES SELECTED TOPICS WERE CHOSEN BY 137 PUBLIC ADMINISTRATION FACULTY MEMBERS FOR INCLUSION IN A REQUIRED COMPUTER COURSE

The title sounds fine, but it can be shortened. A shortened version is:

TOPICS CHOSEN BY 137 PUBLIC ADMINISTRATION FACULTY MEMBERS FOR A REQUIRED COMPUTER COURSE

Note that the number of words was reduced by about half, with essentially no loss in knowing the subject of the data. The body of the table could contain the following information:

Topic	Frequency	Percent
Computer Applications	75	55%
Technical Subjects	39	28
Impact Issues	31	23
Legal Issues	18	13
Policy Issues	10	7
Other	8	6

The heading in the first row gives the main variable of interest, Topic. The two adjacent column headings in this row give the type of statistic that will be reported - Frequency and Percent. The other row headings give the topic categories and the cells contain the statistical information. This is the customary way to arrange a table dealing with a single variable. The row categories are ordered - in this case by the number of times a topic was selected. This is a judgment call since there are alternatives - for example an alphabetical listing. In this case the arrangement is from the highest to lowest frequency which is reasonable given what a reader would most likely be interested in knowing.

Technically we could call the table a **frequency table** since it presents the information in respect to the number of respondents that gave a certain response. It's also important to point out that in addition to the count of people with a given answer, a last column has been added to show the percentage of times a topic was recommended. The addition of percents is helpful since it puts the information in perspective. It's difficult to conceptualize the meaning of 75 out of 137 (the total number of respondents) and the percentage number, 55%, gives the raw numbers added meaning. As a rule of the thumb, it's best to also assume a count will be accompanied by the corresponding percent unless there are good reasons not to do this. Examples of good reasons not to include the percent would be when the total count is small. If there were only 30 respondents in a sample, the percent data would not be very helpful - most people can interpret frequency numbers when the total sample size is 30 or less.

Note that the percent sign (%) appears in the first cell of the last column. Here we are dealing with a matter of taste or personal preference. Even though the column heading states that the cells are percents, the % sign reinforces that specification and is a worthwhile, but not a required addition. Also note that the % sign is not repeated since such repetition would very likely add to the tables bulk without commensurate comprehension, but again that's a matter of personal taste.

You should also realize that the percentages are expressed as whole numbers - i.e. without decimals. The advent of computers has been a great aid in the statistical world, but sometimes we have too much of a good thing. Technically, the 55% shown in the table could come out of a computer as 55.1459 or with even more decimals showing. Again the writer must decide how useful the decimal places are. When showing percents, where most of the numbers are two digit whole numbers, the addition of even one decimal adds almost nothing to a reader's comprehension, but it takes up space and is therefore a liability. Here again one must use his or her own good sense on what is the best practice given the data at hand.

The table we have worked with is a single variable table, in that it dealt only with the single characteristic - topics. A presentation of one variable is sometimes called a **univariate table** or it may be called a **one-way table**. It is also common to show the interrelationship between two variables in tables and in this case the term **bivariate table** may be used. A typical bivariate table might be one that shows the educational level for males and females. Since two variables are involved, educational level and gender, sometimes you will hear people refer to the bivariate table using terms such as a **two-way table**, a **cross-tab table** or a **contingency table**. The table below is an example of a two-way table.

Educational Level by Sex

Sex	Educational Level			
	High School	Bachelor	Master	Total
Female	8	8	6	22
Male	5	5	7	17

Another type of bivariate table might be one that includes a continuous scale variable as one (or both) of the data categories. For example, the table on the next page shows an age distribution based on gender. Note that the column heading, Age, includes the clause "in years" so the reader knows the units of measurement for this variable.

Sex	Age (in years)					
	<60	60-69	70-79	80-89	90-99	Total
Female	4	2	7	5	4	22
Male	2	4	5	3	3	17

Age by Sex

An important decision that needs to be made regarding a table with an interval scale variable is the number of row or column entries to include. In the above table the choice of categories for the gender variable is easy - there are only two, female and male. However, for age a decision is required since there are many ages in a typical data set and they must be grouped into classes before they can be presented. How many categories should there be? A good rule of thumb is to use at least four, but no more than eight categories. However, there are many exceptions to this rule - e.g., too many important details could be lost with only eight categories. Using good judgment is called for.

Not only must the number of categories be determined, but also the size of an interval needs to be decided upon. In our example of age, the situation is simplified by the fact we expect to see age in certain classes - i.e., in decade groups where all those in their 20s are in one category, all those in their 30s in another group, etc. This encourages us to use class sizes of 10-year intervals, and this might work fine with our interest in using four to eight categories. However, if the age range was from newborns through the most elderly we would have over ten categories and would possible want to use a class size of 20 rather than 10. Much would depend on the purpose of the distribution - if decade groups were critical, then one would need to use as many categories as necessary to cover the entire range. On the other hand, if the age distribution were only for background purposes then the use of 20-year classes would be a reasonable choice.

If there is no obvious choice for a class size, then class lengths can be determined by dividing the data range by the number of classes one feels is reasonable. The formula for this calculation is:

$$\text{Interval size} = \text{Data Range/Number of Classes}$$

For example, if the data range is 33 (i.e., 6-39) and it is decided that 5 classes are sufficient then the class interval is 6.6 (33/5). To avoid introducing intervals that require more decimals than the original data set contained, the class size should be rounded to the highest integer. In this case, the next highest integer is 7 which means that seven unique values should be in a category. Hence the classes would be: 6-12, 13-19, 20-26, 27-33 and 34-40.

In respect to a title for a two-way table, the two variables can be linked by the term "by" in the title. In way of illustration, if the table was constructed for the employees of the XYZ organization the title could read "Age Distribution by Sex for XYZ Employees".

In this case, age categories should be presented as the column entries using the female and male titles for the row headings.

Usually the first variable mentioned in the title of a two-way table is the one that is inserted as the column variable. However, the choice of what variable is listed as the column heading and which as the row heading frequently depends on space considerations. Since we typically use 8.5 by 11-inch paper, it's often necessary to have the variable with the longest set of headings introduced as rows. When space is not an issue, many writers prefer to list the dependent variable as the row and the independent variable as the column.

For tables, you also need to decide if you want the marginal totals to be included. Marginal totals refer to the sum of the row (row marginal total) and column (column marginal total). The last two tables had row marginal totals only. Generally speaking, you have to decide how valuable the total will be to the reader. In many cases either a row or column marginal total is useful, unless you are including percents. However there are cases when both row and column totals are informative. The greater the number of categories and the larger the overall sum, the more reason to include totals for rows and columns.

Preparing Figures

The four most popular types of figures are referred to as line, bar, scatter and pie graphs. There are other types and variations on these four major forms as well. For example, there are stacked bar graphs, 3-D pie graphs, etc. Each of the four primary types generally is used for a different purpose. The **line graph** is well suited to show a time trend - say the change in a variable over time. The **bar graph** is preferred when the goal is to show comparisons between quantities. To present the relationship between two variables the **scatter graph** is usually selected. The **pie graph** is used to compare the size of the subparts of a larger unit. Examples of each type are shown on the next page.

Synonyms are frequently associated with these graph types - **polygon** for line, **histogram** for bar, **XY** for scatter, and **circle** for pie. With some computer programs you can easily interchange graph types for the same data set and see how the data will appear with different graphical representations. For example a bar graph can also show trends and is a perfectly acceptable alternative. On the other hand, if there were more than a few trends to show in a figure the line graph may be easier to comprehend than a bar graph representation.

The construction of a graph (with the exception of a pie graph) is in the shape of the letter L. The horizontal part of the L is called the **X-axis** (synonym = **abscissa**) and the vertical part the **Y-axis** (synonym = **ordinate**). For formal presentation there should be a title and the axes should be given headings as well. The same considerations, previously discussed, regarding a title for a table applies to a figure as well.

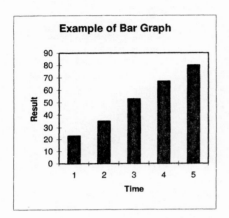

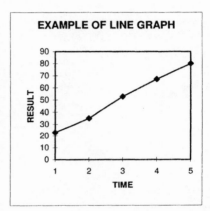

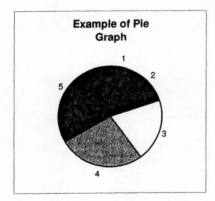

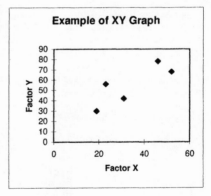

There are special situations that need to be addressed when presenting figures. Graphs are especially prone to misinterpretation. Unfortunately sometimes the misinterpretation is intentional, other times it is inadvertent. In either case, be sensitive to some of the most common problems. The first one we'll examine is the **moving scale** issue. Take a quick

Statistical Methods for Researchers

look at the two graphs below, which present data on profits by quarters for company A and company B. Pick the one you feel shows the best performance in terms of profit gain.

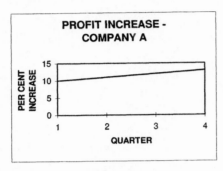

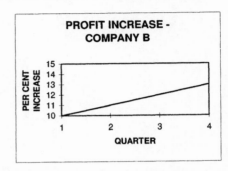

On close examination you will see that both presentations give identical information and one company is not better than the other. Both figures begin with a value of 10 in quarter 1, have a consistent rise and end with a value of 13 at quarter 4. Yet the images appear quite differently. The result for company B looks better than that for company A.

All that's been done in the figures is to simply alter the Y-axis. In the figure for company A the Y-axis begins with 0 and goes to 15. The Y-axis for company B begins with 10 and goes to 15, but note that it occupies the same area as the company A figure. Consequently the Y intervals for company B are much broader than they are for company A. The result is a line with a much sharper angle for company B than company A and the impression of a greater increase in profits.

Bar graphs are also vulnerable to the moving Y scale problem. Costs and benefits are shown in the two graphs below.

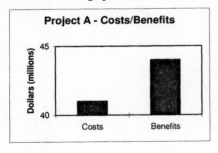

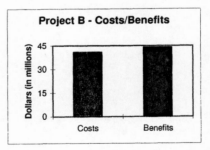

The presentation in the second figure for project B appears to have little difference between costs and benefits, but the one in the first figure for project A has a marked difference. Yet in both cases the costs are 41 million with benefits of 44 million.

It's tempting to recommend that the Y-axis of all graphs begin with 0 to avoid the type of distortion exhibited in these examples. However, there are cases when such a rule would be imprudent because graphs would take up too much space without a commensurate gain in understanding. For example, if we were showing the weight of professional football linemen it would be wasteful to start a scale at 0 when all the weights are over 200 pounds - there is no way the typical weight for a lineman could be any where near 0. Thus, starting the scale at 200 is more space efficient and reasonable, but we still have not eliminated the potential for distortion. An alternative to handle the problem is to show that there is a break in the scale. Start the scale at 0 then add two dashed lines across the Y scale near the 0 point and then pick the next Y point at any number you wish - e.g. 200 for the football linemen example (See the following figure).

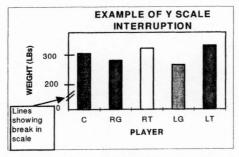

The double break lines alert the reader that there is an interruption in the scale - it serves as a warning and in many cases it is a reasonable compromise. Another problem is encountered when we rely on a pictorial graph to show comparative data (see below)

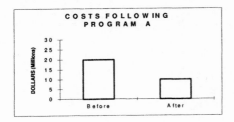

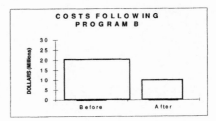

The two figures displayed on the previous page show changes in cost; before and after instituting a program. In both cases the costs go from 20 thousand to 10 thousand. Nonetheless, note that the graphic for program A makes the difference look more impressive. This distortion is because the bars are multidimensional and the total area is exaggerated.

There is nothing wrong with using pictorials in place of bars, but be sure the pictorials do not cause distortions. In the following graph the use of light bulbs is reasonable since the shapes avoid the multidimensional effect.

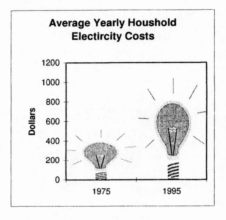

A delightful book on problems with data presentations is by Huff and is included in the bibliography.

General Guidelines

Regardless of whether you're working with tables or figures there are other general recommendations that can be made. One of the most important is not to overlap categories. In other words, the same number should not be in more than one category. In the age example, use categories such as the following:

20-29
30-39
40-49
etc.

as opposed to:

> 20-30
> 30-40
> 40-50
> etc.

In the first case it is clear where a 30 year old will be counted, but in the latter case that is not so. Be sure to include the units of measurement in your headings unless it is perfectly clear what the units are. For instance, all recognize that age is usually collected in terms of years, but there are occasions when other time units (e.g. months) may be used. For other variables there may be competing units of measurement that can be used. For example, weight is usually recorded as pounds in the U.S., but in scientific fields and other countries it is recorded in kilograms. Thus for weight it is important to add the unit of measurement. This can be easily done by following the heading "Weight" with "(in kilograms)" - i.e., place the unit of measurement in parentheses. When dealing with money, dollars can safely be assumed to be the unit of measurement in the U.S. and the term "dollars" is sometimes omitted in a table heading. However, if your audience included non-U.S. citizens you would be wise to include the dollar specification. Furthermore, if you have large sums you may wish to truncate the amounts by reporting information in millions or thousands of dollars. For example you may be dealing with county revenues in the millions of dollars and instead of showing a category as 12,000,000 - 15,000,000 you may want to just record 12-15. In this case you need to add the expression "(in millions)" to the heading. Also remember to define terms that are not understood by your reading audience. Don't use a heading such as "Homicide Rate" without adding the definition for the rate - e.g., "(Deaths per 1,000 residents)".

To avoid inclusion of categories that have low frequencies you can use greater than and less than notation (i.e. the symbols ">" for greater than and "<" for less than). Assume we have an age distribution that has a range of 5 to 92. To use decade intervals requires 10 categories yet there may be very few people under 30 or more than 70. Thus you can truncate categories by having a category "Less than 30" (<30) which includes all those 5 to 29 and a category for all those "70 and over" ($\geq$ 70).

In some cases, especially surveys, you may get no answer to some questions, or responses that indicate the respondents don't know what to respond. We can refer to these kinds of responses as "non-specific". You need to make a judgment on whether the non-specific responses are significant. For instance they would probably be significant if it appeared the respondents were unfamiliar with some issues being examined by the questionnaire. Consequently, you would want to include them as a category in your table or figure. On the other hand these unspecified answers may be immaterial - there may be few of them and the proportion is not unexpected - some people just fail to answer all questions. In this case, they may be omitted as an independent category, but there should be a comment in the exhibit to indicate that the tabulations did not include the non-specific responses.

Integrating Exhibits and Text

If you have chosen to provide an exhibit, then it is fair to assume that the exhibit exists for a purpose. In a report you will want to refer to the exhibit in your text. If you do not intend to comment on the exhibit, then remove it since it serves no purpose. In referring to the exhibit three elements should be covered in the text.

    1. What is the purpose/function of the exhibit?

    2. What did the author conclude from the data?

    3. What evidence is there in the exhibit to support the conclusion?

Frequently, the exhibit is just for descriptive purposes. This purpose can be easily covered in a few sentences such as:

> "The distribution of ages for males and females is shown in Table 1. The data indicate that males tend to be somewhat older than females with a higher proportion of males in the upper age categories."

If the exhibit is included for more substantive purposes a greater amount of attention would need to be devoted to the interpretation and possibly the supporting evidence. On the other hand, remember that you do not need to discuss all the numbers in a display. An attempt to do so is usually a disaster since the reader becomes bored and distracted. You need to discuss only enough numbers so the reader knows why you reached your conclusion.

As an example, if we use the data about faculty member choices for a computer course we could include the following comments in the text:

> "The table gives the results of a questionnaire sent to 137 Public Administration faculty members regarding their choice of topics that should be included in an MIS course. The selections show a clear preference for topics on computer applications. Computer applications were chosen by over 50 percent of the respondents. The topic with the second highest rating, technical subjects, was selected by only 28 percent of the respondents."

Summary

In this chapter the importance of presenting data in a clear, efficient and informative manner was emphasized. Hopefully you realize that there are guidelines and goals to data presentation and observing these recommendations can avoid presenting material that is misleading or poorly understood. Statistical and research methods cover a broad range of topics and many of those topics are difficult to master. However, the ability to display data in an objective and intelligent fashion is a skill that can be mastered by all students.

PROBLEMS

1. Review a newspaper or a magazine and find at least three of the four types of graphs discussed in this chapter. What negative and positive features do you find in each of the graphs?

2. Add titles and headings to the tables you created for problem 6 in chapter 1. Display the data in that table as a bar graph and as a pie graph.

3. Critique the format of the following table:

Response Time for the Metropolitan Fire Department	
Response Time	Number of Calls
<1	2
1-3	12
3-5	14
5-7	21
7-10	16
10-15	21
>15	23

4. For what purpose is the line, bar, pie and XY graph best suited?

5. For the following data set on hourly wages (in dollars) for males and females, construct a table and write a paragraph that could be used in a report on the subject.

	Female Wages				Male Wages	
6.10	6.20	7.40	5.60	7.70	8.60	9.75
8.70	6.00	10.10	5.40	9.00	9.00	8.00
6.30	8.80	7.60	8.20	6.80	9.30	10.20
7.00	5.80	8.30	5.20	8.80	8.10	9.10
9.90	7.80	7.00	7.20	8.90	10.40	
6.80	5.80	8.10	5.20	7.85	9.20	
8.90	6.30	9.20	5.70	5.60	9.15	
5.50	5.40	6.80	5.80	9.40	9.50	

6. Refer to the data below.

STUDENT INITIALS	MID-TERM GRADE	FINAL GRADE
AB	79	83
CD	68	70
EF	91	78
GH	82	75
IJ	88	88
KL	60	73
MN	68	72
OP	82	85
QR	67	72
ST	85	81
UV	74	76
WX	79	78
YZ	81	83

   a. Construct an XY graph.
   b. What can you learn from a review of the data in your XY graph?

7. Refer to the Mid Term grades given in question 6 above.
   a. What is the mean, median and population standard deviation?
   b. If we added one value to the data set so that it increased the mean but did not change the median, what would that value be?

8. Refer to the Final grades given in question 6 above.
   a. What is the range, mode and population variance?
   b. What is the largest value we could add to the data set so that it decreased the mean and lowered the median?

Excel Instruction

Tables

We prepared a one way table for a discrete variable in Chapter 1. What we want to do now is to prepare a two-way table for two discrete variables. To do this we will again use the PivotTable Report procedure. The two variables we will use are Sex and a person's Educational status (i.e., a High School, Bachelor's or Master's degree).

It is best to make a copy of the original database for this exercise. You will need to use the commands discussed in the Preface and Chapter 1. To copy the original database, highlight the complete database (cells A1:N43). Select copy, 🖺; from the Standard Toolbar; click on a new file, 🗋; click on paste, 🖺; and you should have a duplicate of the original database. Using the File /Save As commands give this duplicate file a name – e.g. "Stat3".

To create a two-way table select "Data" from the Main Menu and "PivotTable Report" from the drop down menu that appears. The "PivotTable Wizard - Step 1 of 4" dialog box opens. Be sure the circle for "Microsoft Excel list or database" has a dot in it. Click "Next >" to reach the "PivotTable Wizard - Step 2 of 4" dialog box. The location of the cells that contains the data for our table is requested. In our case we enter B3:N42 in the box for "Range" since this is the range that contains the sex information (Column B) and educational status (Column N) as well as a heading for each variable (SEX and EDUC). There are other variables included (Variables in columns C through M), but their presence won't matter. Click "Next >" and a dialog box for "PivotTable Wizard - Step 3 of 4" appears (see below).

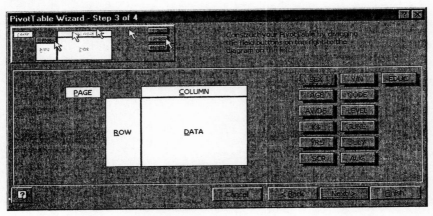

In addition to instructions at the top of the box and a representation of a table layout, there is a series of buttons on the right hand side that correspond to the variables identified in step 2. In this case we only want to work with two of the buttons – SEX and EDUC. We will use the sex categories as rows so place the pointer on the SEX button and drag it to the ROW area of the table. We will use EDUC as the column entry so click on the EDUC button and drag that button to the COLUMN section of the table. These two steps have provided the headings and categories for the rows and columns, but we still need to provide information for the cells of the table. In the dialog box the section called "DATA " is used to identify the statistic that will become the cells of the table. Click on the EDUC button and drag that button to the DATA portion of the box. We want frequency counts and it is important that the gray button in the DATA section states "Count of EDUC". If it doesn't, double click on the gray button in the DATA area and you will be given a choice of statistics that can be calculated. Since we want frequency counts select Count and then click "OK". Your screen should look like this:

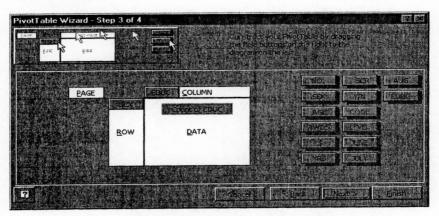

You now click on Next > and you will be at the "PivotTable Wizard - Step 4 of 4". You need to identify a place for your table to appear on your worksheet or have it go onto a new worksheet. Place a dot in the appropriate box (if you chose "Existing worksheet" be sure to enter the cell location). Click "Finish" and there should be this table on the screen:

Count of EDUC	EDUC			
SEX	BS	HS	MS	Grand Total
F	8	8	6	22
M	5	5	7	17
Grand Total	13	13	13	39

To prepare a two way table in which one or both of the variables in the table is a continuous variable you need to convert the raw data of the continuous variable into categories. For example if we want to compare sex and test scores as our fields in a two-way table we will need to convert the test scores to category entries. We will record the converted raw score in column O so we should add a heading in cell O3 for those values. We will give column O the label "CATEGORY" so type that term in cell O3.

Previously we set up categories for test scores - less than 60, 60-69, 70-79, 80-89 and "more" which included scores 90 - 99. We need to assign each raw score a category value. Our first raw score in cell F4 is 55 which fits in the <60 category so type "<60" into cell G4. The score in Cell F5 is 69, which falls into the 60-69 category so type "60-69" in cell G5. You may continue converting all the scores in column F to a category score in column G, but you may prefer a more efficient procedure. If you had many scores to convert it is quicker and more accurate to first sort the scores and then assign categories. To do this, begin by highlighting cells A4 to N42. Next, select the "Data" command on the Main Menu Bar and then "Sort" from the drop down menu that appears. The "Sort" dialog box should be on your screen (see below).

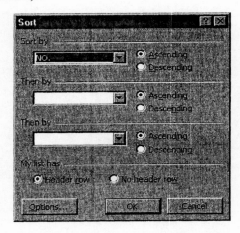

We will want to sort by test score, which is in column F and labeled SCR. At the bottom of the dialog box there is an item that states "My List Has". There should be a dot in the "Header Row" circle and if so, click on the arrow adjacent to the "Sort by" window at the top of the screen. From the listing that appears click on SCR. If, on the other hand, the bottom of the dialog box has a dot in the "No Header Row" circle click on the arrow adjacent to the "Sort by" window. Then select Column G from the options presented. Once "SCR" or "Column G" is indicated in the "Sort by" box, check to be sure the "Ascending"

circle has a dot in it since we will want our sort in ascending order. You execute the sort by selecting the OK button.

With the data set ordered by test score, you can enter the term "<60" in cell O4 and copy it to all the other cells in column G that have a score of <60 (i.e. cells O5 to O9). For the next score in the series, 63.4, enter "60-69" in the CATEGORY column. Copy that expression for all other scores in the 60s. In a similar fashion, you enter "70-79" for scores in the 70s, "80-89" for scores in the 80s and "90-99" for scores in the 90s.

When you have finished assigning the test scores to an appropriate category name, you may begin creating the pivot table by selecting "Data" from the main menu and "PivotTable Report" from the drop down menu. In step 1 of the PivotTable Wizard be sure only the "Microsoft Excel List or Database" circle has a dot in it. In step 2, set the range as A3:O42. In step 3 select SEX as the Row variable and CATEGORY as the Column variable. Also use CATEGORY as the "Data" variable. Be sure that in the DATA area the entry reads "Count of CATEGORY". At this point, your screen should look like this:

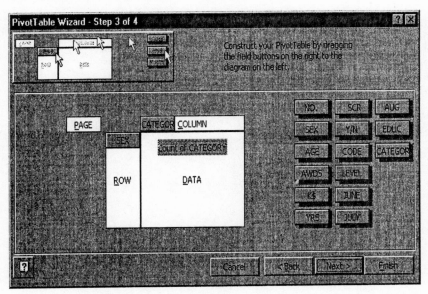

Click the "Next >" button to proceed to step 4. Identify a location for the table (e.g. B46 of the existing worksheet would be a suitable place) and click on "Finish". You should have a display that begins in cell B46 and looks like this:

Count of Category	Category					
SEX	<60	60-69	70-79	80-89	90-99	Grand Total
F	4	2	7	5	4	22
M	2	4	5	3	3	17
Grand Total	6	6	12	8	7	39

To restore the data set we used above to the original ID NO order, highlight cells A4 to O42. Click on "Data" from the Main Menu and select "Sort" from the drop down menu. For the "Sort by" box select either "NO." or "Column A" depending on how the "Header Row" circle is set. Indicate that the sort should be ascending and execute the sort by selecting OK. The data set should now be in order of ID NO.

Figures

To prepare a figure it may be necessary to first organize the data you want to display. An efficient way to organize the data is by first creating tables. We made a pivot table for a single variable, SEX, in chapter 1. Either locate that table or repeat the steps to create a new pivot table for SEX. In our figure, we will want to show the number of females and males so highlight the two cells that contain the appropriate counts (the counts are 22 Females and 17 Males). We will assume for purposes of this explanation that the labels (F and M) are in cells A3 and A4 and that the frequencies are in columns B3 and B4).

We first should identify the data we want presented in the chart. For our example, the data for the sex distribution chart are in cells A3:B4, so highlight those cells and go to the Standard Toolbar and click on the ChartWizard icon, ▉. The first ChartWizard screen, "Step 1 or 4 – Chart Type" shown below should appear).

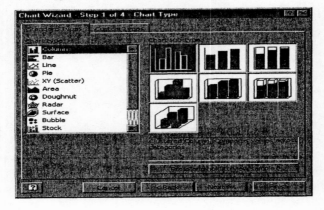

When you clicked on the chart icon you may have gotten the Office Assistant box which is designed to help you develop charts. This explanation will not rely on Office Assistant so you may close the Office Assistant box by clicking the "x" button on the screen.

We use step 1 to choose the type of chart we want. For our purposes a column chart would be appropriate. As shown above you have the choice of different types. If you want to see the form of another type, move the cursor to that type (e.g. Pie) and click. Samples of different pie charts will appear in the Chart sub-type window, which is on the right hand side of the screen. In our case a column chart seems fine, but we still need to select a subtype. The initial type, which is described as "Clustered Column" is suitable for our purposes and we can see how our data will appear on that type of chart by clicking and holding the button that reads "Press and hold to view sample". On the assumption we are satisfied with our choice, we hit the "Next >" button to move to the second step (see below).

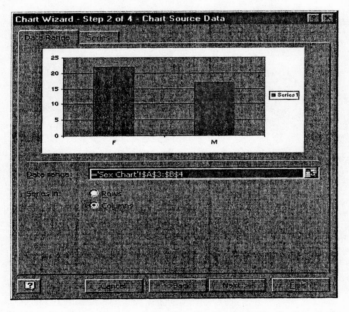

At step 2 we again have an opportunity to identify the range of cells for our chart. If you started the process without selecting that range, this step gives you the opportunity to identify the range. Since we have the correct cell range (i.e. A3:B4) we may proceed to step 3 by hitting the "Next >" button. The other terms and $ symbols in the "Data range" box were automatically created by Excel because we highlighted the cells A3 to B4 before we selected the chart icon. If we had failed to pre-select cells A3:B4, we could, at this stage, simple type in those cell locations.

Step 3 gives us additional charting options (see below). The options for the chart are pre-set by Excel and those settings, referred to as "defaults", produce the chart shown below. However, the defaults may be modified by using the 6 different tabs shown at the top of the step 3 screen (e.g. Titles, etc.)

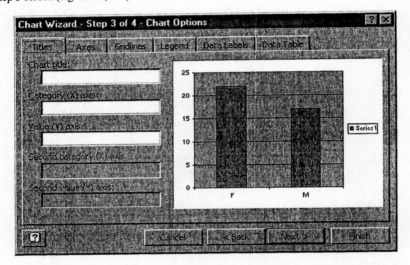

To add a chart title, the heading for the X-axis and the heading for the Y-axis click on the tab "Titles". Type in the appropriate box the title, the Category (X) axis (i.e., the horizontal axis) label, or the Value (Y) axis (i.e. the vertical axis) label. In our case the title could be "Sex Distribution", the X-axis heading "Sex", and the Y-axis heading "Number of Employees".

We will briefly describe what the other tabs are for, but will only show the screens for those tabs that we need to change. You advance to a tab by clicking on the tab name (e.g. for the Axes tab we would click on that tab "Axes" shown in the above screen).

The "Axes" tab allows you to suppress or display the scales for the X and Y-axes. For example if we didn't want the frequency categories (e.g. 0, 5, 10 etc) to appear on the chart, we could delete them using this tab.

The "Gridlines" tab allows you to add or delete the horizontal and vertical gridlines on a chart. For example, the gridlines at the frequency counts of 5, 10 etc. are probably unnecessary and can be eliminated. To delete the horizontal gridlines we click on the "Gridline" tab and the Gridline dialog box appears (see below).

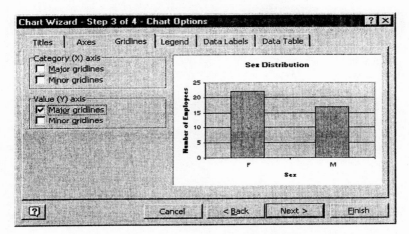

Go to the section labeled "Value (Y) axis". Click on the "Major gridlines" box with a check mark (✓) in it, eliminating the ✓. The gridlines are then removed from the chart.

The "Legend" tab allows you to add, delete or reposition the legend (i.e. the "Series 1" box on the right). When you select the Legend tab, the following screen appears:

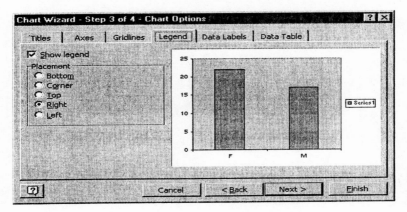

We do not need a legend for our chart so we would click on the box for "Show legend" to change the check mark to a blank box. This action will eliminate the legend from our chart.

The "Data Labels" option allows you to display the value of a data point on the chart (e.g. showing the frequency counts of 22 and 17 above the female and male bars of the chart).

The "Data Table" tab allows you to have the data, used to create the chart, appear immediately below the chart area.

To go to the next step hit "Next >". The "Chart Location" dialog box appears which allows you to place the chart on the current worksheet or on a new worksheet (see below).

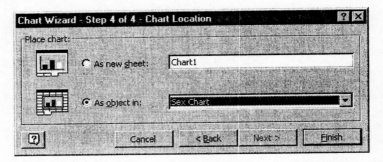

In our case we have plenty of room on the current worksheet so we will place the chart there. Make sure there is a dot in the "As object in" circle and then complete the operation by hitting "Finish". The final chart (see below) will then appear on your worksheet.

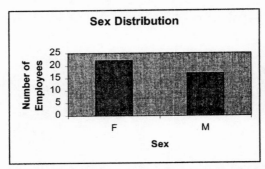

The chart, when it first appears, will have small handles on its edges. You can use these handles to re-size the graph. Place your pointer on one of the handles (the pointer becomes a doubled headed arrow) and you can now move the pointer to change the dimensions of the chart. To reposition the whole chart, you place the pointer on an interior portion other than the plot area. The plot area is the gray section of the chart that contains the two bars. Hold down the left mouse button and your pointer will change to a four-arrow configuration. You can then drag the chart to a new location.

To see how our sex distribution would look like as a pie chart, click on the chart so the handles appear. Then click on the ChartWizard icon on the Standard Toolbar and from the "Chart Type" options, select "Pie". Pick the second chart in the first row as the sub-type and close this screen by hitting "Next >". You should now be at step 2. We should be satisfied with the information on this screen so hit "Next >" again and we'll be at step 3. Use the "Legend" tab to eliminate the legend if it appears on the chart. The only other option we need to change is to add the M and F labels to our chart so click on the "Data Labels" tab. Change the Data Labels option from "None" to "Show Label" (see below).

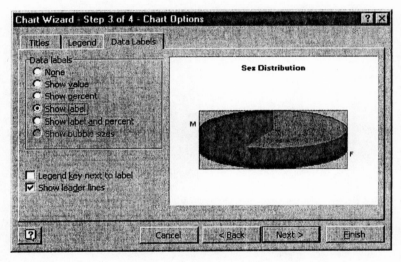

We can now complete our chart by hitting the "Finish" button and the chart shown in the above screen will appear on our worksheet.

We can also prepare line graphs. The PC Proficiency scores would be suitable for a line graph so copy the scores form the original data set to a new worksheet (Columns A, B and C would be a reasonable location for the June, July and August data on the new worksheet).

A graph of the mean proficiency scores for each month would be a good way to display the data. We need to first calculate the means. We learned how to calculate a mean in the last chapter so for the June scores type "=average (A4:A42)" in cell A45. To get means for July and August copy the A45 cell formula to cells B45 and C45. We will want to use the labels June, July and August on our chart so either copy or type those terms in cells A44, B44 and C44 respectively. Next highlight cells A44:C45 (i.e. the 6 cells with the month label and the month means). Next select ChartWizard from the Standard Toolbar. For chart type select Line and for the sub-type select the first chart in the second row. Click "Next >" to go to step 2 and since we do not need to modify the "Data Range" again click on "Next >" to go to step 3. At this screen we should add a title and a heading for the X and Y-axes. "PC Proficiency" for the title, "Month" for the X-axis heading and "Score" for the Y-axis heading are reasonable titles so type those terms in the appropriate boxes. The only other thing we need to do is eliminate the gridlines and the "Series 1" notation. Repeat the actions described for the "Gridline" and "Legend" tabs in the description of how to prepare the Sex Distribution chart and then hit "Finish". You should have the following chart on your worksheet.

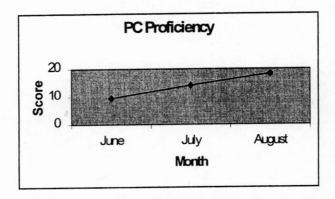

Another type of chart we should know how to produce is the XY or scatter chart. An interesting relationship we might want to look at is the connection between age and pay. To do this, copy the two variables onto a new worksheet. Copy the Age data including the headings to Column A and the PAY data, with its heading, to column B. The "K$" notation in cell B3 should be changed to "PAY" since "K$" will not be a meaningful heading in our chart. Highlight cells A3:B42. Click on the ChartWizard icon. At Step 1 select "XY (Scatter)" as the chart type and for the sub-type select the only chart in row 1. At step 2 be sure the "Data Range" is set at A3:B42. The sequence of the input variables is important (in our example column A is for the age variable and column B contains the pay information). The first variable you enter (Age in our example) will become the X variable - the variable for the horizontal axis. The second variable (in our example the amount an employee is paid) becomes the Y variable - the variable for the vertical axis. At step 3 add a title (e.g.

"Relationship Between Age and Pay", the X-axis heading (e.g. "Age") and the Y-axis heading (e.g. "Pay -in thousands"). We'll retain the gridlines for this chart, but use the "Legend" tab do delete the Pay legend and then click on "Finish". The chart displayed below should appear on your worksheet.

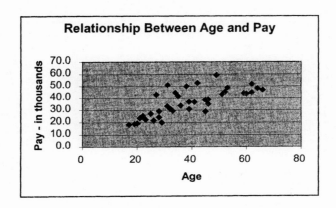

Here are a few exercises for you to practice on using your database.
1. Prepare a table showing sex and advancement potential.
2. Prepare a table showing educational level and the July PC proficiency scores.
3. Represent educational levels as a bar graph and as a pie chart.
4. Prepare an XY graph of test scores and pay.

Chapter 3

Probability

Introduction

You can easily take a statistics course, never deal with probability theory and survive quite nicely. On the other hand, there are programs where entire courses are devoted to the topic. For our purposes it will be useful to spend only a small amount of time on the subject. However, we will accomplish a great deal - we will learn how to interpret a probability, the basic properties associated with a probability, the rules to execute the most common types of probability problems and what an expected value is.

Appreciating probabilities is important because they are frequently the main output of a statistical analysis. As a matter of fact, statistical hypothesis testing is built around making decisions based on probabilities. We will deal with hypothesis testing in a later chapter.

Definition

Perhaps the easiest way to define a **probability** is to say it is the chance that an event will happen. We can expand that notion by adding that the probability of an event is the proportion of times the event will occur in a very large series of identical repeated trials. "Very large" means hundreds or thousands or even millions of repetitions.

Framed as a formula, we get the following proportion:

$$Probability = \frac{Number\ of\ times\ the\ event\ A\ occurs}{Number\ of\ times\ the\ trial\ is\ repeated}$$

Event A can be almost anything - getting a head on a coin toss, the Chicago Cubs winning a World Series, a snow storm occurring on Christmas day. Event A is used in the formula to represent any event that we're interested in - the event for which we want to determine the probability of its occurrence. The word "trial" is a generic term that refers to the activity that produced the event. If the event were getting a head on a coin toss, then the coin toss is the trial. If the event were drawing a white ball from an urn, then the trial is the drawing of the ball.

What if I asked students: What is the probability a 3 will result when we roll one die? Almost all would come up with the right answer - it's 1/6. If we try to use our formal probability formula to explain why students know the answer we would conclude that each student threw a die many many times and found that 1 out of 6 times a 3 turned up. Of course, that explanation is absurd. Even if people threw a die 600 times it's most unlikely that exactly 1/6th of the throws would turn out to be a 3. After 600 throws they might have a result around 100, but there is only a .02 probability that they'd get exactly 100 3s.

So how is it so many people know the correct answer? We know the correct answer by our ability to reason - we use our knowledge of the structure of a die. A die has six sides and only one of those sides is a 3. We can call this an **a priori probability** - we know it in advance. In addition, there is also a class of probabilities that can be called **posterior probabilities**. Posterior probabilities rely on past experience to determine the likelihood of an event. For example, insurance companies keep track of how long people live and they can therefore calculate the probability of death for different groups of people. Clearly the premium an insurance company charges customers, for the amount of life insurance the customers want, is dependent on the probability of their dying within a certain number of years.

Probability Properties

There are two important properties of a probability - learn them and they'll keep you from making some mistakes, or at least they will warn you that you have an incorrect answer.

1. The numeric limits of a probability are 0 to 1.
    A probability of 0 means the event never occurs. A probability of 1 means the event always occur. An example of a 0 probability could be the event your next class will be taught by Abraham Lincoln, the ex-President. An example of a probability of 1 is the likelihood that the sun will rise tomorrow morning.
2. The sum of all the probabilities of all possible events is equal to 1.
    There are six events or outcomes possible when we roll a die. The probability of each outcome is 1/6. The sum of the probabilities of the six events is 1.
    There is a valuable corollary to this rule, which applies to two-event situations. A two-event situation means that there can only be two outcomes - e.g. a head or a

tail with a coin toss. A 3 or not a 3 when rolling a die is another example. The corollary to property 2 says that, when faced with a two-event situation, once you know one of the probabilities, you automatically know the other probability. All you have to do is subtract the known probability from 1 to get the unknown probability. In term of a formula we have:

Probability of event B = 1 - Probability of event A

If the probability of failing an exam is .23 than the probability of passing the exam is 1-.23 or .77. The value of this corollary may seem insignificant, especially when dealing with events like a coin toss where it's easy to figure out the probability of a head and a tail turning up. However, it may be useful to quickly determine the probability of not getting a 3 on a roll of the die. We could add the probability for the other five outcomes to get that answer, but it's quicker just to subtract 1/6 from 1 to find the answer of 5/6.

Calculating Probabilities

There are four types of probability problems that will be presented and for each problem type there is a way to set-up the required calculations. We can think of these set-ups as rules. For each rule there is a formula which provides a solution to a problem. One set of problems involve **OR probabilities** - What is the probability of event A or B happening? When referring to this class of probabilities the letters, "OR" will be capitalized for emphasis. The term "or" means "either one" - i.e., either event A happens or event B happens. Just the addition of the word "either" helps reinforce what is wanted when presented with an OR problem - What is the probability of getting either event A or event B?

In writing a formula the following notation will be used:

P(A) = Probability of the event A
P(B) = Probability of the event B

Thus the expression P(A or B) means the probability of either the event A or the event B occurring.

Simple Addition Rule

The simple addition rule says that the probability that either event A or event B occurs is equal to the probability of A plus the probability of B. However, there is an additional clause which states that you may simply sum the two probabilities providing the events are mutually exclusive. The term **"mutually exclusive"** needs to be defined. It means that the events stated in the problem can not occur together. You can not be at home and driving your car at the same time - those events can not exist together and are mutually exclusive. But you can be at home and sleeping in a bedroom since those events are not mutually exclusive. Both those events can exist together - you can be in both places at the same time. The formula for the simple addition procedure is:

$$P(A \text{ or } B) = P(A) + P(B)$$

Assume that the probability of your being in New York at a particular point in time is 5/24 and the probability of your being in Detroit is 3/24. A question could come up - what's the probability you're in New York or Detroit? To apply the simple addition rule the events must be mutually exclusive. Can you be in Detroit and New York at the same time? Since the answer is "no" the events are mutually exclusive and you can apply the simple addition rule:

$$P(A \text{ or } B) = \frac{5}{24} + \frac{3}{24} = \frac{8}{24}$$

The fraction 8/24 simplifies to 1/3 which can be converted to the proportion .33. When working with probabilities that are expressed as a fraction, you'll have maximum accuracy if you retain the fractions in the calculation. The exception would be when the fraction easily converts to a specific proportion. For example, 1/2 converts to exactly .50, but 1/3 has no precise equivalent as a proportion. In general, it's also more accurate to express the answer as a fraction as well. However, sometimes a fraction is harder to interpret than a proportion. For instance, the fraction 352/551 would be more readily understood as its equivalent proportion of .64. There is no hard and fast rule on when to choose the fraction or the proportion format for your answer. It probably is best to decide on the basis of what form is more understandable to your reader.

If you select the proportion format, a question often comes up on how many decimal places to show in your answer. Check to see how many decimal places are in the set of numbers you began with. Your answer may then contain one more decimal place. However, you still need to apply common sense - if the answer is very small (e.g. less than .00) you may need to add more decimal places so the reader has a better idea of the significance of the result. On the other hand if the number is very large (e.g. 2,567,328.3) there is no need to include the decimal place.

You can extend the simple addition rule to cover three events by just adding the third event to the equation. Assume the probability of your being in Chicago is 2/24. We'll call your presence in Chicago event C. We also know that event C is mutually exclusive of both event A and event B. We can apply the simple addition rule and, after incorporating event C, we get:

$$P(A \text{ or } B \text{ or } C) = P(A) + P(B) + P(C)$$

Our answer is the sum of the three probabilities:

$$P(A \text{ or } B \text{ or } C) = \frac{5}{24} + \frac{3}{24} + \frac{2}{24} = \frac{10}{24} = .42$$

General Addition Rule

The simple addition rule works fine as long as the events are mutually exclusive. But what do we do if the events are not mutually exclusive? Let's examine a problem. What's the probability of selecting a student at random from the university's enrollment list who is either a male or a graduate student? Assume the probabilities are .50 for male and .60 for graduate student. If we applied the simple addition rule we'd add the two probabilities and get 1.10. This obviously is wrong since it violates the property that the sum of the probabilities can not exceed 1. The problem is that being male and being a graduate student are not mutually exclusive. A person can be both male and a graduate student. We need a rule for problems that involve OR events that are not mutually exclusive. The procedure that fits this situation is called the general addition rule. The formula is:

$$P(A \text{ or } B) = P(A) + P(B) - [P(A) * P(B)]$$

The * symbol is used to indicate multiplication - i.e., multiply the probability of A times the probability of B.

Note that what's happened is that a new term, $[P(A) * P(B)]$, was introduced into the simple addition formula. Mathematically the general addition rule says add the probabilities of the A and B events and then subtract the product of the two probabilities. If we add the probability for events A and B (.50+.60) we get 1.10. The product of the two events is .30 (.60*.50) and when we subtract that value from 1.10 the result is .80. The answer to the question "what's the probability of selecting a student, at random, from the university's enrollment list who is either a male or a graduate student", is .80.

If we take all the people classified as male and add them to all the people who are classified as graduate student, we count the people who are both male and graduate students twice. They were counted when we brought in the male group and they were counted a second time when the graduate student group was included. The sum of the two events, male and graduate student, is 1.10. We need to reduce that figure by those who were counted twice. The .30 is the proportion of people counted twice so that value is deducted from the 1.10 figure. The result (1.10 - .30) gives us the proportion who are either male or a graduate student (.80).

To see if you have understood how to use the probability procedures discussed above, answer the following questions.

Suppose you work on the second floor of the building that houses your organization. There is one copying machine on the first floor and one on the third floor. The number of people and the functions performed on each floor differ so the probability that a copier is free varies. The probability that a machine is free is:

Floor	Probability
1st	.42
3rd	.30

What is the probability that you will find a free copier on either the first or third floor?
(The space below is provided for you to record your answer)

Your answer should be .594. This is an OR probability question so we can use one of the addition probability rules. To determine whether to use the simple or general rule, you must decide if the events are mutually exclusive. Since you can have a copier free on any floor, the events are not mutually exclusive and you must use the general addition formula:

$$P(A \text{ or } B) = P(A) + P(B) - [P(A) * P(B)]$$

If A is the event: there is a copier free on 1st floor, and B is the event: there is a copier free on 3rd floor, the formula becomes:

$$P(A \text{ or } B) = .42 + .30 - [.42 * .30]$$
$$P(A \text{ or } B) = .72 - .126$$
$$P(A \text{ or } B) = .594$$

Here is another problem for you to solve:
      Assume that the probabilities in the table above refer to the chance that your boss is visiting staff on the different floors. What is the probability that you will find your boss on the first or the third floor?
In this case your answer should be .72. Again this is an OR probability question, but the two events are mutually exclusive. Your boss can not be on the first and third floors at the same time. Thus, you need to use the simple addition formula to solve this problem:

$$P(A \text{ or } B) = P(A) + P(B)$$

If A is the event, boss on 1st floor, and B is the event, boss on 3rd floor, then the formula becomes:

$$P(A \text{ or } B) = .42 + .30$$
$$P(A \text{ or } B) = .72$$

Simple Multiplication Rule

We have so far worked with OR events. It is also important to be able to calculate AND probabilities. An **AND probability** means all events specified must occur - i.e., event A must happen <u>and</u> event B must happen. To solve this class of problems we use the simple multiplication rule <u>providing</u> the events are independent. To be **independent** the outcome of event A can not influence the event B outcome. In other words the occurrence of event A does not change the probability of event B occurring.

For example, what's the probability of getting two heads when we flip a coin twice. Most students know the answer - it's ¼, or .25 when expressed as a proportion rather than a fraction. How do they know that? In some cases it's just part of a person's memory - the answer was learned at some point in time and it stuck.

A probability tree is a good way to understand what is happening. A probability tree for the problem we're looking at is shown below.

<u>Flip 1</u>	<u>Flip 2</u>	<u>Result</u>
	H	HH
H		
	T	HT
	H	TH
T		
	T	TT

Note that on the first flip there can be only two outcomes - a head or a tail. If the first flip resulted in a head, then on the second flip there can only be another head or a tail. If, on the other hand, the first flip turned out to be a tail, then the second flip can be either a head or a second tail. We see that there can only be four possible results - two heads, a head followed by a tail, a tail followed by a head or two tails. The result we are interested in is HH (two heads) which is one of the four possibilities - hence we say the probability of HH is 1/4 or .25.

There is another way to get the answer to our problem - we can use the formula for the simple multiplication rule. That formula is:

$$P(A \text{ and } B) = P(A) * P(B)$$

Remember, this formula is only useable when the events A and B are independent. In our example the probability of a head on the first toss is 1/2 or .5 and that is also the probability of a head when we toss the coin a second time. Substituting the probabilities into our formula gives us .5 * .5 = .25. Note that the two events are independent. Getting a head on the first toss has no influence on our second toss - there is still a .5 chance of getting a head.

An example of two events that lack independence is drawing balls from an urn when the balls are removed after they are drawn. Imagine an urn with five white and five black balls. We want to draw a first ball (event A) and then a second ball (event B). We will not return the first ball to the urn. How would we go about determining if selecting a white ball on the first draw influenced picking a black ball on a second draw? The probability of getting a white ball on the first draw is 5/10 or .50. There are 5 white balls out of a total of 10. Now when we go to make our second draw the urn contains only 9 balls - four white and five black. The first draw has influenced our second draw by reducing the number of white balls from five to four. In this case we say we lack independence since the first event has influenced the second event. Note that had we replaced the first ball before drawing the second ball, we would have had independence. The urn, under that arrangement, would contain the original five white and five black balls. In a sense we restored the urn to its original state by replacing the white ball.

We can extend the simple multiplication rule to more than two events by just including the new events into the equation. The new events must be independent of each other and the original events. For example what's the probability of flipping a coin three times and getting three heads. We'll let the third toss represent event C. We need to expand our formula by adding event C and we get:

$$P(A \text{ and } B \text{ and } C) = P(A) * P(B) * P(C)$$

Substituting the .50 probability for getting a head our formula reads:

$$P(A \text{ and } B \text{ and } C) = .50 * .50 * .50 = .125$$

We can represent .125 in its fractional form of 1/8. If you expand our probability tree for one more event (flip 3) you will also see that a HHH result is one of the eight outcomes that occurs.

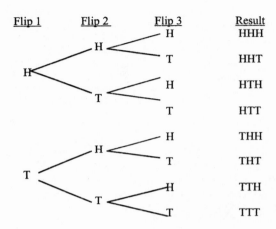

Flip 1	Flip 2	Flip 3	Result
	H	H	HHH
		T	HHT
H		H	HTH
	T	T	HTT
	H	H	THH
		T	THT
T		H	TTH
	T	T	TTT

Sources of Confusion

What's not always easy to understand for some people is the fact that a HHH sequence has the same probability as a HHT sequence. They both occur 1/8 of the time. The expanded probability tree shows the 8 outcomes that occur when a coin is flipped three times. Note that the sequence HHT is one of the eight possibilities just as HHH is one of the eight possibilities. Hence both outcomes have the same 1/8 probability.

You can determine the size of a probability tree, and hence the number of possible outcomes, by a simple calculation. By size, I mean, how many different kinds of results do you end up with in your tree. In the tree above we have 8 different results. To find the size of a tree first determine how many outcomes the variable you're interested in has. In the coin flip above there are 2 possible outcomes (a head or a tail). Then note the number of replications you're using. Replications means the number of times you're repeating the trial. In the example above we had 3 trials - i.e., we repeated the coin flip 3 times By convention the number of outcomes is called n and the number of trials or replications is called r. The formula for the size of a probability tree is:

$$\text{Number of Possible Outcomes} = n^r$$

The superscript r is called an exponent and means you start with the base number n and you multiply it by itself r times - i.e. n*n*n when r = 3. In our example the base number n is 2 (the 2 outcomes) and the r is 3 (the 3 replications) so our calculation becomes:

$$2^3 = 2*2*2 = 8$$

Frequently the expression "n raised to the power r" is used for the notation n^r.

We can take another example of how this elementary formula can be useful. If we had a red, white and blue ball in each of two urns and wanted to know how large the probability tree would be if we drew a ball from each urn we'd have n=3 (the number of possible outcomes) and r=2 (the number of replications) so our answer would be:

$$3^2 = 3*3 = 9$$

A word of caution - n and r will be used in other formulae that will be presented in the next chapter. The n and r may have slightly different definitions. It's a bit confusing, but it's better to use the conventional notation - there are already too many variations in notation used in statistical texts, which only make understanding the discipline more difficult.

To avoid another frequent source of confusion, visualize two coins - coin A and coin B. Coin A has never been flipped and coin B has been flipped seven times in a row and each of those seven flips has produced a head. What is the probability of a head when coin A is tossed? It's 1/2 or .5. What's the probability of a head when coin B is tossed? It's also 1/2 or .5. Coins have no memory - previous outcomes are irrelevant when flipping a coin. Examine the first two results in our probability tree - the HHH and HHT outcomes. For these results the first and second coin tosses were both heads giving us an HH sequence at that point in the exercise. Now the third toss can be either a head or a tail and they are equally likely. If coins could remember their history, things might be different, but coins don't recall what they did in the past. The expression gambler's fallacy is used for people who believe that a tail is more likely than a head after a series of coin flips has ended up heads. These people think that a change is more likely to balance out the 50/50 outcome inherent in a coin toss.

It is true that a long run of heads (or tails) is very unlikely before the fact, but in that case we are dealing with a series of tosses, not just one toss. In working on probability problems it's important to clarify the difference between a single event and a series of events. Note that when we ask for the probability of a single event, say the third coin toss, the probability of a head is 1/2. Even when the question is phrased - what is the probability of a head on a third coin toss after the first two coin tosses ended up heads - we are only interested in a single event; getting a head on the last coin toss. When we ask what is the probability of getting three heads in a row, a HHH series, we ask a different question. The probability of the sequence HHH is 1/8. We've asked two different probability questions and quite properly have two different answers. In the first question the first two tosses were history - they don't influence the third toss and therefore we ignore them when we calculate our answer. We need only to calculate what is the probability of a head on a single toss. In the second circumstance we are asking a question involving three tosses. We have to look at the whole probability tree rather than just isolate out the last toss.

To work probability problems you must carefully clarify the question for which an answer is sought. In our examples above, we talk about 3 tosses in both cases. However, in the first case we're only interested in the probability of a head on toss number 3. The first two tosses are <u>not</u> relevant parts of the probability question. In the second case we ask - what is the probability of getting 3 heads in a row. All 3 coin tosses are part of the probability question.

In this particular example, you may see the formula P^n in place of the simple multiplication formula where P stands for the probability of the event and n for the number of repetitions. What is the probability of getting four heads in a row? Using this formula, we know the probability of a head is .5 so that value becomes P. Since there are to be 4 repetitions, n becomes 4. Substituting in our formula gives us:

$$.5^4 = .5*.5*.5*.5 = 0625.$$

The probability of getting 4 heads in a row is .0625.

General Multiplication Rule

We now turn to our last rule - the general multiplication rule. The general multiplication procedure is used for AND probability problems in which the events lack independence. The formula is:

$$P(A \text{ and } B) = P(A) * P(B \mid A)$$

The expression "P(B|A)" means the probability of event B, given that event A has occurred. In other words, what is the probability of the second event (i.e. B) after the first event (A) has occurred. This expression is also considered a **conditional probability** since the outcome B is dependent on what happens with event A. Take an urn example in which there are 5 white and 5 black balls. Balls are not replaced after they have been drawn from the urn. Our goal is to determine the probability of getting a white ball followed by a black ball. We can rephrase that goal as follows: What is the probability of 1] picking a white ball and 2] then picking a black ball, remembering that the white ball was removed from the urn? We arrive at the probability for picking the black ball by noting that there are nine balls left in the urn after the first white ball is picked. More precisely, there are four white balls and five black balls left in the urn. The chances of getting a black ball on the second draw therefore becomes 5/9. That fraction, 5/9, is the probability of B given A. The answer to our question, about the probability of pulling a

Statistical Methods for Researchers

white ball and then a black ball from an urn containing five white and five black balls, requires the general multiplication formula presented above because the events are <u>not</u> independent. Event A changes the probability of event B. Before event A takes place the probability of event B, drawing a black ball, is 5/10. However, after event A occurs the probability of event B changes to 5/9.

The probability of event A, pulling a white ball from the earn containing the ten original balls, is 5/10. As discussed above, the probability of the second event (pulling a black ball) after a white ball was taken on the first draw is now 5/9. Thus our probability calculation is:

$$P(A \ and \ B) = \frac{5}{10} * \frac{5}{9} = \frac{25}{90} = .28$$

We can take a new urn problem and use it to show another example of how the general multiplication rule works. Let's place two red, three white and five blue balls in an urn. What's the probability that we will draw a red ball and then draw a blue ball? The probability is to be determined with the understanding that the first red ball is not returned to the urn. Since the first ball was withheld, we have a situation that lacks independence and we must use the general multiplication rule.

Event A, the probability of picking a red ball, is 2/10 - there are two red balls and ten balls altogether. The probability of event B|A is 5/9 - there are five blue balls, three white balls and a red ball still in the urn. Since we removed a red ball on the first draw only nine balls are left in the urn. We now multiply the two probabilities to get our answer:

$$P(A \ and \ B) = \frac{2}{10} * \frac{5}{9} = \frac{10}{90} = .11$$

It's also possible to see what is happening when we have an AND probability problem where independence is lacking by use of data presented in the following table:

Age	Gender		
	Male	Female	Total
<20	15	35	50
21 - 40	85	65	150
>40	30	20	50
Total	130	120	250

Assume the question to be answered is: Based on the data in the table, what is the probability of selecting a person who is less than 20 years old and is a male? We'll define event A as being less than 20 years old and event B as being male. The problem calls for the general multiplication procedure and requires us to first determine the probability of event A - the probability that you select someone who is less than 20. That probability is 50/250 or .20. There are 50 people less than 20, out of the 250 people in our data base. The conditional probability, B|A, is the chance of selecting a male given that you have already selected a person in the less than 20 year age category. The word "given" in this example means restrict your area of interest to the data in the <20 row of the table. That portion of the table is given below.

Age	Gender		
	Male	Female	Total
<20	15	35	50

Note that there are 15 males among the 50 people who are less than 20. The probability for B|A is therefore 15/50 or .30. Using the general multiplication rule and substituting the probabilities we just derived gives us an answer of:

$$P(A \text{ and } B) = .20 * .30 = .06$$

It doesn't matter what we define as the A or B event - we still get the same answer. If the A event is selecting a male and the B event is being less than 20, we find that the probability of selecting a male is 130/250. The probability of selecting a person less than 20, given that the person is a male, confines us to the second column of the original table - the column showing the distribution of males.

Age	Male
<20	15
21 - 40	85
>40	30
Total	130

What is the probability of selecting a male less than 20 years of age? As shown above that probability is 15/130 and is our B|A probability. Thus using the general multiplication formula, and substituting the probabilities we just identified, gives us an answer of:

$$P(A \text{ and } B) = \frac{130}{250} * \frac{15}{130} = \frac{15}{250} = .06$$

It may be worthwhile to point out that there is a link between the name of the rule and the kind of arithmetic involved. Both rules that have the term "multiplication" in their

Statistical Methods for Researchers

title use that operation (i.e. multiplication) in their execution. To a large extent the same is true for the first two rules we examined which have the term "addition" in their name. The simple addition rule uses only addition and the general addition rule uses addition, but also subtraction and multiplication.

As a check on how well you understand calculating probabilities using the multiplication rules we'll work a few problems.

Assume in your neighborhood that there is a raffle for free dinners at a local restaurant. Two free dinners are awarded. You buy 20 tickets and a total of 250 are sold. The raffle is conducted so that after the first winning ticket is selected, it is discarded and then the second drawing takes place. What's the probability that you will win both free dinners (i.e., one of your 20 tickets is picked on the first drawing and one is picked on the second drawing)?

Your answer should be 380/62,250 or .0061. This is an AND probability situation since you need to have one of your tickets picked at the first drawing *and* a second ticket picked at the second drawing. You now must decide if the two events (first drawing and second drawing) are independent. Does the outcome of the first drawing affect the second drawing? Yes it does. Note that the first winning ticket is discarded. That means the second drawing will have one less ticket and the odds have to change. Thus, this situation lacks independence. In this case you need to use the general multiplication rule. Event A is drawing one of your tickets on the first drawing. The probability of that occurring is 20/250. You bought 20 tickets and 250 were sold. Event B is drawing one of your tickets in the second drawing. What you need to determine is the probability of B given A (i.e. P(B|A).

Event A has reduced the number of your tickets by 1 and the total number of tickets by 1 as well. Consequently you now have only 19 tickets in drawing 2 and there is a total of 249 tickets in that drawing. Thus, the B|A probability is 19/249. Substituting the correct probabilities into the formula gives us:

$$P(A \text{ and } B) = \frac{20}{250} * \frac{19}{249} = \frac{380}{62,250} = .0061$$

What if the people running the raffle change the rules - they allow the first winning ticket to be replaced prior to the second drawing. Now what is your probability of winning two dinners?

If your answer is 400/62,500 (or .0064) you've done very well. In this latter example note that replacing the ticket means that the first drawing and second drawing are identical - there are the same number of winning tickets for you in both drawings and the total number of tickets in each drawing also remains unchanged. The first drawing now has no affect on the second drawing - the events are independent. We therefore can use the simple multiplication formula which is:

$$P(A \text{ and } B) = P(A) * P(B)$$

The probability of event A, drawing a winning ticket the first time is 20/250. The probability of event B is also 20/250 so our solution becomes:

$$P(A \text{ and } B) = \frac{20}{250} * \frac{20}{250} = \frac{400}{62,500} = .0064$$

Decision Tree for Selecting AND and OR Probabilities

A decision tree will serve as a good summary of our exploration into probability theory. When presented with a probability problem, two decisions are required:
1] Is the problem an OR problem or is it an AND problem?
2] The second decision looks at the relationship between the events. In an OR problem, are the events mutually exclusive? In an AND problem, are the events independent?

By answering these questions we are led to the correct rule (and formula). The corresponding decision tree is presented below:

Type of problem	Relationship between events	Procedure	Formula
OR	Mutually exclusive	Simple Addition	$P(A)+P(B)$
	Not mutually exclusive	General Addition	$P(A)+P(B) - [P(A)*P(B)]$
AND	Independent	Simple Multiplication	$P(A)*P(B)$
	Not Independent	General Multiplication	$P(A)*P(B|A)$

Word Problems

Many people who are very good at solving problems using formulae still have a fear of word problems. That fear is reasonable - converting the words so the problem can be described in the form of a mathematical equation can be difficult. At least in the text book world, always remember that the word problem you are given fits one of the formulae you have learned. The trick of course is to discover what formula to use. For example we have learned to solve an OR probability as well as an AND probability problem. We have also worked with word problems and converted the words into the appropriate mathematical problem. The decision tree given above helps us through the decision making process. Nevertheless, the word problem can be presented a bit more obscurely than the problems we have worked on thus far. Suppose the problem is worded as follows:

What is the probability of getting two heads when a coin is flipped two times?

Note that neither of the key words, "or" nor "and", appear in the problem statement. Thus it is up to you to figure out what kind of problem it is. Visualize the outcome - it's the sequence: Head-Head. It means a Head followed by a Head. This latter statement could also be expressed as a Head on the first flip <u>and</u> a Head on the second flip. The translation tells us we have an AND probability problem which we can now solve. What is the answer?

Your answer should be 1/4 or .25 since this is an AND probability with independence (what happens on the first flip has no impact on what will happen on the second flip) and the formula is:

$$P(A \text{ and } B) = P(A) * P(B)$$

Substituting 1/2 for P (A) and 1/2 for P (B) gives us the answer which is 1/4.

Word problems can also combine two or more rules into one problem. Here's a typical example:

What is the probability of getting either a 1 or a 2 on the first roll of a die and a 5 or 6 on the second roll?

Work through these types of problems slowly. Begin by looking for the individual parts that make up the problem. The first part (a 1 or 2) is clearly an OR probability, but note there is also a second part (a 5 or 6) which is also an OR probability. Also observe that the two parts are joined by the word "and" making the whole problem an AND probability. The procedure to solve the problem requires that we begin by determining the probability for each part first. The probability for a 1 or 2 is an OR probability in which the events are mutually exclusive so we'd use the formula:

$$P(A \text{ or } B) = P(A) + P(B)$$

Since the chance of getting a 1 is 1/6 and the chance of getting a 2 is 1/6 we substitute 1/6 for P (A) and 1/6 for P (B). The result is 2/6 or 1/3. Why don't you determine the probability for the second part - a 5 or 6.

You should also have gotten 1/3 as your answer since the logic is the same for a 5 or 6 as it is for a 1 or 2.

The last step requires us to use the AND probability formula for independent events because what we get on the first roll of a die is independent of what we get on the second roll. The formula is:

$$P(A \text{ and } B) = P(A) * P(B)$$

and substituting 1/3 for P (A) and 1/3 for P (B) gives us the answer 1/9 or .11.

It should also be pointed out that the four procedures we learned are very useful, but they do not cover all types of probability problems. We learned how to use the four types to solve a problem if the specific sequence was defined. For example if the question was

What is the probability of getting 2 heads followed by a tail when a coin is flipped 3 times?

We know the outcome is Head-Head-Tail, a specific sequence. In this case we apply the AND probability formula for independent events to get our answer (the answer is 1/8). But what if the question was:

What is the probability of getting 2 heads and 1 tail when a coin is flipped 3 times?

Note that this latter question does not specify a sequence. There has to be only 1 tail, but it can occur on the first, second or third flip. The order was specified in the first question - the tail occurred on the third flip. But order doesn't matter in the second question - the tail can occur on any flip. To solve this latter problem we need to learn a few more procedures which will be covered in chapter 4.

Expected Values

An interesting application of probability theory is the notion of an expected value. To understand this concept we will use an illustration all people are familiar with. In a game of chance there are multiple outcomes: you place your bet on one of the options and if it occurs, you win a certain amount of money. If the outcome you picked doesn't occur, you lose your wager. For example, we can make coin flipping a game of chance. You can bet on a head or tail coming up, and if you are right, you win some money. If you're wrong, you will lose whatever you bet. For instance, the game could be constructed so that it costs $1 to play and, if you guess the right outcome, you receive $2. If you make the wrong choice, you receive nothing. An **expected value** is the average value of the option you select when the game is repeated indefinitely. The coin flipping game where we bet on a head occurring can be presented in the following manner:

Outcome	Probability	Payoff
H	.5	$2
T	.5	0

The **outcome value** of each option is its probability times its payoff amount. This relationship, expressed as an equation, is:

$$\text{Outcome Value} = \text{Probability} * \text{Payoff}$$

The sum of the outcome values is the expected value of the wager. This relationship, expressed as an equation, is:

$$\text{Expected Value} = \Sigma \, (\text{Outcome Values})$$

Expanding our table, we can add the outcome values and obtain the expected value (see next page).

The sum of the outcome values is $1 and that is our expected value of the game. In our example, since it costs $1 to play the game, and the expected value is $1, we have a **break-even** situation. In the long run you will end up with no profit or loss in this game.

Outcome	Probability	Payoff	Outcome Value
H	.5	$2	$1 (.5*2=1)
T	.5	0	0 (.5*0=0)
Sum	1.0		$1 ◄————— Expected Value

We can be a bit more realistic since there would be little reason to sponsor or organize such a game - the organizers would show no profit for their troubles. Let's set up a game where the digits 0 to 9 are written on tennis balls and one ball is then randomly chosen. If you guess the number of the ball chosen you win, otherwise you lose. It costs $10 to play this game, but if you win you collect $90. Let's say you pick the number 5 as the likely winning number. Diagrammatically our game is displayed below:

Outcome	Probability	Payoff	Outcome Value
1	.1	$0	$0 (.1*0=0)
2	.1	0	0 (.1*0=0)
3	.1	0	0 (.1*0=0)
4	.1	0	0 (.1*0=0)
5	.1	90	9 (.1*90=9)
6	.1	0	0 (.1*0=0)
7	.1	0	0 (.1*0=0)
8	.1	0	0 (.1*0=0)
9	.1	0	0 (.1*0=0)
0	.1	0	0 (.1*0=0)
Sum:	1.0		$9 ◄————— Expected Value

The expected value in this game is $9. Note that the expected value is a special kind of average - it is what happens after many repetitions. In spite of the name, the expected value does not refer to what we expect in any one game. In no single game do you win $9 - you either win $90 or you win nothing. The expected value refers to what we will get, on average, after many many games. Furthermore, since a game costs $10 but the expected value is only $9, we can see that in the long run we lose money playing this game. On average the loss is $1 per game played. This example resembles a very simplified lottery. Lotteries have become a very popular endeavor among a number of states. The state lotteries are more complex, but the approach is essentially what we've described here.

We could ask the question - why do people play a game that in the long run results in a loss? The explanation is based on how people perceive the value of money. Imagine a lottery where the cost of a wager is $1 and there is a single winner who receives one million dollars. However, two million people participate and under that circumstance the expected value of your $1 wager is 50 cents (Since two million people participate you have a .0000005 chance of winning. You win one million dollars and one million times .0000005 = .50.) In the long run you lose - in fact your expected loss is 50 cents every time you play. So why would people play? For many people $1 is expendable - lose it and it has no effect on their standard of living. However, winning one million dollars would be a fantastic event in their lives and even though it is very improbable, they are willing to take the chance.

We can use our knowledge of expected value to deal with more pragmatic issues. The issue could involve such things as selling off government properties to private investors. Let's look at a hypothetical case. A decision is required on whether to invest $12,000 in order to proceed with the sale of four properties owned by the city. How many properties will be sold is problematical, but with the assistance of reliable real estate investors the likelihood of the sales is obtained. The gain or loss from the sales, excluding the up-front $12,000 investment, is also determined and is shown in the table below.

Properties Sold	Probability	Net Gain/Loss
0	.10	-34,000
1	.30	-16,000
2	.40	+15,000
3	.15	+47,000
4	.05	+86,000

Although the properties are identical, the basic costs of selling means there could be a loss if too few are sold. Should you proceed with the $12,000 investment at this point or discontinue the effort to sell the properties?

An expected value table can be constructed to help with this decision. Converting the above data to an expected value table format results in the following table.

Properties Sold	Probability	Net Gain/Loss	Outcome Value	
0	.10	$-34,000	$-3,400	
1	.30	-16,000	-4,800	
2	.40	+15,000	+6,000	
3	.15	+47,000	+7,050	
4	.05	+86,000	+4,300	
Sum	1.00		$+9,150	◄── Expected Value

Note that based on this analysis the investment of $12,000 would appear unwise because the expected payoff is only $9,150.

What if we eliminate the need to invest the $12,000; how does that change our decision? In that case all we need to do is look at the expected value - it's plus $9,150 and rationality says go ahead with the sales. However, even in that case, some people may decide not to proceed. Note that 40 per cent of the time you will suffer a loss. If you sell no property or only one property the payoff is negative (-34,000 and -16,000 respectively). For some investors the risk of losing money, especially money that has been given to them as a public trust, is so frightening that they will not take that risk in spite of the long term expected gain. The term **risk averse** is used to refer to people who act in this fashion.

This latter scenario is useful to reinforce the fact that an expected value is a quantitative issue. In an earlier example, we illustrated a case where people may participate in a lottery even though the quantitative answer shows it is a bad choice. Here we may have a case of refusing to participate in a program where the mathematical results say it's a good investment. Our behavior in real life may not be consistent with quantitative rationality, and that's all right. However, we should be aware of the quantitative facts before we make our decisions.

Understanding the quantitative elements of a decision gives us a more informed and therefore stronger position when we eventually decide what to do. If we decide to go against the quantitative answer we need to develop a rationale for the apparent contrary stand we are taking. In our last example, it would probably be ineffective to argue we were not supporting the property sales because we are risk averse. The public embarrassment of losing money 40 per cent of the time might make a more tolerable argument. However, in this problem and in many other problems such as this, there are uncertainty elements, which are represented as definitive data. For example, the likelihood values cited are probably not that precise. If we simply reduce the probability for selling 3 properties by 5% (i.e. from .15 to .10) and 4 properties by 5% (i.e. from .05 to 0.0, and add 5% to the probability of selling 0 or 1 property (so they become .15 and .35 respectively), we get the results shown in the next table.

Properties Sold	Probability	Net Gain/Loss	Outcome Value
0	.15	$-34,000	-5,100
1	.35	-16,000	-5,600
2	.40	+15,000	+6,000
3	.10	+47,000	+4,700
4	.00	+86,000	0
Sum	1.00		0 ← Expected Value

Observe that the $9,150 gain is lost, after making only a rather small change in the figures.

The purpose here is not to discredit using a quantitative approach because of uncertainty. Instead we should be stimulated to do better at getting reliable information and be willing to introduce different scenarios to see how the answer changes. To see how well we understand expected values find the answer to the following problem.

You are a supervisor in an agency, which is undergoing an audit. A sample of cases is to be drawn by selecting one of the case workers who work for you and reviewing all the cases in that person's case load. There is a 50 per cent chance of picking your newest worker, Case Worker A. For all your other workers (Case workers B to F) there is a .10 probability of selection. The relevant case load data are presented in the table below:

Case Worker	Load
A	52
B	47
C	32
D	69
E	61
F	41

What is the expected number of cases that will be audited?

Your answer should be 51. Remember an expected value requires the calculation of outcome values. An outcome value is the probability that the outcome occurs times its payoff amount. The sum of the outcome values is the expected value. What we need to do is convert the information in the above table into an expected value table. The table below shows an expected value table for our data:

Outcome	Probability	Payoff	Outcome Value
A	.50	52	26.0
B	.10	47	4.7
C	.10	32	3.2
D	.10	69	6.9
E	.10	61	6.1
F	.10	41	4.1
Sum	1.0		51.0

Expected Value

This example illustrates that an expected value need not be a monetary value - it can be any kind of numeric variable. In this case the expected value is expressed in terms of the number of cases that could be audited.

Summary

In this chapter we learned that a probability is a number between 0 and 1 and represents the relative frequency of an outcome in a large series of trials. Furthermore, the sum of the probabilities of all possible outcomes is equal to 1. We also discovered that probability problems can be divided into two main categories. The OR set seeks to find the probability for event A or event B. If events A and B are mutually exclusive the simple addition rule is used. If events A and B are not mutually exclusive, the general addition rule is used. The other problem category is for AND problems - determining the likelihood for events A and B. If events A and B are independent the simple multiplication rule is used. If events A and B are not independent, the general multiplication rule is used. We also studied expected values and found that an expected value is a special kind of average - it is the average that is reached after many replications of an event.

PROBLEMS

1. In a single roll of a die what is the probability of getting:
   a. A 5 or 6
   b. An odd number

2. Assume the organization you work at has 37% Democrats, 31% Republicans, 18% who belong to other political parties and 14% without a political preference. Each month an employee's name is drawn from a hat and is featured in the organization's newsletter. A person may be a repeat winner. What is the chance that in a 3 month period the newsletter will feature:
   a. Only Republicans
   b. No Republicans
   c. Only Democrats or Republicans
   d. A Republican in month 1, a Democrat in month 2 and a person without a political preference in month 3?

3. At your school all students must eventually take a set of core courses. The registrar reports the proportion of freshman enrolled in the core courses as follows:

Course	Enrollment
English	.72
Philosophy	.30
History	.46
Mathematics	.28
Psychology	.49

   a. What's the probability that a freshman student chosen at random is not taking psychology?
   b. What's the probability that a freshman student chosen at random is taking English or History.
   c. What's the probability that a freshman student chosen at random is taking English and History.

4. The town you live in has the following composition:
   20% are overweight
   16% are underweight
   50% are female
   25% have lived in the town less than 5 years
   75% have lived in the town 5 years or longer.
   What is the probability of choosing a person with the following traits: (Assume weight, sex and length of residency are independent).
   a. Either overweight or female?
   b. A male who has lived in the town less than 5 years and is underweight.

5. In an urn there are 25 balls - 6 white, 8 black, and 11 red. When balls are removed they are not replaced. Give the probability of:
   a. Selecting a red or white ball on the first pick
   b. Selecting 2 black balls in a row
   c. Selecting a white ball and then either a black or red ball

6. Redo problem 5, but this time assume the balls are replaced after each draw.

7. If you toss a coin what is the probability of a tail under the following circumstances?
   a. The flip follows 8 consecutive tails
   b. The flip follows a head
   c. It is the first flip for the coin

8. In a game of chance you can bet on different colored balls being picked at random. There are an equal number of white, blue, green and yellow balls. It costs $2.50 to play. If a white ball is chosen you get nothing back, if a blue ball is chosen you get $1, if a green ball is chosen you get your $2.50 back and if a yellow ball is chosen you get $5.
   a. What is the expected value of this game?
   b. Assume you are a perfectly rational person, should you play this game?

9. In the above game suppose you find out that there are twice as many blue balls as the other colored balls. Assume you win $10 when a yellow ball is picked and that the same cash awards as stated in problem 8 apply to all the other balls.
   a. What is the expected value for the game now?
   b. Assume you are a perfectly rational person, should you play this game?

10. What will be the size of a probability tree for the following conditions:
    a. A coin toss that has 5 replications
    b. A throw of a die that is repeated once. Essentially this is the same result you get when you throw a pair of dice.

11. Draw the probability tree for problem 10b (throwing a pair of dice). Note that if you add the face value of the two die together, a lot of the results have the same numeric value.

12. Based on the tree you drew in problem 11, determine the probability of having the numeric value of the dice total:
    a. 2
    b. 11
    c. 7
    d. 7 or 11

13. If you throw a white and a black die, what is the probability that the white die will be a 1 or a 6, and the black die will be a 6?

14. Imagine a bowl that has 2 red, 3 white and 4 blue balloons. What is the probability of the following events:
    a.   Obtaining a blue balloon, breaking it and then obtaining a red or white balloon
    b.   Obtaining a blue or red balloon, replacing it, then obtaining a white balloon which you break, and then obtaining a red balloon

15. If a value equal to the mean is added to a data set do the following statistics increase, decrease, remain the same, or can't you tell.
    a. Standard deviation
    b. Range
    c. Median
    d. Mode
    e. Variance

16. A county agency wants to know how tax rates for commercial buildings affect the sales of those properties. A random sample of properties in different communities, which have a different tax rate, produces the following results:

Community	Per Cent Sold	Tax Rate (Per $1,000 assessed value)
A	78	3.00
B	63	1.50
C	73	2.20
D	88	2.85
E	70	3.15
F	82	1.40
G	66	3.25
H	90	2.10
I	92	2.65

    a. What is the best way to display the results as a figure?
    b. Does there appear to be a relationship between tax rates and salability?

Excel Instruction

Most all of the calculations presented in chapter 3 deal with the basic mathematical operations of addition, subtraction, multiplication or division. Techniques to perform these operations were covered in a previous chapter so there is no need to repeat them here. However, note that Excel performs calculations from the left to the right, but it does all the multiplication and division functions before it does any addition or subtraction. If you need to do an addition/subtraction operation before a multiplication/division operation is carried out, place the addition/subtraction function in parentheses and it will be done first. For instance if you want to add 5 and 18 and then multiply the product by 3 you enter "=(5+18)*3", and Excel returns 69 as the answer. If you simply entered "=5+18*3" Excel would first multiply 18 * 3 and then add the 5 to the product giving an answer of 59.

You will also need to use the parentheses to separate major portions of more complex formulae. For example, suppose you want to perform the following calculation:

$$\frac{43-8}{4+3}$$

The answer is:

$$\frac{35}{7} = 5$$

However if we write this in Excel as "=43-8/4+3" Excel would return an answer of 44. This happens because Excel first does the division (8/4) and gets 2. Excel next subtracts the 2 from the 43 which results in the value 41. Finally, Excel adds 3 to 41 and ends up with the answer, 44. However, if we enclose both the numerator operation and denominator operations in parentheses, Excel will perform those operations independently. The correct Excel format is "=(43-8)/(4+3)". When this expression is used Excel will return the correct answer, 5.

A new operation, introduced in chapter 3, was the use of an exponent. To calculate 3^3 in Excel, you could write "=3*3*3". However, what if the calculation were 3^{13}? It would be rather laborious to find the answer in Excel by doing 13 multiplications. A better way to find the answer is to use the following command "=3^13". In Excel the ^ sign means "raise to the power" and this expression is a more efficient way to arrive at our answer, 1,594,323.

You can also do a square root using Excel. Suppose you want to know the square root of 160. In this case you'd write "=SQRT(160)" and Excel returns the correct value, 12.65. If the value 160 were already in a cell (e.g. cell G8) and you wanted to determine the square root, you would simply write "=SQRT(G8)" in an empty cell.

Here are problems you can practice on.
1. From your database perform the following calculations for the age variable.

a. X-µ for each employee

b $(X-\mu)^2$ for each employee

c. $\Sigma(X_i - \mu)^2$

d. $\dfrac{\Sigma(X - \mu)^2}{N}$

e. $\sqrt{\dfrac{\Sigma(X - \mu)^2}{N}}$

f. What do you call the statistic calculated in 1.d and 1.e?

2. Perform the following calculations using Excel:

a. $(5)^7 / (10.5)^3$

b. $\sqrt{5645.8} * (4.2)^3$

c. $\dfrac{10 * 5^2}{25}$

d. $\dfrac{(8 * 6)^2}{25}$

e. $\dfrac{10 + 5^2}{25} + \dfrac{(8 + 6)^2}{30}$

3. Write equations for each of the following problems in a single cell (i.e., write an expression, which includes all the operations in a single command])

a. Problem 2.a
b. Problem 2.b
c. Problem 2.e

Chapter 4

Permutations, Combinations and Binomial Probabilities

Introduction

A valuable area of statistics deals with permutations and combinations. Essentially permutations and combinations allow you to figure out the number of ways an event can occur. In this chapter we will examine three basic techniques that give us valuable formulae from which we calculate probabilities. We begin with a discussion of the statistical process called permutations. We then study the combination procedure. We close by examining a special distribution called the binomial probability distribution.

Permutations

Permutations are used when the goal is to know the number of ways objects can be selected from a data set and the order of the items selected is important. The clause "order of the items selected" means we have an interest in what object is picked first, second, third, etc. When this interest is present we will say we want to use an ordered selection process. We will see in a moment that there are situations in which all that matters is which object is chosen. In that case, whether an object is picked first, second or third, etc. is immaterial. This latter situation will be called an unordered selection process.

To appreciate the difference between an ordered and an unordered selection process, consider the following example. Three people love the ballet, but only 2 tickets for the performance Saturday night are available. How many ways are there to select the two people who will be given the tickets? We'll identify the people as A, B and C. The ways 2 people can be picked from this group of 3 is displayed below:

| 1st | 2nd | |
Pick	Pick	Sequence
A	B	1
A	C	2
B	A	3
B	C	4
C	A	5
C	B	6

We will call this kind of diagram an enumeration display. As shown in the enumeration display there are exactly 6 possible ways to pick the 2 people - no more, no less. Starting at the top of the display, we see that in the first sequence, A is picked followed by B. In sequence 2, A is picked followed by C. Note that sequence 3 picks B then A. However, A and B were also selected in sequence 1. The only difference between sequence 1 and 3, is the order of selection. If order doesn't matter we would not need to distinguish between sequence 1 and 3, nor between sequence 2 and 5 (in both sequences A and C are chosen), nor between sequence 4 and 6 (in both sequences B and C are chosen). If it's not important to identify each unique order of selection then we have only 3, not 6, ways to make our selections. Those 3 results consist of choosing A and B, A and C and B and C. As indicated above, we call the method of distributing the tickets an unordered selection. Furthermore, we conclude that there are 3 ways to distribute the 2 ballet tickets among the 3 people. However, what if one of the seats was a better seat than the other one (e.g. an orchestra seat versus a seat in the balcony). We'll revise our selection plan so that the first person picked receives the orchestra seat and the second person chosen is given the balcony ticket. In this situation the order of selection is important - in sequence 1, person A gets the orchestra seat and B gets the balcony seat. In sequence 3, it is B who gets the orchestra seat and A ends up with the balcony seat. We need to use an ordered selection process when there is a difference in who gets picked first and who gets picked second.

To sum up - when you are selecting items and you want an ordered selection, you use a procedure called a **permutation**. On the other hand, when you're not interested in the selection order (i.e., we want an unordered selection) you use a procedure called a **combination**. We will discuss how to solve a combination problem after we complete our discussion on permutations.

The problem we presented above involved selecting 2 objects from a set of 3. We found that there were 6 ways that that could be done when the order of selection was important. Let's look at another permutation problem. What if we want to know how many different ways 4 students could be ranked from high to low? With a data set of 4, it is apparent that to list all the different sequences could be a time consuming task if we used an enumeration display. Fortunately, to get the answer we can use a formula. The formula, called the **Complete Permutation Formula**, is:

$$W = n!$$

where W is the number of ways to make a selection and n the number of items in the data set.

The term "complete" is used in the name to emphasize that you are selecting the order for all 4 students - the <u>complete</u> data set. In our first example we did not select the whole set — the set contained 3 items and we selected 2. Later we will see that we need to use a different formula when we want to select only a subset of the total set.

The n in the formula for a complete permutation refers to the total number of objects in the data set. In our example, we have 4 people so we can substitute that number for n in our formula. The ! after a number means factorial. To perform a factorial operation you start with the number represented by n and multiply that number by the next lower integer (n-1). You continue introducing the next lower integer until you reach 1. In our case we start with 4 (n), the next lower integer after 4 is 3 (n-1) and the next one after that is 2 (n-2), and finally we reach 1 (n-3). We stop here because the last number included in a factorial is always 1. Thus 4! = 4*3*2*1 which is equal to 24. There are 24 different ways we can rank 4 students from highest to lowest.

If we want to know what the 24 arrangements look like we will need to create the enumeration display. In the process of producing the display we will also see why the formula, n!, is appropriate. First we will identify our 4 students as A, B, C and D.

We can pick either A, or B or C or D on our first selection - and at this stage our enumeration display looks likes this:

<u>1st Pick</u>
A
B
C
D
Sequence Calculator = 4

Note that there are 4 choices for our first pick. We will keep track of what we're doing by using a "sequence calculator". The purpose of the sequence calculator is to record the total number of sequences identified after each selection step. Since, after the first pick, there are the 4 sequences shown above, we just have the number 4 in the sequence calculator.

For the second selection there are 3 options for each of our initial picks. If our first pick was A then we can only select B, C or D at pick 2. For the B selection at the first pick, we can chose either A, C or D as our second pick. There are also only 3 options when C is chosen as the first pick and 3 options when D is chosen at the first pick. Our display now looks like this:

1st Pick	2nd Pick	1st Pick	2nd Pick
A	B	C	A
A	C	C	B
A	D	C	D
B	A	D	A
B	C	D	B
B	D	D	C

Sequence Calculator: 4*3 = 12

Note that our sequence calculator has the entry "4*3" for a total of 12 sequences enumerated at this stage. The number of sequences is obtained by multiplying the number of options at the first pick (4) by the number of options each of those selection had at the second pick (3). We obtain the 3 for the second pick because after picking A on the first pick, we can pick B, C or D on the second pick – 3 choices. For our other choices, there are also 3 choices at the second pick. Thus, after pick 2, the sequence calculator contains this entry: 4*3=12.

At the third pick, each of the 12 sequences developed so far has 2 options. For example, the sequence A then B can only have as the third pick, a C or a D. A two choice option is also available for all of the other sequences as well. The choices at the third pick are shown in the next enumeration display.

1st Pick	2nd Pick	3rd Pick	1st Pick	2nd Pick	3rd Pick
A	B	C	C	A	B
A	B	D	C	A	D
A	C	B	C	B	A
A	C	D	C	B	D
A	D	B	C	D	A
A	D	C	C	D	B
B	A	C	D	A	B
B	A	D	D	A	C
B	C	A	D	B	A
B	C	D	D	B	C
B	D	A	D	C	A
B	D	B	D	C	B

Sequence Calculator: 4*3*2 = 24

When we count the number of sequences in the above table we find that there are a total of 24. Arithmetically what we did was to take our 12 options from pick 2 and multiply those 12 sequences by 2 - the number of option we have for each sequence at the third pick. Consequently, after pick 3 our sequence calculator has this entry: 4*3*2 = 24.

We finally come to the fourth and last pick. Note that for each of the 24 sequences there is only 1 choice for the last pick. In the sequence ABC we can only chose D. Having only a single choice is true for all other sequences as well. Our enumeration display now looks like this:

1st Pick	2nd Pick	3rd Pick	4th Pick		1st Pick	2nd Pick	3rd Pick	4th Pick
A	B	C	D		C	A	B	D
A	B	D	C		C	A	D	B
A	C	B	D		C	B	A	D
A	C	D	B		C	B	D	A
A	D	B	C		C	D	A	B
A	D	C	B		C	D	B	A
B	A	C	D		D	A	B	C
B	A	D	C		D	A	C	B
B	C	A	A		D	B	A	C
B	C	D	D		D	B	C	A
B	D	A	C		D	C	A	B
B	D	B	A		D	C	B	A

Sequence Calculator: 4*3*2*1 = 24

When we count the number of sequences in the above table we find there are still 24. Arithmetically, what we did, was to take our 24 options from pick 3 and multiply those 24 sequences by 1 - the number of options we have at the fourth pick for each sequence. Consequently, after pick 4 our sequence calculator has the following entry: 4*3*2*1 = 24. There are 24 different ways to arrange 4 students using the enumeration method. This answer is the same as the one we obtained by using the complete permutation formula, n!.

If you look back at the complete permutation formula, n!, and we plug in 4 for n, we get 4*3*2*1 which equals 24. This is the identical result we obtained with the sequence calculator. The point of this rather lengthy exercise is to show that the complete permutation formula does in fact explain just what is happening when we enumerate all possible sequences.

If we had asked how many ways are there to rank 17 students from high to low, we would need to calculate 17!. The result is a very large number - 355,687,428,096,000. However, what if we only want to know how many ways the top 2 students, out of the class of 17, could be selected - i.e., we want to know the number of ways that a valedictorian and salutatorian could be selected for this class. Note that order matters in this case. We want a specific person in each position. We won't even try to determine this answer by an enumeration display, but go right to a formula to provide the answer. The formula we need is called the **Partial Permutation Formula** and takes on this form:

$$W = \frac{n!}{(n-r)!}$$

where W is the number of ways to make a slection, n is the total number of objects in the data set and r is the number of objects we want to select. In our example n is 17 and r is 2 so we need to calculate:

$$\frac{17!}{(17-2)!} = \frac{17!}{15!}$$

Our calculation effort is lessened if the factorial, 15!, in the denominator is reduced to 1. Dividing both the numerator and denominator by 15! leaves us with the following calculation:

$$\frac{17*16}{1} = 272$$

There are 272 different ways we can select a valedictorian and salutatorian from a group of 17 students.

If we look at this calculation, it says multiply 17 by 16 to find the number of ways you can pick a person for valedictorian and a different person for salutatorian out of a class of 17 students. We did not do an enumeration display for this case, but hopefully you can see that to do the display we would have 17 options for the first pick. In other words, all 17 students would be eligible for selection as valedictorian. For the second selection, there would be 16 students eligible to be the salutatorian. Of the 17 students, we eliminated the one chosen as valedictorian since it is obvious that the same person can not be valedictorian and salutatorian at the same time. The sequence calculator would record that there are 17*16 or 272 unique sequences at the end of pick number 2. And here is where we would stop - since only 2 students are to be selected. It's important to realize that the formula we use for a partial permutation is directly related to what we do when we prepare an enumeration display. The presence of factorials in the formula of a partial permutation may look strange, but once we appreciate what a factorial means, we see that the formula communicates exactly what needs to be done to get the right answer.

Practical Problems

Before we move on to the next topic we'll test your knowledge of permutations by presenting a couple of problems.

You are in charge of candidates night and need to establish the seating arrangement at the head table for the 7 candidates. No one but candidates will be seated at the head table. How many different seating arrangements are there?

Your answer should be 5,040. This is a complete permutation problem since the order of selection is important and we need to consider all possible seating arrangements for the entire set of 7 candidates. It matters to the politicians who sits in the first seat, second seat, etc. The solution requires the complete permutation formula because the entire data set of 7 people is involved. The formula is n! and since n is 7 the answer is 7! or 7*6*5*4*3*2*1. When we multiply this out we get 5,040.

We'll try another problem. The professional group you belong to has 50 members and needs to appoint a delegate and an alternate. How many different ways are there for the membership to fill the 2 officer positions?

Your answer should be 2,450. This is also a permutation problem because there will be a specific person chosen as delegate and another specific person selected as alternate. However, in this case we are only selecting a subset of the total group so we will want to use the partial permutation formula, n!/(n-r)!. There are 50 members so the n is 50. We want to select 2 people so 2 is our r. Substituting these values in the formula gives us:

$$\frac{50!}{(50-2)!} = \frac{50!}{48!} = 50 * 49 = 2,450$$

Combinations

A combination is used for a slightly different situation than a permutation. In the examples above, we had an interest in the order in which the selections were made. What if we are satisfied with an unordered selection - we don't care who's picked first, or second, or third, etc. All we want to do is to select a certain number of people from a larger group and the order of selection is immaterial. A **combination** tells us how many ways there are to select objects from a set when order is not of interest. Here's an example: Suppose terrorists are holding 5 hostages and agree to free 2. How many groups of 2 are possible? Note that the order of selection is not critical in this problem. Whether you are picked first or second makes no difference, in either case you will be freed.

If our 5 hostages are coded A, B, C, D and E we can start with person A and then pair that person with each other person. We begin by pairing A with B to make our first group of 2. Then A could be paired with C for our second grouping, etc. When we've finished with A, there would be 4 groups identified (AB, AC, AD and AE). We'd move to B and pair B with C, then B with D and finally B with E. We'd now have 3 more groups. Next we'd pair C with D and then C with E and we'd have as additional 2 groups. Finally our last pair would be D with E. The full set of pairs resulting from this exercise is: AB, AC, AD, AE, BC, BD, BE, CD, CE, DE - a total of 10. Again we can identify the number of possible pairs by a formula. The **Combination Formula** is:

$$W = \frac{n!}{r!*(n-r)!}$$

where W is the number of ways to make a selection, n is the total number in the data set and r is the number to be selected. For our hostage example the total number of hostages is 5. The number to be selected is:

$$\frac{5!}{2!*(5-2)!} = \frac{5!}{2!*3!}$$

since there are a total of 5 people (n) and 2 (r) are to be selected.

If we convert the factorials to their multiplication status we get:

$$\frac{5*4*3*2*1}{(2*1)*(3*2*1)}$$

Canceling (3*2*1) in the numerator and denominator gives us:

$$\frac{5*4}{2*1} = \frac{20}{2} = 10$$

There are 10 possible ways to select the hostages, which is the same answer we obtained when we enumerated each possible pair.

If we examine the combination formula we will see that it is very similar to the formula for a partial permutation. Refer to the partial permutation formula and you will see that the only difference is the presence of r! in the denominator of the combination formula. Remember when we started this chapter we determined the number of ways we could select 2 people to attend a ballet from a group of 3. When this was an ordered selection problem, 6 unique sequences were identified. We found the 6 sequences by the enumeration method, but had we used the partial permutation formula, we would have also found that there were 6 ordered sequences. In way of illustration, when we substitute n = 3 and r = 2 into the partial permutation formula we get:

$$\frac{3!}{(3-2)!} = \frac{3!}{1!} = 6$$

However, we also said that if we weren't interested in order, there would be only 3 sequences. The combination formula is used for unordered selections and when applied to this situation it does, in fact, give us an answer of 3:

$$\frac{3!}{(2!)(3-2)!} = \frac{3!}{2!*1!} = 3$$

When our interest switched from an ordered to an unordered selection situation, we lost 3 or half of the 6 sequences. When order no longer mattered, the AB sequence and the BA sequence were redundant - we could eliminate one of them. This was also true for the other redundant sequences involving AC and BC. What happened then was that the 6 sequences were halved - i.e., the number 6 was divided by 2 to give us the 3 sequences for an unordered selection.

The addition of r! to the denominator of the combination formula performs this halving operation. In the example r = 2 and therefore r! = 2! or 2*1 which is equal to 2. Thus, adding r! to the denominator of the combination formula reduced the result from the partial permutation formula by half. What happened is that the addition of r! to the denominator of the combination formula eliminated the redundant sequences.

Hopefully you can see that the combination formula is not a magical equation that has no interpretative value. It, just as the other formulae we've studied, is a statement that gives us a common sense explanation of how to solve a problem.

Using Permutations and Combinations to Solve Probability Problems

We have used permutations and combinations to determine the number of ways objects can be selected. We can also use this knowledge to calculate probabilities. A probability can be defined as follows:

$$\text{Probability} = \frac{\text{Number of Times an Event of Interest Occurs}}{\text{Total Number of Possible Events}}$$

In the most popular form of poker, 5-card draw, each player is dealt 5 cards. We could ask what is the chance you will be dealt a club flush (i.e., all 5 of the cards you are dealt are clubs). To find the answer we need to determine 1] the total number of possible 5 card hands there are and 2] the number of hands in which all the cards are clubs. The total number of 5 card hands becomes the denominator and the numerator is the number of hands containing only clubs.

How many poker hands are there? Based on what you've learned so far you should be able to arrive at this answer - try it.

The answer is 2,598,960. What you first needed to do was to decide if order mattered. In dealing a hand of 5-card draw poker, what card you get first, second, etc. is immaterial - you just worry about the set of cards and the order they come to you is insignificant. Since the question involves an unordered selection we use the combination formula:

$$\frac{n!}{r!*(n-r)!}$$

There are a total of 52 cards and 5 cards constitute a poker hand. Therefore n = 52 and r = 5. Substituting these values in the formula gives us:

$$\frac{52!}{5!*(52-5)!}$$

We won't show the rest of the calculations since the work is straightforward and the answer is given above.

We still have to calculate our numerator - the number of hands that will contain 5 clubs. Try solving this question.

The answer you should have gotten is 1,287. Again order is irrelevant and therefore we use the combination formula. There are 13 clubs in a deck so 13 is our n. All the cards must be clubs and we are selecting 5 cards so 5 is our r. Substituting the 13 and 5 in our combination formula gives us:

$$\frac{13!}{5!*(13-5)!}$$

We won't show the rest of the calculations. The most important part of a mathematical problem is the set-up, with computers the calculations should be a mere formality.

Since we know the values for the numerator and denominator all we have to do is divide to get the probability we wanted:

$$\frac{1,287}{2,598,960} = .0005$$

There are approximately 5 chances out of 10,000 or 1 chance in 2,000 to be dealt a club flush - it's not too likely.

Probability Distributions

If we drew a graph in which we identified all the possible outcomes of a trial on the X-axis and used as the Y-axis, the probability of the outcome occurring, we'd have what statisticians call a **probability distribution**. Note that a probability distribution is similar to a frequency distribution, but the probability distribution gives the probability for each outcome rather than the frequency. The probability distribution of a single coin toss consists of two possible outcomes - a head and a tail with a .50 probability for each outcome. The probability distribution for a single toss of a coin then is pretty uninteresting. When the results are graphed, we' have two bars of equal height (see below).

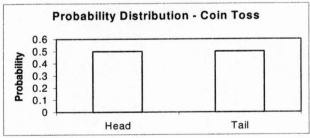

In the last chapter we listed all the possible outcomes from flipping a coin three times along with their probabilities. After 3 tosses the possible outcomes were 0, 1, 2 or 3

Statistical Methods for Researchers

heads. For each outcome there was a corresponding probability of its occurrence. The three coin toss results produces a probability distribution in which there are 8 outcomes and each outcome has the same 1/8 probability of occurrence (see below).

Diagram of a Probability Distribution

Flip 1	Flip 2	Flip 3	Result	Probability
		H	HHH	1/8
	H			
		T	HHT	1/8
H				
		H	HTH	1/8
	T			
		T	HTT	1/8
		H	THH	1/8
	H			
		T	THT	1/8
T				
		H	TTH	1/8
	T			
		T	TTT	1/8

A graph of this probability distribution is shown below.

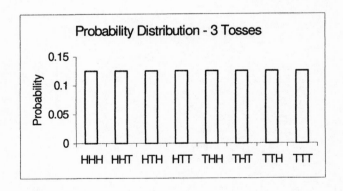

The Binomial Probability Distribution

As we know, a coin toss can end up with only 2 discrete outcomes - a head or a tail. A coin toss belongs to a special type of probability distribution called the **binomial probability distribution**. A binomial probability distribution uses only discrete variables. A different kind of distribution is used for continuous variables and will be examined in the next chapter. Remember a discrete variable can take on only a limited number of values and is used to classify people or objects. However, a binomial distribution adds a further restriction on discrete variables. As the prefix bi suggests - the variable can have only two states. Gender satisfies this requirement since people can be only male or female. The toss of a die is not a binomial variable since there are 6 possible outcomes of equal probability.

The two states of a binomial variable also need to be independent of each other. In the last chapter, we stated that independence means that the occurrence of one of the outcomes can not influence the probability of the other outcome.

A binomial probability distribution often involves doing repeated trials with a variable. Statisticians have an interest in mapping the outcomes of such variables and they refer to the pattern of outcomes as following a **Bernoulli Process**. However, in addition to the conditions of 1] a two-event discrete variable and 2] independent outcomes, a third requirement is added to a Bernoulli Process. The third requirement is that time must not influence the outcome probabilities. For example, consider the operation of a roulette wheel. With each use, the outcome (e.g. an odd number) could be an independent event at the start, but over time it could change due to machine deterioration. Variables in which the probability of the outcomes changes over time should not be used to create binomial probability distributions.

We can determine the probability distribution for a variable that satisfies the three conditions described above. Recall that a probability distribution is the pattern of probabilities for the different outcomes that can occur for a variable.

We'll now develop a probability distribution for a more complex situation than a single coin toss. Assume that we know that the presence of a harmless bacterium occurs in 1 out of 10 people. If we also assume that the test is 100% accurate, when we test a random group of people we have a .10 probability of finding a positive test for the bacterium. We could ask - If we test three people, how many tests (0, 1, 2, or 3) tests will yield a positive result?

There are a number of ways we can solve this problem. One thing we can do if to draw a probability tree. In the tree we'll use the symbol "+" for a positive result and the symbol "-" for a negative result.

As shown in Table 4-A on the next page, there are 8 possible outcomes or sequences that can occur.

Statistical Methods for Researchers

Table 4-A. Probability Tree for Three Tests

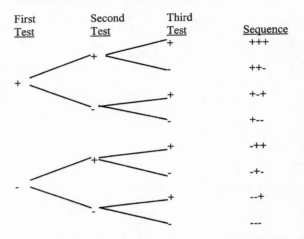

First Test	Second Test	Third Test	Sequence

We should add another column to Table 4-A to record the probability of each outcome. To determine the probability of the first case, the +++ sequence, we begin by recalling that the probability of a + is .1. Since this is an AND probability with the independence requirement satisfied, we can use the simple multiplication rule. We multiply the three events together to obtain our answer:

$$.1*.1*.1 = .001$$

For the next sequence, ++-, we need to know the probability for a - result. Most all know that's it's .9, but if we want authoritative support for our answer we can call on the second corollary to probability property described in chapter 3. That corollary says that in a two-event trial, the probability of an unspecified event is equal to: 1 - the specified probability. In our example the stated probability for a + test is .1. Thus to get the probability of a - test we subtract .1 from 1 and get .9. We can now use the simple multiplication rule to calculate the probability for the second sequence:

$$.1*.1*.9 = .009$$

There is a .009 probability for a ++- sequence. We continue this process until the probabilities for all 8 outcomes have been determined (see Table 4-B).

In Table 4-B, we added a column for the number of + tests and row number - reference to the row number will help to explain some concepts shortly.

Table 4-B. Probability Tree for Three Tests - With Additional Information

First Test	Second Test	Third Test	Sequence	Probability	No. + Tests	Row
		+	+++	.1*.1*.1=.001	3	1
	+		++-	.1*.1*.9=.009	2	2
+		+	+-+	.1*.9*.1=.009	2	3
	-		+--	.1*.9*.9=.081	1	4
		+	-++	.9*.1*.1=.009	2	5
	+		-+-	.9*.1*.9=.081	1	6
-		+	--+	.9*.9*.1=.081	1	7
	-		---	.9*.9*.9=.729	0	8
Sum:				1.0		

When we are done with our tree we can check to see if we made a mistake by calculating the sum of the probabilities. Probability property 2 says the sum should be 1.0 and if that is not our answer we know there must be a mistake somewhere. In our case the sum is 1.0 so we're OK at this point.

Even though we've done all this work, we still need to do a bit more since we haven't answered the original question - we need to determine the probability of obtaining 0, 1, 2 and 3 positive tests. We need to do some summarization to finish the task. For the probability of 0 + tests, note that there is only one sequence that results in 0 + tests (the one given in row 8) and the corresponding probability is .729. Observe that there are three sequences that result in just one + test (See rows 4, 6 and 7). The accumulative probabilities for these three rows is .243 (.081 + .081 + .081 = .243). In a similar fashion we can arrive at the probabilities for 2 and 3 + tests. Table 4-C on the next page summarizes these results.

A graph (Figure 4-A) of the data in table 4-C is shown right below the table. Figure 4-A is simply a graphic depiction of the probability distribution. The probability of 3 positive results (.001) is so small it may appear to be missing from Figure 4-A, but it's there as a very fine line.

Table 4-C. Binomial Probability Distribution for Three Tests

Number of Tests	Probability
0	.729
1	.243
2	.027
3	.001
Sum:	1.000

Figure 4-A

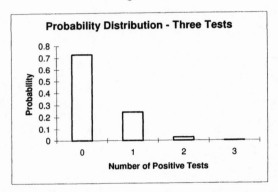

We obtained the results in Table 4-C by painstakingly making a probability tree and then summarizing the results in a table. As you may suspect there is a quicker and easier way to obtain the information then creating probability trees. We can use a formula - it's called the **Binomial Formula** and is shown below. An apology is in order for the complicated appearance of the formula, but it's not as complex as it seems.

$$\text{Binomial Formula} = \frac{n!}{r! * (n-r)!} * p^r * (1-p)^{(n-r)}$$

First we need to define the algebraic symbols and we'll start with the last portion of the equation - the part beginning with p. The p represents the probability of obtaining a positive result. Usually the generic term "success" is used for p, but since our interest is in positive tests we will use that term rather than success. The (1-p) is the probability of not getting a positive result. In our example (1-p) is equal to .9 (1-.1). The symbol for the probability of a positive result is p and (1-p) is the symbol for a negative result. Some texts prefer to use the notation q for (1-p) so if you see that notation don't be surprised. In addition, the generic term for (1-p) is a failure since, as noted above, the generic term for p is a success. The n and r have similar definitions to the ones we used before. The total number of trials involved is n. In our case, a trial is a test so n is 3. The r is the number of trials that have the characteristic we are looking for, which in our example means r is the number of positive test results. In the first row of Table 4-C our interest is in the probability of getting no positive tests so r for that situation is 0. The whole expression p^r $*(1\text{-}p)^{(n\text{-}r)}$ is the probability of getting a specific sequence. For the first sequence presented in Table 4-C (sequence +++), the values of p, (1-p), n and r are: .1, .9, 3 and 3 respectively. The calculation, for the probability of obtaining the +++ sequence, is:

$$p^r * (1 - p)^{(n-r)} = .1^3 * .9^{(3-3)} = .1^3 * .9^0$$

The expression $.9^0$ will look strange to non-mathematicians. Whenever you have an exponent of 0 such as $.9^0$ the result is always 1. It doesn't matter what the base number is. A base number of .9, 99 or .01, when raised to 0 power, results in a value of 1. Consequently, the equation above simplifies to: $.1^3 * 1 = .001$. In chapter 3 we learned another way to find the probability of a +++ sequence. We used the formula for an AND probability in which there was independence (the simple multiplication rule). Employing that approach we get:

$$.1 * .1 * .1 = .001$$

This answer is identical to the one we obtained using the binomial formula (as well it should be). Later in this chapter we will describe the circumstance when the binomial formula, and not the simple multiplication formula, must be used.

Now for the first part of the formula — if you isolate that portion you will see that it looks like this:

$$\frac{n!}{r!*(n\text{-}r)!}$$

This portion of the formula is an old friend - it's the combination formula we used to determine how many ways there were to select hostages for release. In this example, this portion of the binomial formula will tell us the number of ways we can get one positive test. It's appropriate to use the binomial formula here, since there is more than one sequence that results in 1 positive result. Note that 1 positive test result occurs in rows 4, 6 and 7 of Table 4-B. Each of these rows has a single positive result, but the + appears in a different position each time. The string of test results in row 4 (+--) has the + result in the first position - i.e., in the initial test. In row 6 the string is -+- and in row 7 it is --+. These 3 rows represent the number of different ways that we can select 1 positive test from a group of 3 tests. It's the combination issue all over again. The incorporation of the combination formula, as part of the binomial formula, tells us how many ways we can obtain 1 positive result when 3 tests are run. Remember that the last part of the equation, $p^r*(1-p)^{(n-r)}$, tells us how likely it is that we will have a sequence with a given r in it. A given r in our example could be either 0, 1, 2, or 3.

Let's pause and see where we are. What we have found so far, in words not in mathematical notation, is that the binomial formula tells us that the probability of getting r success in n trials is equal to the number of sequences that have r success times the probability of obtaining r successes:

$$\left(\begin{array}{c} Probability\ of\ r \\ successes\ in\ n\ trials \end{array}\right) = \left(\begin{array}{c} Number\ of\ sequences \\ with\ r\ successes \end{array}\right) * \left(\begin{array}{c} Probability\ of \\ an\ r\ success \end{array}\right)$$

With reference to our bacterium example, we could use the above formula to state the probability of obtaining exactly 1 positive result (r) among 3 tests (n). Here is what we would say:

$$\left(\begin{array}{c} Probability\ of\ 1\ positive \\ result\ in\ 3\ tests \end{array}\right) = \left(\begin{array}{c} Number\ of\ sequences \\ with\ 1\ positive\ result \end{array}\right) * \left(\begin{array}{c} Probability\ of\ 1 \\ positive\ result \end{array}\right)$$

We can now use the mathematical notation to obtain our answers. The first part of the binomial formula will give us the number of sequences with one positive result:

$$\left(\begin{array}{c} Number\ of\ sequences \\ with\ 1\ postive\ result \end{array}\right) = \frac{n!}{r!*(n-r)!} = \frac{3!}{1!*(3-1)!} = 3$$

The answer, 3, is the number of sequences that contain a single positive result.

The second part of the binomial formula will tell us the probability of obtaining a single positive result. Recall that the probability of a positive result (p) is 0.1 and the probability of a negative result (1-p) is 0.9. When we substitute these values we obtain the following answer:

$$\left(\begin{array}{c} Probability\ of\ 1 \\ positive\ result \end{array}\right) = p^r*(1-p)^{(n-r)} = .1^1 * .9^{(3-1)} = 0.081$$

Note that the probability, 0.081, is the same value that we obtained from the probability tree (Table 4-B) for the sequences with 1 positive result. Finally we just need to multiply the two components of the binomial formula together to get our answer:

$$3*0.081 = 0.243$$

The answer, 0.243, is the same as that shown in Table 4-C. The other three results can be calculated in the same way we arrived at the answer for 1 positive result. We only need to change r to match the number of positive results specified in the table. We won't show the work involved, but you may want to try doing the calculations - the correct answers are in Table 4-C.

In this chapter we presented additional rules, formulae, procedures etc. associated with probabilities. There are even more sophisticated approaches than these, but the most important ones related to statistics have been covered.

Decision Tree for Selecting a Permutation or Combination

The secret in working problems using the techniques learned in this chapter is to first decide what kind of a problem you have. We have presented procedures to figure out the number of ways an event can occur. We learned we can use permutations or combinations for these kind of problems. A decision tree to distinguish between the methods used to find the number of ways to select elements is shown below.

Type of Sequence	Will Total Group be Selected?	Procedure	Formula
Ordered	No	Complete Permutation	$n!$
	Yes	Partial Permutation	$\dfrac{n!}{(n-r)!}$
Unordered		Combination	$\dfrac{n!}{r!*(n-r)!}$

Note that to differentiate between a permutation and a combination you first have to decide if order matters in the selection process. If order does matter, you have a permutation problem. If order doesn't matter, you have a combination problem.

There are two types of permutation problems - one involves finding the ways to select a subset of the total group. For these problems, you use the partial permutation formula. If you are looking at ways in which the whole group can be selected, you use the complete permutation formula.

Remember that the following symbols are used in these formulae:

n The total number in a group

r The number in the group who are to be selected

In addition to these counting techniques we also learned when and how to use the binomial probability procedure. The binomial probability function gives us an additional tool to the procedures we learned in Chapter 3. You can use the binomial probability formula when the problem involves a binomial variable in which the probability of an outcome remains constant over time and is unaffected by the outcome of a previous trial (i.e. it is independent). With a binomial probability problem it is important to remember that there is potentially more than one sequence that will satisfy the conditions stated in the problem. In chapter 3, we solved problems that dealt with a specific sequence and, in that case, the simple multiplication rule was appropriate. In the example we worked in this chapter, there was more than one way to get a success (e.g. a positive test result). A positive result could occur on the first, second or third trial. Had we asked what is the probability of a positive result at the second trial with failures at trials 1 and 3, then the problem involves only a single sequence (i.e. -+-). Thus, the simple multiplication procedure could be used to gives us our answer:

$$.9*.1*.9 = .081$$

However, when we ask what is the probability of a single positive result in a series of three trials, without specifying which trials produce the positive result, then we need to apply the binomial formula:

$$\frac{n!}{r!*(n-r)!} * p^r * (1-p)^{(n-r)}$$

and in that case our answer is .243 because there are 3 different sequences that can produce one positive result and the probability of a positive test is .081.

Summary

In this chapter we learned how to determine the number of ways items can be selected from a data set. If the order of selection is important, and we want to select all items in the data set, we use a process called a complete permutation. If order is important, but we only need to select a portion of the items in the data set, the process is called a partial permutation. For selections in which the order of item selection is immaterial, the process is called a combination. A formula to calculate the number of possible ways a selection could occur was provided for each type of process.

We also studied the behavior of variables in which there are only two possible outcomes and those outcomes are independent of each other. After doing repeated trials and recording the outcomes, the distribution of outcomes can be developed for the variable. The distribution is called a binomial probability distribution. We also found that we can determine the exact probability for the elements that make up such a distribution by using the binomial formula. For some of the elements, it is possible to obtain the answer using the simple multiplication procedure. A decision tree to help us decide when we use the binomial procedure rather than the simple multiplication method is shown below.

Decision Tree for Selecting the Simple Multiplication or Binomial Procedure

Desired outcome comes from only one sequence	Procedure	Formula
Yes ———————	Simple Multiplication	$P(A)*P(B)$
No ———————	Binomial	$\dfrac{n!}{r!*(n-r)!} * p^r * (1-p)^{(n-r)}$

Statistical Methods for Researchers

PROBLEMS

1. If n is 5 and r is 2 what is the answer for the following problems:
   a. Partial Permutation problem
   b. Combination problem
   c. Assume n is 6 - what is the answer for a Complete Permutation problem?

2. Four contestants (Al, Bob, Cal and Don) will all compete in a free throw shooting contest. The first one who makes a free throw will win the contest so the order in which they shoot matters. We want to determine the number of ways to select the contestants.
   a. Is this a partial permutation, complete permutation or a combination problem?
   b. What is n?
   c. How many ways can the shooting sequence be arranged?
   d. Draw the enumeration display for this problem.

3. Assume the order of selection is important and we want to select 4 objects from a set of 7.
   a. How many ways can we do that?
   b. If the order of selection isn't important and we want to select 4 objects from this set of 7, how may ways can we do that?

4. You need to transfer 4 people from the home office in Washington DC. One must go to Chicago, another to New York, another to Los Angeles and the fourth to Atlanta. The person chosen first will go to Chicago, the second selection to New York, the third to Los Angeles and the fourth to Atlanta. How many ways are there for the four staff members to be assigned to the four cities?

5. From 6 satellite offices at your agency, 4 offices are to be chosen at random for an in-depth safety inspection. How many ways can the 4 offices be chosen?

6. How many different five letter word combinations can be created from the word FOCUS? The letter combinations do not have to result in recognizable words.

7. Assume you have 5 people who are eligible to go to a special training class, but only 2 may go - how many different ways could you select the 2 people?

8. Suppose you have 10 distinct jobs that can be done by 10 different workers. Each worker is capable of doing each job. How many ways could you make the 10 assignments?

9. You belong to a 30 member professional group that needs to elect a President. Two nominees are to be selected to compete in the presidential election. How many ways are there to select the two nominees?

10. What is the probability for the following events when you throw a die two times?
    a. Getting 2 ones?
    b. Getting a one on the first or second throw?
    c. Draw an enumeration display for this problem and check your answers to parts a and b.

11. In a group of 10 volunteers, 5 are to be selected to serve as the control subjects in a study. How many different control groups is it possible to create?

12. In a multiple choice test with 4 options per question, what is the likelihood of getting the first questions correct by guessing?

13. In a multiple choice test with 4 options per question, what is the likelihood of getting the first three questions correct by guessing?

14. In a multiple choice test with 4 options per question, what is the likelihood of getting just one of the first three questions correct by guessing (the question you get correct can be no. 1, 2 or 3)?

15. In a multiple choice test with 4 options per question, what is the likelihood of getting each of the first four questions wrong by guessing?

16. In a multiple choice test with 4 options per question, what is the likelihood of getting 8 out of 10 questions wrong by guessing?

17. If the probability of employee A being absent is .1 and the probability of employee B being absent is .15 then what is the probability that:
    a. They will both be absent tomorrow
    b. Employee A will be present and Employee B will be absent tomorrow
    c. They will both be present tomorrow

18. Answer the following questions regarding bridge hands (In bridge each player is dealt 13 cards).
    a. How many possible bridge hands are there?
    b. How many ways are there to be dealt only spades in a bridge hand?
    c. What is the probability of being dealt only spades in a bridge hand?

19. Suppose you can send 3 of your staff on a special mission. There are 8 qualified people - 3 women and 5 men. You decide to pick the 3 names out of a hat so there is no appearance of a preference on your part. The hat picking ends up with all 3 of the women being chosen.
   a. What is the probability that a women is picked?
   b. Does picking a woman as the first selection affect the probability of the second pick?
   c. What's the probability that the 3 selections ended up being all women?

20. There are 8 conservatives and 5 liberals on your regional governing body. By tradition a parliamentarian and a sergeant at arms are chosen at random from among these 13 representatives.
   a. What is the probability that the parliamentarian will be from the conservative party?
   b. What is the probability that the parliamentarian and sergeant at arms will both be from the conservative party?
   c. What is the probability the parliamentarian and sergeant at arms will be from the same party?

21. In a field there are 10 white, 8 black and 7 brown cows. What is the probability of randomly selecting for milking a white cow, then either a black or brown cow and finally another white cow on the third selection? In each selection the cow is removed from the field and taken to the milking barn where she remain until all cows are milked.

22. As a member of the local school board you are asked to vote on a proposal to expand the school facilities. Two options are offered. Option A is designed for a major facilities increase and option B offers a more modest plan to increase facilities. Consultants have conducted cost-benefit analyses for both options. The cost benefit analysis assumes four scenarios concerning population (and therefore student) growth. The chances that each scenario will occur are also provided. The table below summarizes the consultant's report:

		Cost-Benefit [in Millions]	
Population Change	Probability	Option A	Option B
1. Decrease	.05	-40.4	-12.8
2. Low growth	.30	-11.3	0.4
3. Moderate growth	.45	6.1	5.0
4. High growth	.20	26.2	3.5

   a. What option will give you the best financial return?
   b. If you have a risk averse personality what option would you choose?
   c. What is the best scenario regarding population change under option A?
   d. What is the best scenario regarding population change under option B?

Excel Instruction

The type of calculations we did in chapter 4 can be easily done in Excel. One thing we did involved the calculation of a factorial. There is a function in Excel for factorials and it is named FACT. Its format is:

=FACT(Number of the factorial)

To obtain the value for 3!, you'd find an empty cell, type =FACT(3) and finish by hitting the enter key. The answer 6 appears in the cell into which you wrote the instruction.

Excel uses what is known as scientific notation - it's a popular compression technique used by computers programs when a value is very large - i.e. takes up a lot of space. For example 17! may give you the following answer in your spreadsheet: 3.56E+14. Obviously we know enough about factorials to realize 17! has to be a lot larger than 3.56. How much larger is the question. To determine how much larger, we focus on the E+14 portion of the value. E+14 means that the preceding number, 3.56, needs to be increased by multiplying it by 100,000,000,000,000. Note that the number we will multiply by has a 1 followed by 14 zeros. In practice you don't really have to multiply 3.56 by 100,000,000,000,000 - just move the decimal point 14 places to the right - that gives the same result as you get by multiplying by 100,000,000,000,000. The 3.56 has been rounded so you do lose some specificity in the scientific notation answer. 17! with all its precision retained is what was given in the text (355,687,428,096,000). You can get that level of precision to appear on you spreadsheet if you are willing to expand the cell size so there is room for all the digits and you reformat the cell (See an Excel manual on formatting cells).

Incidentally, if a number is larger than the cell width allocated Excel places a series of # symbols in the cell. You need to expand the cell width until the full number will fit.

We also calculated permutations in chapter 4. Again Excel has a function for permutations - its name is PERMUT. The format is:

=PERMUT(Total number of objects in group, Number of objects to be selected)

Excel uses the term "objects" as a generic expression to refer to widgets, people, etc. PERMUT can solve partial permutation problems. For example, to find the number of ways we can pick 2 students (e.g. a valedictorian and salutatorian) from a group of 17 we would write =PERMUT(17,2) in an empty cell. After hitting Enter your cell would contain the correct answer (272).

You can also use the PERMUT function for complete permutation problems. For example to find how many ways to arrange 5 flags on a display using PERMUT we'd make n=5 and r=5. In other words you would type =PERMUT(5,5). The answer will appear as 120. Since the formula for the complete permutation case is n! we could just use the factorial function as described above. With either method you arrive at the same answer.

To calculate combinations you also use a built-in Excel function called COMBIN. The format is:

=COMBIN(Total number of objects in group, Number of objects to be selected)

For the number of ways 2 hostages can be picked from a group of 5, we go to an empty cell, type =COMBIN(5,2), hit Enter and the result, 10, will appear in the cell.

The last type of calculations we performed were binomial probabilities. Once again there is an Excel function available for this calculation which is called BINOMDIST. The format is:

=BINOMDIST(Number of successes, Number of trials,

Probability of success, True/False option).

In our coin toss work, if we wanted to know the probability of getting only 1 head in 3 tosses we would write:

=BINOMDIST(1,3,0.5,False)

Excel uses the generic terminology mentioned in the text so the number of success refers to the number of heads (1) and the number of trials refers to the number of tosses (3). The probability of a success or a head is 0.5. The last element you need to specify is a little ambiguous. You type either TRUE or FALSE. If you type FALSE the probability calculated is for the exact number of success you specified. If you type TRUE you will get the accumulative probabilities up to and including the number of success specified. For example, if you type TRUE and enter 1 as your number of successes, you will get the sum of the probabilities for 0 successes plus 1 success. W we want the probability of getting exactly 1 head so we would type "FALSE". Our cell entry would be:

=BINOMDIST(1,3,0.5,FALSE)

and the answer you obtain is .0375.

The TRUE alternative is a convenient option since many times one is interested in the chances of having an event occur at least so many times or at most so many times. For example, what's the chance of getting at most 2 heads when a coin is flipped 5 times? For this question the phrase "at most" means getting 0, 1 or 2 heads. You could find the probability of exactly 0 heads, then the probability of exactly 1 head and finally the probability of exactly 2 heads. The three probabilities you obtained could then be added to together to answer the question. Using this approach would obviously be time consuming. By using the function:

=BINOMDIST(2,4,0.5,TRUE)

you could get your answer (.6875) in a single step.

If the question you had to answer involved getting four or more heads (i.e. getting at least 4 heads) when a coin is flipped 6 times you could find the probability for 4 heads, 5 heads and 6 heads and then add the probabilities together. You could also save yourself some work by finding the probability for up to 3 heads using this expression

=BINOMDIST(3,6,0.5,TRUE)

and then subtracting that probability from 1 to get your answer. The expression we would write is:

=1-BINOMDIST(3,6,0.5,TRUE)

and Excel returns .344. Note that there is a minus sign between the 1 and BINOMDIST.

In this latter case we are using the corollary to probability property 2. We create a two event situation - getting up to 3 heads vs. getting 4 or more heads. If we know one probability we can subtract that probability from 1 to get the other probability. In this case the probability of getting up to 3 heads is .656 so the probability of 4 or more heads is .344 (1-.656).

Here are a few problems for you to practice on to see how well you understand how to use Excel. For some problems the answer is given in brackets at the end of the question.

1. For the following problems in this chapter use Excel to obtain the correct answer:
   Problems, 1a, 3a, 3b, 7, 8, 9, 11.
2. Use Excel to obtain the 4 probabilities used in Table 4-C of this chapter.
3. Use Excel to determine the probability of the following coin tosses:
   a. Exactly 5 heads in 10 tosses [.246]
   b. Three of fewer tails in 12 tosses [.073]
   c. Six or more heads in 15 tosses [.849]
4. In a 10 item True/False test, use Excel to find the probability of getting a score <u>over</u> 60% when you have to guess at each question. [.172]
5. Assume there are 27 blue, 39 red and 52 white balls in a barrel. Use Excel to find the probability of the following sequences:
   a. Red, White and then Blue when only the white ball is removed. [.034]
   b. Red, White, Blue, Blue, Red and then White when only the red balls are removed. [.00114]
6. In a 20 item multiple choice test (5 options per question) use excel to find the probability of scoring 50% or worse when you have to guess at each answer. [.999]
7. Use Excel to calculate 11!/50000
8. Use Excel to calculate: 21!/(17!*5!)

Chapter 5

Normal Curve and Z Values

Introduction

In the last two chapters we studied probabilities for discrete variables. We used that knowledge to tell us how unusual a certain event was. As we saw in chapter 3, flipping a coin four times and getting all heads is reasonably rare ($.5^4 = .0625$). In chapter 4 we found that if the probability of not getting a positive test was .9, then it is quite likely (probability = .729) that there would be no positive result after even three tests ($.9^3 = .729$). On the other hand it would be very unusual to get 3 positive tests (probability = $.1^3$ or .001). As a matter of fact, if we were responsible for a quality control operation at the laboratory that performed the tests and found 3 consecutive positive results, we might wonder if something had gone wrong with the testing procedure.

We have seen how the binomial distribution allows us to calculate probabilities for discrete data. What we will learn in this chapter is a comparable procedure that we can apply to find probabilities associated with continuous data. To accomplish our goal we will identify a special distribution called the normal distribution. We will also develop a powerful statisitcs called a Z score.

Recall that the probability distribution for a discrete variable is the probability assigned to each of the outcomes that the discrete variable can assume. Each of those outcomes is discrete: 2 heads, 3 positive tests, etc. You don't have 2.33 heads or 1.52 positive tests. When we plot the distribution each category on the X-axis is distinct from the other categories - a 0 or a 1 or a 2, etc. The distribution for a continuous variable presents a more complex situation. When dealing with continuous data you really shouldn't talk about the number of items in a <u>specific</u> category. Nevertheless, we often do this. For instance we report the heights of men who are 70 inches tall, 71 inches tall etc. However some of the men placed in a category may be a bit over or under the inches designated. For example those in the 72 inch category may be a tad over 72 inches and some may be a little under 72 inches. We lump them altogether in the 72 inch category even though we know they are not exactly 72 inches tall. If we were willing to define height with more precision (i.e. use a better measuring device) we could use tenth of an inch categories rather than whole inch categories.

Because of the difference between discrete and continuous variables, the procedures we have learned for finding the probability apply nicely to discrete variables, but don't work very well for continuous variables. We, therefore, need to develop a different approach that we can use with continuous variables. That approach utilizes a distribution called the normal distribution.

Discrete and Continuous Distributions

In earlier chapters we observed that a **discrete distribution** typically involves variables that can be counted. We can count people by their sex or we can enumerate the outcomes of a die (e.g. the outcomes of 1,2,3,4,5,6). We can therefore associate probabilities with each distinct outcome - the probability that the next person we meet is a male or the probability of getting a 3 when we toss a die. The binomial distribution discussed in chapter 4 is an example of a discrete distribution.

In theory, a distinct number can not represent a **continuous variable**. Any number we would assign can be altered by performing a more precise measurement. For this reason we can not assign a probability to a distinct value of a continuous variable. What we can do however, is assign a probability to a range of values for a continuous variable. For example, although we can not assign a probability to a weight of exactly 150 pounds we can determine the probability that a person will weigh between 149.5 and 150.5 pounds.

Normal Distribution

For a great many continuous variables a distribution called the normal distribution can be used to determine probabilities. A graphic representation of a normal distribution follows a pattern called the normal curve. The **normal curve** is bell shaped, peaked in the middle with long tapering tails. In a normal distribution the mean, median and mode are all right in the middle of the distribution. It is also symmetrical (i.e., the two sides are mirror images of each other). An example of a normal distribution is shown as Figure 5-A. There is a formula for a normal curve, but it's very complex and not important for our level of understanding.

Figure 5-A Normal Curve

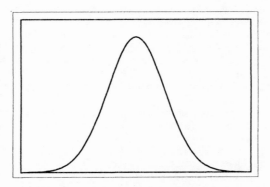

A data distribution is called **skewed** if its peak is more on one side of the distribution as opposed to being in the center. When the longer tail is on the left (Figure 5-B) the distribution is referred to as having a negative skew and when its on the right side the distribution is said to have a positive skew (Figure 5-C).

Figure 5-B Negative Skew   Figure 5-C Positive Skew

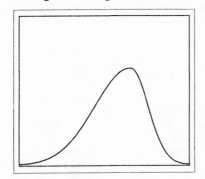 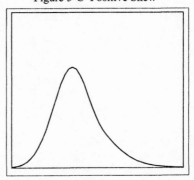

The distribution of people's weight tends to produce a normal distribution. Many other phenomena such as amount of rainfall, size of fish, cost of a pound of apples also take on

the appearance of a normal curve when graphed. The use of the term "normal" is possibly misleading - there is no necessity in nature that a continuous variables take on the shape of the normal curve. A continuous variable that takes on a different shape isn't abnormal. The word "normal" is used in the sense of being typical rather than having any connotation of violating an acceptable standard.

As an aside there is an alternative name for the normal curve - it's the **Gaussian curve** named for Karl Gauss, a German mathematician who did a great deal of the early theoretical work associated with the normal curve. However, Gauss was not the first person to discover the curve - that honor belongs to a French mathematician, Abraham De Moivre.

The normal curve is extremely useful because it allows us to determine probabilities for continuous variables. Test scores are usually considered continuous variables. We'll use test scores in our understanding of how the normal curve can be used to tell us when we have an unusual result. For instance, we have a good idea of what are unusual test scores - scores 98 or better don't occur that often - fortunately neither do scores below 60. In a graduate program we expect most test scores to be in the low 80s to the mid 90s. Since we have an impression of a test score distribution, we have a sense of what scores are peculiarly high or low.

To determine what's unusual in data sets, when we have little knowledge about the data, we would need to gather a massive set of measurements, graph them and then find out what values are extreme and therefore unusual. However, it would be better if we could gain a sense of what's unusual in a more efficient manner. An efficient way to obtain the probability of values in a data set is to use two statistical characteristics of the data set and make an assumption about the shape of the distribution. The statistics that we need to know are the mean and the standard deviation. The assumption we make is that the variable has a normal distribution.

What makes the normal distribution so valuable is that, as you move away from the middle of the distribution, a precise proportion of the values are included as you go to the left or right. Since the mean is at the mid-point and the distribution is symmetrical, we know that 50% of the values are below the mean and 50% above it. However, what is unique about the normal curve is that if you have a point 1 standard deviation to the left of the mean then approximately 34% of all the scores of the distribution are included between that point and the mean. Likewise due to the symmetry of a normal distribution, if the point were 1 standard deviation to the right of the mean, about 34% of all the scores of the distribution would be included in the interval between that point and the mean. By moving out 2 standard deviations to the left of the mean you'd include roughly 48% of the scores and moving 2 standard deviations to the right would also include about 48% of the scores. This means that around 96% of the values will fall within an interval that is 2 standard deviation below the mean to two standard deviations above the mean. This relationship between standard deviation distances from the mean and proportion of values included in the interval is a very powerful tool. This means that we can tell how unusual any value is <u>providing</u> we know the population mean, the population standard deviation and have reason to believe the data set is normally distributed. A diagram of the normal curve with the percentage of scores included for various standard deviation units is shown as Figure 5-D.

Figure 5-D

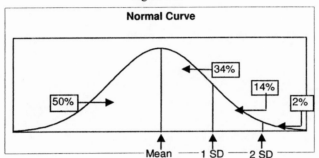

The normal distribution is a powerful tool in statistics. Although there are an enormous number of variables that have a normal distribution, you can describe the shape of any one with only two numbers – the mean and standard deviation. Knowing the mean and standard deviation allows you to know the shape of the entire distribution.

As an example, think of an SAT test which has a mean of 1000 and a standard deviation of 100. Also assume that the distribution of the test scores is normal in shape. A graph of that relationship is shown as Figure 5-E on the next page. Let's see how much we can say about this distribution just by using the mean, standard deviation plus the normality assumption.

We know that 50% of the scores will be less than 1000 and 50% will be greater than 1000. We also are aware that about 34% of the scores will fall between 1000 and 1100 because one standard deviation equals 100 points, the distance between 1000 and 1100. We can then figure out that scoring above 1100 occurs only 16% of the time because 50% of the scores are less than 1000 and 34% are between 1000 and 1100 leaving the remainder or 16% above 1100.

Remember that the standard deviation is 100 and the mean is 1000. With this information we can determine that 48% of the scores fall between 800 and 1000 because 2 standard deviations equals 200 points, the distance between 800 and 1000. We also know how unique scores below 800 are. A score of 800 is 2 standard deviations below the mean. Since 48% of the scores are contained in the interval from the mean to 2 standard deviations below it, that leaves 2% of the scores under 800.

Figure 5-E

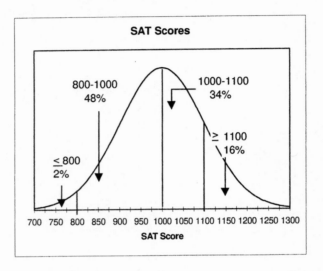

In Figure 5-F (see next page) we show a hypothetical normal distribution for men from outer space where the mean height is 40 inches with a standard deviation of 7 inches. By using the same approach as used above we'd be able to say a lot about what constitutes tall or small men from outer space.

We'd also be able to easily pick out obese or thin wombats (an Australian mammal) if we were told the species has a mean weight of 40 grams, a standard deviation of 3 grams and the weights are normally distributed. The normal distribution for adult wombats is presented in Figure 5-G. Both of these variables (men from outer space and wombats) have the same mean (40), but the distributions look somewhat differently since each variable has a different standard deviation. They are also in different measurement units - inches for outer space men and grams for wombats.

Z Scores

Even though the means in Figure 5-F and Figure 5-G are the same, the difference in the standard deviations creates different looking curves. The mathematical calculations needed to produce each curve is tedious because of the complexity of the normal curve equation.

Figure 5-F   Figure 5-G

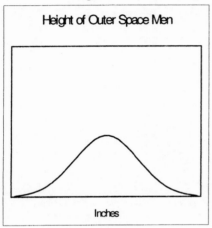

Height of Outer Space Men

Inches

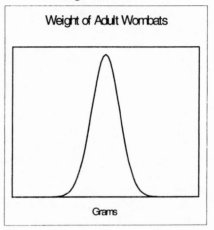

Weight of Adult Wombats

Grams

To create a unique relationship for each data set would be a monumental task. Fortunately by standardizing the data, one curve can be prepared that can be applied to all data sets. The goal of standardization is to transform all normally distributed variables to a distribution that has a mean of 0 and a standard deviation of 1. We carry out the transformation by creating Z scores. The formula for a Z score is:

$$z = \frac{x - \mu}{\sigma}$$

where **Z** is the **standardized normal value**, X the value of an element in the original data set, μ the population mean of the data set and σ the population standard deviation of the data set. Frequently the standardized normal value is called a **Z score** or a **Z value**.

The mean of the transformed data set is always 0 and the standard deviation will always be 1. In other words no matter what kind of variable you begin with (as long as it is normally distributed), after you convert each raw data elements (X) to a Z score, the converted distribution will have a mean of 0 and a standard deviation of 1. We can show what occurs using the following sequence:

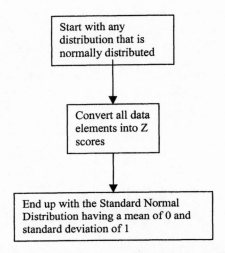

Consequently the transformation to the Z scores allows one curve to handle all variables that are normally distributed. The name of the curve using Z values is the **standard normal curve** (See Figure 5-H).

Figure 5-H

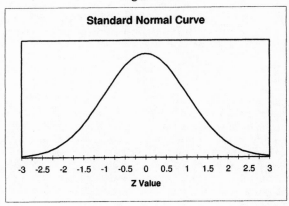

Note that the X axis of a standard normal distribution is composed of Z values (i.e. the transformed values). Z scores are also frequently called **standard scores**.

The total area under the standard normal curve is assigned the value 1. Assigning the whole area below the curve a value of 1 makes a lot of sense. Think of all the area involved as 100%. There is 50% of the area to the left of the mean (remember the mean is at Z = 0) and 50% to the right. There is a mathematical relationship between percentages and proportions. A percent divided by 100 equals a proportion. The value 50% is equal to the proportion .50. All we do to convert a percent to a proportion is to divide by 100. Ten percent is equal to the proportion .10, 20% equals the proportion .20 etc. Consequently 100% is the same as the proportion 1.00.

The advantage of thinking of the area as a proportion rather than as a percent is that proportions are the way we talk about probabilities. The probability .25 means the proportion of times an event occurs is .25 or, stated as a percent, it means the event occurs 25% of the time. Thus, areas between any two Z values are essentially probabilities. The area from the extreme right hand side of figure 5-H to the point where Z is 0 (the mid point of the scale) is .50. We can also say that the probability that someone scores above the mean is .50.

To summarize, Z values make up the X-axis scale of the standard normal curve. A Z value is the number of standard deviations a raw score is from the mean. The Z values are scaled differently than we usually find for X-scale variables. A Z of 0 corresponds to the location of the mean of the distribution. Also note that Z = 0 is at the center of the distribution rather than the left hand side. Why this is done will be clear in a moment.

In our SAT example, A SAT score of 1100 is 1 standard deviation above the mean since we defined the mean as 1000 and the standard deviation as 100. Let us return to the formula for a Z score. It is:

$$Z = \frac{X - \mu}{\sigma}$$

In the formula X represents the SAT value of interest (1100) - it is any value in the data set in which we have an interest. μ is the mean SAT (1000) and σ is the standard deviation (100). The Z score corresponding to a SAT value of 1100 is:

$$Z = \frac{1100 - 1000}{100} = \frac{100}{100} = +1.0$$

The + sign is intentionally included in the answer. As we can see in Figure 5-H, negative Z scores are frequent (they make up the whole left-hand side) and it's best to reinforce the positive value of Z by including the + sign. Negative Z score should always be preceded by a minus sign.

As noted above, the Z score corresponding to the mean SAT value is zero and we can verify this fact by substituting the mean SAT score of 1000 into the X formula. Our result is Z = 0 (see below).

$$Z = \frac{1000 - 1000}{100} = \frac{0}{100} = 0$$

We can use the standard normal curve and Z values because there are fixed areas under the standard normal curve between any two Z values. For example, between Z=0 and Z=+1, lies .34 of the total area of a normal curve. In our example, we have found that a SAT score of 1000 corresponds to Z=0 and a SAT score of 1100 corresponds to Z=+1. Since areas of the normal curve correspond to probabilities and we know .34 of the area is between a Z of 0 and a Z of +1.00, we could say that there is a .34 probability of someone scoring between 1000 and 1100. Between Z=+1 and Z=+2 lies an additional .14 of the area and we could say that the probability of someone scoring between 1100 and 1200 is .14.

Since the standard normal curve is symmetrical, the area between Z = 0 (the mean SAT score of 1000) and Z = -1 (a SAT score of 900) is also .34. Thus, the probability of someone scoring between 900 and 1000 is also .34.

So far we have been dealing with Z values that turn out to be whole numbers such as 1 or 2. What about finding the area corresponding to a Z of +1.25 which will occur with a SAT score of 1125

$$Z = \frac{1125 - 1000}{100} = \frac{125}{100} = 1.25$$

To find the answer we need more information about the distribution of the standard normal curve.

Areas Under the Normal Curve

Tables have been constructed giving the areas that correspond to various values of Z. The tables are usually titled "Areas Under the Normal Curve". A portion of a typical table is shown on the next page.

A complete Area Under the Normal Curve Table is provided as Table 1 in the Appendix. The abbreviation AUNC (Area Under the Normal Curve) will be used to refer to this table.

Areas Under the Normal Curve

Z	Area in left tail	Area in right tail	Z	Area in left tail	Area in right tail	Z	Area in left tail	Area in right tail
0.0	0.5000	0.5000	**0.5**	0.6915	0.3085	**1.0**	0.8413	0.1587
0.05	0.5199	0.4801	**0.55**	0.7088	0.2912	**1.05**	0.8531	0.1469
0.1	0.5398	0.4602	**0.6**	0.7257	0.2743	**1.1**	0.8643	0.1357
0.15	0.5596	0.4404	**0.65**	0.7422	0.2578	**1.15**	0.8749	0.1251
0.2	0.5793	0.4207	**0.7**	0.7580	0.2420	**1.2**	0.8849	0.1151
0.25	0.5987	0.4013	**0.75**	0.7734	0.2266	*1.25*	*0.8944*	*0.1056*
0.3	0.6179	0.3821	**0.8**	0.7881	0.2119	**1.3**	0.9032	0.0968
0.35	0.6368	0.3632	**0.85**	0.8023	0.1977	**1.35**	0.9115	0.0885
0.4	0.6554	0.3446	**0.9**	0.8159	0.1841	**1.4**	0.9192	0.0808
0.45	0.6736	0.3264	**0.95**	0.8289	0.1711	**1.45**	0.9265	0.0735

There are columns in the AUNC table for Z values. Locate 1.25 in the Z column. The cell is italicized for purposes of this demonstration. The column to the right of the Z column is labeled "Area in left tail" and the number in that column is .8944. This number, .8944, is the area of the curve which lies to the left of the Z value +1.25 - i.e. it is the area in the left hand tail of the curve. The next column titled "Area in right tail" gives the area of the curve which lies to the right of the Z value of +1.25. The area between 1.25 and the end of the curve on the right side is .1056. The two regions identified above are shown in Figure 5-I below.

Figure 5-I

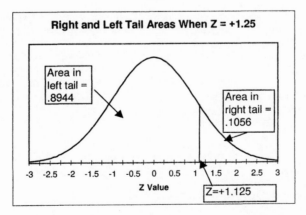

Note that the AUNC table in the Appendix (Table 1) contains negative and positive values. As shown in Figure 5-I, the negative values are all on the left hand side of the curve and the positive values are all on the right hand side. Also note that due the symmetry of the curve, the area of the curve which lies between the center (the point at which Z = 0) and the calculated Z score of -1.0 is the same as the area between the center and a Z=+1.0 (they are both .3413). After you reach the center of the curve (where Z = 0), and start moving to the right-hand side, the Z values are all positive. Also note that if go to the beginning of Table 1 of the Appendix and you look at the values in the column titled "Area in left tail", they begin with values near zero (when Z = -3.0) and approach 1.0 when you are at the end of the table (Z = +3.0). Also observe that the entries in the column titled "Area in left tail" are cumulative areas as you move from left to right along the axis.

If we go back to the SAT data and think in terms of probabilities we can re-interpret what we said earlier. To guide us through this exercise we can use the following three step procedure:

1. Convert the value of interest in the original data set (denoted as X) to a Z score using the data set mean (μ) and standard deviation (σ). The conversion formula is:

$$Z = \frac{X - \mu}{\sigma}$$

2. Look up the Z score in the AUNC table (Table 1 of the Appendix) and find the area to the left of the Z score (using the Area in left tail column) and the area to the right of the Z score (using the Area in right tail column).
3. Interpret the area in the left tail as the probability of obtaining values less than X. Interpret the area in the right tail as the probability of obtaining values greater than X.

Below is a demonstration on using the 3-step procedure. In the SAT example, assume our value of interest is 1125 so we:

1. Convert X=1125 to a Z score using the Z formula where μ = 1000 and σ = 100:

$$Z = \frac{1125 - 1000}{100} = \frac{125}{100} = +1.25$$

2. The area in the left tail for Z=1.25 is .8944 and the area in the right tail for Z=1.25 is .1056.
3. We conclude that the probability of scoring less than 1125 on the SAT is .8944 and the probability of scoring more than 1125 is .1056.

We can also use the AUNC table to find the probability someone will have a SAT score that is in a given range. Suppose we want to know the probability someone will score between 1000 and 1125. We already know the probability of scoring less than 1125 – it's .8944. However, we need to know the probability of scoring less than 1000. Using our 3 step procedure we get:

1.   Convert X=1000 to a Z score where μ = 1000 and σ = 100 using the Z formula:

$$Z = \frac{1000 - 1000}{100} = \frac{0}{100} = 0.0$$

2.   The area in the left tail for Z = 0.0 is .5000 and the area in the right tail for Z = 0.0 is .5000.
3.   We conclude that the probability of scoring less than 1000 on the SAT is .5000 and the probability of scoring more than 1000 is .5000.

We can now take the difference in the probability of scoring less than 1125 and subtract off the probability of scoring less than 1000. The difference between the two probabilities is .3944 (.8944-.5000). We conclude that the probability of scoring between 1000 and 1125 is .3944. Figure 5-J shows this probability diagrammatically (see next page).

Figure 5-J

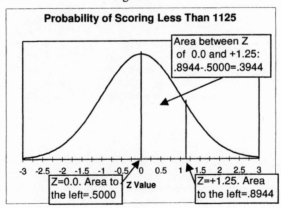

We will take a moment to show how the 3 step procedure works for values that result in negative Z scores. Suppose we'd like to know the probability associated with a SAT score of 850.

1.   Convert X = 850 to a Z score where μ = 1000 and σ = 100 using the Z formula:

$$Z = \frac{850 - 1000}{100} = \frac{-150}{100} = -1.50$$

2.   The area in the left tail for Z = -1.50 is .0668 and the area in the right tail for Z = -1.50 is .9332.
3.   We conclude that the probability of scoring less than 850 on the SAT is .0668 and the probability of scoring more than 850 is .9332.

It is important to appreciate what is going on when we used our 3 step procedure. We began with an observed value (in this case an SAT score), we converted it to a statistic (in this case the Z score) and then used a table (in this case an AUNC table) to tell us how unusual the original value was (i.e. the probability of a obtaining the observed SAT score). Almost all our statistical testing is based on a similar process. When we encounter other statistical methods (e.g. t tests, Chi Square tests, etc.) they will also involve calculating a statistic and relying on tables to determine probabilities. The practice of using a table to determine probabilities was the common approach until the advent of computers. In the computer age we can often omit tables since the computer can provide the probability information directly.

When working with the standard normal curve there is a peculiarity that should be mentioned. There is no probability associated with a specific value for a continuous variable. For example we do not obtain a probability for scoring exactly 1000 on the SAT. As noted earlier, continuous variables do not have precise values such as 1000. If we had a better measuring instrument the value would be a little larger or smaller than 1000. Consequently, when we calculate probabilities with the standard normal curve, we always find a probability for a range of values. Now if we rephrased the question to ask what is the probability of scoring between 999.5 and 1000.5 then you can calculate a probability. In this case it would be infinitesimally small, but you could calculate it.

Outlier

Statisticians are very concerned about recognizing values in a data set that are aberrant. These atypical values are referred to as **outliers** because they appear not to belong to the main body of data points. Outliers that are truly non-members of a data set can cause a large distortion in the mean and standard deviation of the data set. Although there is no hard and fast rule on what an outlier is, some statisticians will inspect any value that is more than three standard devations above or below the mean. Most AUNC tables include area between −3.00 and +3.00, but very small areas are associated with more extreme Z values. If the inspection shows that the data element is erroneous, then it usually will be discarded. However, if the inspection does not clearly show that the data element is an invalid member, it should be retained.

Practical Problems

Some of the kinds of problems we can solve by using the AUNC table are given below. We'll stay with our SAT example where $\mu = 1000$ and $\sigma = 100$.
1.   What score do you need to be in the top 10% of those taking the SAT test?
2.   What score cuts off the lowest 20% of the scores?

How you go about answering each question is given below. A picture is constructed for each problem and it is recommended that you always prepare a figure when you work with AUNC problems. The figure will keep you from making mistakes and the visual reinforcement will enhance your understanding.

Problem 1. What score do you need to be in the top 10% of those taking the SAT test? First of all, realize that for this problem we need to find a SAT score (i.e. an X value) rather than an area. In this case, we are given the area under the curve - it's 10%. As a result, we next need to find the Z value matching that 10% area. Then we'll use the Z formula to solve for X (the SAT score). We know that 10% is equivalent to .10 when expressed as a proportion and, since the problem asks for the "top" 10%, we need to focus on the right hand side or positive side of the curve. We look through the column in the AUNC table headed "Area in right tail" till we find the value .10. There is no value exactly matching .10, but .1003 comes very close so we will use it. The Z value corresponding to the .1003 area is 1.28. We now substitute that Z value in our formula and solve for X. We already know the mean (1000) and standard deviation (100) so X is our only unknown. Our formula looks like this:

$$1.28 = \frac{X - 1000}{100}$$

If you forgot your algebra you first multiply both sides of the equation by 100. That eliminates the fraction on the right hand side and you get:

$$128 = X - 1000$$

You now want to isolate the X so add 1000 to both sides of the equation and you get:

$$1128 = X$$

That's the answer - a score of 1128 cuts off the top 10% of the scores. We can write a single expression when we want to solve for the X in the Z formula. The equation we use is:

$$X = (Z * \sigma) + \mu$$

and after substituting the appropriate values in this formula you'll get your answer.

$$X = (1.28*100) + 1000$$
$$X = 128 + 1000$$
$$X = 1128$$

Your answer is that a score of 1128 cuts off the top 10% of the SAT scores. Draw the figure that shows the area of the standard normal curve matching our answer below.

The figure you drew should look like that shown in figure 5-K.

Figure 5-K

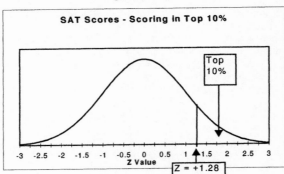

Problem 2. The score that cuts off the lowest 20% of those taking the test is found by following the steps used in problem 1, but remember now we're on the left (i.e. the lower or negative side of the curve). We again begin by finding a Z value - in this case one that matches a .20 area in the "Area in left tail". We don't have an exact match, but .2005 is very close to .20 so we will use it. The Z value corresponding to the .2005 area is a minus 0.84 (i.e. −0.84). We now substitute that Z value in our formula and solve for X. We already know the mean and standard deviation so X is our only unknown. Our calculations are:

$$X = (-0.84*100) + 1000$$
$$X = -84 + 1000$$
$$X = 916$$

We have our answer - a score of 916 cuts off the lowest 20% of the scores. Draw the figure that shows the area of the standard normal curve that matches our answer below:

Your drawing should look like Figure 5-L.

Figure 5-L

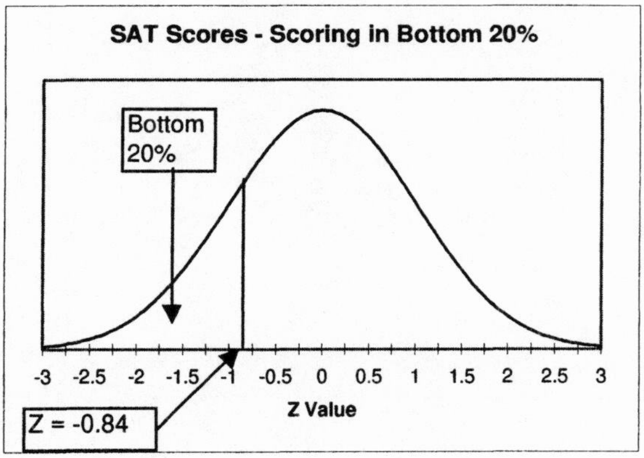

Using the AUNC Table for Binomial Variables

Up to this point we have maintained a separation between distributions for discrete and continuous variables. More precisely, we've used the binomial distribution for discrete variables which have only two outcomes. We'll use the expression "binomial variable" for this class of variables. We've also used the normal distribution for continuous variables. It would be useful to be able to use Z scores and the AUNC properties for binomial variables as well as continuous variables. Fortunately, there are circumstances when this can be done.

When we looked at the binomial distribution in the last chapter (see Figure 4-A) we saw that it didn't look like a normal distribution. It had a step ladder appearance - as you go from one discrete outcome to the next there is a abrupt change in the bar. In chapter 4 we tended to deal with a relatively small number of trials. In addition, for the laboratory test example, the probability for a positive test was fairly rare (i.e. a .1 probability for a positive test). As noted in chapter 4, the generic term for the positive result can be defined as a "success". What happens to the distribution when we have probabilities of successes and failures around .5 and there are a large number of trials? Let's examine a coin toss since the probabilities for heads and tails are .5. What the distribution looks like when we do only 5 tosses (i.e. trials) is shown in Figure 5-M below.

Figure 5-M

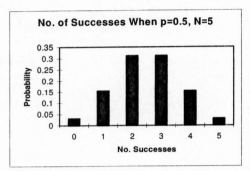

If we do 20 tests our picture looks like that shown in Figure 5-N below. It should be evident that as we increase the number of trials we come closer to creating a distribution that looks like a normal distribution.

Figure 5-N

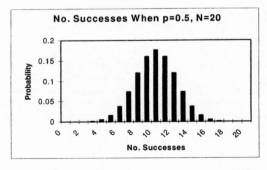

What about the case of small probabilities? If we use a die and define a success as a throw which produces a 3, then there is 1 chance in 6 of having a success. The alternative state is not getting a 3 and that has a 5/6 chance of occurring. As can we seen in Figures 5-O, 5-P, 5-Q and 5-R as we increase the number of trials (the number of trials is given as N in the title of each figure) from 1 to 30 there is a tendency for the distributions to begin to look more like a normal distribution.

Figure 5-O

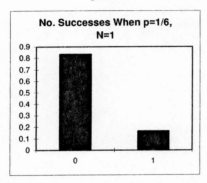

Figure 5-P

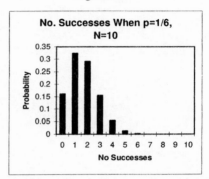

Figure 5-Q

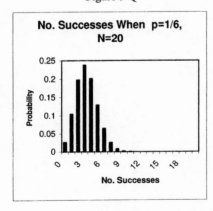

Figure 5-R

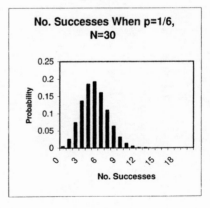

The point is that when we have a large enough N and a p around .5 a binomial distribution will resemble a normal distribution well enough so we can take advantage of the normal curve methodology for determining a probability.

The rule of thumb as to when n and p are the right size is that n*p and n*(1-p) must both be greater than 10 for you to assume that a binomial variable has a normal distribution. As a result 1] small sample sizes and 2] extremely large or small values of p may rule out using the normal curve assumption when dealing with binomial variables. Note, that we could not use the normal distribution for the conditions set forth in Figure 5-R because n*p = 30*1/6 = 5 and that value is less than the required value of 10.

When the required values of 10 are met, we can use the normal curve approximation. You need to determine the expected value of the mean and the standard deviation of the binomial variable. The formula for the expected value of the mean for a binomial variable is:

$$\mu = n * p$$

The corresponding formula for the standard deviation is:

$$\sigma = \sqrt{n * p * (1 - p)}$$

To appreciate what is meant by the expected value for a binomial variable, visualize a coin toss which you repeat 36 times. At the end, you record the number of heads that turned up during the 36 tosses - it could be 15 or 18 or 20 or whatever. If you then repeat the series of 36 tosses you again will have a number of tosses that turn up heads. If you continue doing this you will create a data set containing counts such as 18, 15, 17, 21 etc. Assume we collected a massive number of results and formed them into a data set. From the data set you can calculate a mean and a standard deviation. The mean of the data set is the expected value for the binomial variable. The standard deviation of the data set is the standard deviation for that mean.

The above formulae for μ and σ allow us to directly obtain the mean and standard deviation. If we apply the formulae using .5 as p (the probability of a head) and an n of 36 we get:

$$\mu = 36 * .5 = 18$$

and:

$$\sigma = \sqrt{36 * .5 * .5} = 3$$

When we flip a coin 36 times the expected number of heads is 18 with a standard deviation of 3.

We now will work an example to show how we can use the Z formula for a binomial variable. In our laboratory test illustration, there is a .1 chance of a positive result. What if we have a new lab technician and, after his first 100 tests, it is found that 15 of the tests had a positive result. Is there the possibility that something is wrong - is he getting too many positive tests because of poor technique or is this result fairly common and there is no need for alarm? It's clear that it would be helpful to know how often we might find at least 15 positive tests when the average chance of getting a positive result is .1. To arrive at an answer we should write down what we know: n = 100; p = .1 and 1-p = .9. We would like to solve this problem using the normal curve properties so first we'll check to see if n*p and n*(1-p) are 10 or more. The n*p value is 10 (100*.1) and the n*(1-p) value is 90 (100*.9) so we may proceed.

We'll now calculate the expected mean and the standard deviation. To calculate the mean we use the formula given earlier and after substituting values for n and p we get:

$$\mu = 100 * .1 = 10$$

Substituting the n, p and (1-p) values in the standard deviation formula we get:

$$\sigma = \sqrt{100 * .1 * .9}$$
$$\sigma = \sqrt{9}$$
$$\sigma = 3$$

We next calculate a Z value using the Z formula:

$$Z = \frac{X - \mu}{\sigma}$$

Our X is the number of tests that are positive, which is given as 15. As noted above μ = 10 and σ = 3 so we now have all the numbers we need to find Z and:

$$Z = \frac{15 - 10}{3} = \frac{5}{3} = +1.67$$

Before we look up Z=+1.67 in the AUNC table we should recall that there is no probability associated with a specific Z value. Instead we must work with ranges rather than specific values.

For this problem we do not ask what is the probability of getting exactly 15 positive tests. What we are interested in is: how many times will we get 15 or more positive tests.

If getting 15 or more is fairly common we should not be concerned about the new technician. To find the probability of 15 or more positive tests we find the row for Z=+1.67 in the AUNC table. We note that the "Area in right tail" is listed as .0475. This is the probability of getting 15 or more positive tests. Should we be concerned? At this point we must rely on our own judgment. Some people may be willing to accept the .0475 probability as reasonable and do nothing. Others may be very alarmed and want to remove the technician. Most, I suspect, would treat the result as a warning - they might want to review the new technician's technique to see if it is faulty or they might want an experienced technician to re-run all the test to see if the experienced person gets the same results. We've used statistics to tell us how rare an event is. What we should do with the information is a matter of individual choice. We'll see later where rules of statistical decision making can be invoked, but at this point we are left to rely on our own judgment regarding a course of action.

Remember that the use of the standardized normal curve for a binomial variable gives an approximate and not a precise result. For example in the above problem had we used the binomial formula to arrive at the answer we would have found the probability to be .0327 rather than .0475. The approximation, using the standardized normal curve approach, can be improved by making an adjustment to the X value (in our example the X is 15). We will not explain how the adjustment works. In the computer age one can quickly obtain the precise answer using computer programs that calculate the binomial formula. As a result, trying to learn how to improve a result that gives us a better approximation doesn't seem worthwhile. However, it is important to emphasize that even binomial variables can use the standardized normal curve under certain conditions, but the result one obtains is only an approximate value.

Summary

In this chapter we found that when we have a data set for a continuous variable we can determine how unusual values within that set are, if we know the mean and standard deviation and can assume that the variable is normally distributed,. We convert the individual values to a Z score and then use the AUNC table to give us the corresponding probability of occurrence. We also learned that we could adapt the process and find a specific value that cuts off a given proportion of the distribution. In addition, we discovered that under special circumstances, we can apply the Z score approach to binomial variables as well as continuous variables.

PROBLEMS

1. Draw four graphs that represent the standard normal cure (See Figure 5-6). A free hand drawing is fine, but try to make sure your X axis has properly spaced Z values on it from -3 to +3. On the graphs you prepared, shade in the area that contains the areas specified below:
   a. The upper 16%
   b. The central 84%
   c. The lower 20%
   d. The central 50%

2. Based on question 1, write down next to each graph the upper and lower Z value that corresponds to your area (Use Z =+∞ for the extreme right-hand side of the normal curve and Z = -∞ for the extreme left-hand side. The ∞ symbol means infinity).

3. Assume a variable, which has a normal distribution, has a mean of 70 and a standard deviation of 20. For the following scores, what is the corresponding Z value?
   a. 70
   b. 80
   c. 65
   d. 20

4. What Z value on a standard normal curve corresponds to the following percentile scores: (The 30th percentile score means 30% of the other scores are below it and 70% are above it.)
   a. 30th
   b. 50th
   c. 95th
   d. 15th

5. What is the difference in the X axis for a normal distribution and a standard normal distribution?

6. To obtain a higher position in government, it is often necessary to take a civil service test. Assume a test has a normal distribution, a mean of 150 and a standard deviation of 20. Determine the proportion of test scores that will be:
   a. Below 125 (You can round all answers to 2 decimal places)
   b. Between 120 and 160:
   c. Below the mean
   d. Higher than 200

7. The mean score for a normally distributed aptitude test is 24 with a standard deviation of 4. Find the aptitude score that will be used to cut off the:
   a. Top 10% (You can round all answers to 1 decimal place)
   b. Lowest 10%
   c. Top 75%
   d. Lowest 33%

8. Assume there is a normal distribution for a variable with a mean of 500 and a standard deviation of 40. What is the probability that a person will score:
   a. 600 or higher (You can round all answers to 3 decimal places)
   b. 440 or less
   c. Between 580 and 620
   d. Between 500 and 400
   e. Between 450 and 550
   f. Part d and e both include a spread of 100, why don't they give the same answer?

9. How much area falls below the following Z values:
   a. 0.45
   b. -0.45
   c. 2.40
   d. How much area is between Z=-1.05 and Z=.70?

10. In roulette there are 38 outcomes possible and include the numbers 1 - 36 plus a 0 and a 00 outcome. In addition, in the set of numbers 1-36 half (18) of the numbers are red and half (18) are black. The 0 and 00 outcomes are not colored red or black.
    a. What is the probability of getting a red number?
    b. What is the probability of getting a red or an even number?
    c. What is the probability of getting a red number on one spin of the roulette wheel and an odd number on a second spin?
    d. What is the probability of getting two black numbers on consecutive spins of the roulette wheel?

11 Assume you bet a $1 chip on one spin of the roulette wheel and your bet is that an even number will come up. You will collect $2 if an even number turns up, any other result means you lose your bet.
    a. What is the expected value for this situation?
    b. If you were a completely rational person would you play the game of roulette described above?

12 Ten individuals are selected to have an interview with the bureau chief.

    a. How many ways can the order of the interviews be arranged?

    b. If 4 of the 10 will be selected for a special development course, how many ways are there to select them?

The leadership role for the group of 10 individuals is chosen at random and the leadership role changes every month. If 6 of the 10 people are minorities what's the probability of the following events?

    c. A minority person is chosen as leader only once in the first 3 months?

    d. A non-minority person is chosen as leader three times in a 4 month period?

    e. At most 2 (i.e. 0, 1 or 2) individuals who are selected as leader are non-minorities over the first 4 months.

    f. At least 2 (i.e. 2, 3 4, 5 or 6) of the individuals selected as leader is a minority in a 6 month interval?

Excel Instruction

There are a number of convenient statistical functions in Excel that can help us do the exercises discussed in chapter 5. Be aware that in the examples, the answer will be expressed to two decimal places. In actuality your answer may contain a different number of places depending on how your format specification is set.

The first function we'll examine is called STANDARDIZE. STANDARDIZE refers to the standardized Z value we calculated in the first part of the chapter. This function computes the Z value when you supply the appropriate X, μ, and σ values. The format of the function is:

$$=STANDARDIZE(X,\mu,\sigma)$$

To find the Z value for a SAT problem where the X is 1100, μ is 1000 and σ is 100, we would enter:

$$=STANDARDIZE(1100,1000,100)$$

and the computed Z value (1.0) would be provided.

Another function in Excel that we can use is called NORMDIST which stands for Normal Distribution. Do not confuse this function with a similarly named function called NORMSDIST (i.e., with an S as the fifth letter) which we will discuss next. With NORMDIST you can use the appropriate X, μ, and σ values to obtain the area under the normal curve directly. The format for NORMDIST is:

$$=NORMDIST(X,\mu,\sigma,TRUE)$$

For example. this function can give us area in the left-hand tail corresponding to an X = 950 when μ = 1000 and σ = 100 . The last entry "TRUE" is also required in the expression in order to return the proper area. In our example, we would enter the following:

$$=NORMDIST(950,1000,100,TRUE)$$

and Excel will return the area as .31. We conclude that there is a .31 probability that a person would score 950 or less.

A related function is called NORMSDIST and in this case you obtain the area by directly entering the Z value rather than X, μ, and σ. The S in the name signifies standardized, and indicates that you will be working with Z values. The format is:

$$=NORMSDIST(Z)$$

If we determined that Z was -.50, we would enter the following expression:

$$=NORMSDIST(-.50)$$

in a cell and Excel would return the value .31. Excel only returns the left-hand area of a tail. We need to do some simple arithmetic to get the value for the right-hand tail. If we wanted to know how much area is in the right-hand tail we need to subtract the area NORMDIST gives us from 1. In Excel we would enter the following expression:

$$=1-NORMDIST(950,1000,100,TRUE)$$

The result Excel produces is .69. There is a .69 probability that a person scores 950 or more on the SAT.

The last function we can use is called NORMSINV which computes the Z value corresponding to a given area under the normal curve. The INV in the title indicates that the inverse relationship will be calculated. Thus we can enter an area and find the corresponding Z value with NORMSINV. The format is:

$$=NORMSINV(AREA<X)$$

The AREA entry refers to the area of the normal curve that is less than X. To illustrate the use of this function, we will use the data for our SAT example and ask what score cut off the upper 10 per cent of SAT scores. First we must find the point on the Z scale that has .90 of the area below it and .10 of the area above it. By using =NORMSINV(.90), Excel would respond with the correct Z value (1.28). To determine the actual score we would need to use the Z formula and solve for X. This could be done by entering the following command in a cell:

$$=1.28*100 + 1000$$

and you would get 1128 as your answer.

You can also access each of the functions described above through the function icon (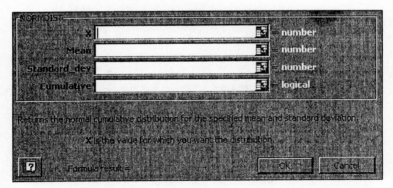.) Click on the function icon and you'll be at a screen titled "Paste Function". Select "Statistical" from the Function Category listing and you will have a listing of functions in the right hand portion of the dialog box (i.e. the section titled "Function Name"). You will find each of the functions discussed above in that listing.

As an example of how to use the Function command click on the Function icon and select "Statistical" from the Function Category listing. Click on NORMDIST from the Function Name list and then hit OK. You will be at a screen titled "NORMDIST" which contains a series of boxes to complete (see below).

We're familiar with the SAT test probabilities so we'll use SAT scores for our example. Take the case where the SAT scores have a mean of 1000 and a standard deviation of 100 and we want to know how many scores will be 950 or less.

The first box on the screen is labeled "X" and corresponds to the X we've used in used in our formulae. In our example X is 950. Type in 950 and then click on the next box titled "Mean". Again we know what the mean value is - it's 1000 so enter that value in the mean box. Next click on the box titled "Standard_dev". This is also a familiar term so type in 100 which is the standard deviation for our problem. The last entry, titled "Cumulative", essentially asks a question - do you want the cumulative area between the far left point of the normal curve to the point identified as X? Always type in "True" for this item. If your screen looks like the one below, hit the OK button and your answer will appear on the worksheet.

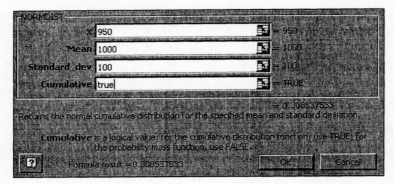

In this case the answer shown is .31. the probability of scoring 950 or less is .31.

The screens you will have for the other functions, Standardize, Normsdist and Normsinv, are shown below.

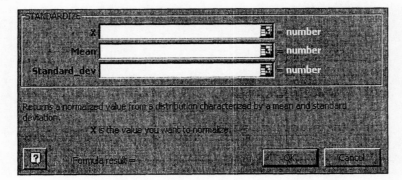

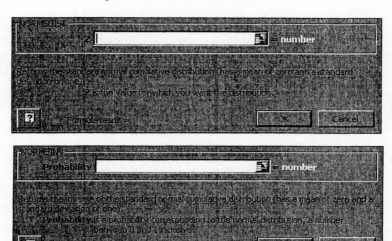

The only other new technique presented in Chapter 5 was the drawing function to add information to a graph. Although the use of the drawing tools in Excel are not necessarily statistical attributes they can be useful and are worth a brief explanation. One of the options on the Standard Toolbar is the drawing tool. The icon for this feature is located on the right hand side and is a picture of a number of different shapes - a triangle, square and circle [▓].

In Figure 5-I, comments and arrows were added to the graph of the normal distribution of SAT scores to note the amount of area associated with different portions of the curve. Comments such as those shown in Figure 5-I are added by selecting the drawing icon from the main toolbox row. A new toolbar with the title Drawing will appear on the screen. From the drawing toolbar click on the text box button - the text box button is an icon that looks like a sheet of paper with the letter A in the upper right hand corner, ▓. After you click on the text box button the pointer will become a cross which you then use to outline the size of the written instruction you wish. You hold down the left mouse key and drag the mouse to form the shape you desire. When you have finished outlining the box you release the mouse key. In the left-hand corner of the box a flashing marker will appear to mark the location where you can enter text. Type the text you wish using the keyboard in the normal fashion. Note that the text box has a box around it with small squares (i.e. handles) on the borders. You can use the squares to resize the box. By moving the pointer to a border, you can then drag the entire box to a new location. When you have completed typing the text, and are satisfied with the size and location of the box, move the pointer out of the box and hit the left mouse key. Your text box will now be a part of the display.

To add a directional arrow, locate the arrow tool which is on the top row of the drawing toolbox - it looks like an arrow (). Select it by left clicking the mouse. The pointer will again become a cross. Move the mouse to the place you want your arrow to begin and then, holding down the left key, move the pointer in the direction you want it to go. When your arrow is in the correct location let go off the left-hand key and the arrow will appear. To delete a drawing object, click on the object and when the handles appear use the delete key to eliminate the object. To delete the Drawing Toolbar go to the Standard Toolbar, click on the drawing toolbar icon and that toolbar will disappear.

There are many more things that can be drawn and there are a large number of formatting options. You may wish to refer to an Excel manual to learn about these features. More detailed instructions on how to revise drawn objects are also given in the manuals.

To practice the techniques described, try the following problems:

1. Refer to chapter 5 and use the Excel function STANDARDIZE to calculate the Z value in problem 3.
2. Refer to problem 6 and use the NORMDIST function to calculate the answers to these problems.
3. Refer to problem 7 and use NORMSINV to calculate the answers to these problems.
4. Refer to problem 9 and use NORMSDIST to calculate the answer to these problems.
5. The precise probability of 15 or more positive test is .0399.
   a. Use Excel to confirm that that is the correct answer.
   b. What is the probability of finding 92 or more negative tests using the standardized normal curve approximation?
6. Create a pie graph for the creativity codes in your database and with the drawing object add a text box by each segment of the pie and in the box include the full name of the creativity code (e.g., insert the term "Very low level" in the box for code 1). Add arrows that connect the text boxes to their respective area of the pie.

Chapter 6

Measurements and Obstacles to Research

Introduction

It's been said that measurement is the foundation of all scientific effort. **Measurement** refers to the rules that are used to assign values to variables. In a scientific project, we frequently introduce an agent (e.g. a new way to treat hypertension) and evaluate an outcome variable (e.g., a person's blood pressure). When we evaluate a variable we take measurements of it. There are a variety of ways to measure a variable. We can measure whether a variable is present, its size, its area, its mass, etc. The measurement may be expressed as micrograms of a substance, absence of a disease, amount of money spent, number of people unemployed, time to run a maze, degree of support for a new health care program, millimeters of blood pressure, etc.

We are also free to select the setting in which we take a measurement. The measurement may be taken in a tightly controlled environment or one with loose controls. By way of illustration, we can record a person's hearing acuity in a sound booth or on a busy city street. It's true that one environment may be preferable to another but we, nevertheless, have a choice.

The point of these examples is to demonstrate that measurements have different characteristics. What we measure and how we perform the measurement are critically important to research projects. In this chapter we will see that measurements can be classified and evaluated in different ways. For instance, we can classify them in respect to the type of scale used to record the measurement. Researchers have established categories based on the way the units of measurement are designated. We will also study measurements in terms of their reliability and validity. Finally we will identify common measurement problems that we may run into during a research study.

145

Scales

A scale refers to how the units of measurement are assigned. The simplest system is based on the ability to just count items. The more advanced scales have the ability to place uniform numeric values along the scale. Normally statisticians refer to four types of measurement scales - nominal, ordinal, interval and ratio. We will treat interval and ratio scales as one type and refer to them as interval scales. A **ratio scale** is an interval scale that has a true zero point. The adjective "true" means there is an actual 0 point. Sometimes we have what could be called a constructed 0 point as opposed to an actual 0 point. In the constructed situation there is a scale with 0 in it because of the way we define 0. As an example, a 0 Fahrenheit temperature reading is based on a definition. We could define 0 differently as is done in the centigrade scale. For centigrade temperature 0 degrees is defined as the point at which water freezes. In contrast to the 0 point in temperatures, weight is a variable in which there is a true 0 point. For weight there is a null state - a point at which an object is weightless. The distinction between ratio and interval scales, however, has no bearing on what we will do, and we will therefore work with three rather than four measurement scales.

When statisticians talk about scales of measurement, they refer to the relationship between the phenomena being studied and the manner in which we express observations about that phenomenon. To illustrate, we may be looking at the phenomenon, attitudes toward the death penalty. We elect to express those attitudes on a scale. The scale may have just two points - favor or oppose. However, it could have more than two points by introducing categories such as extremely favorable, somewhat favorable, indifferent, etc. Although the same phenomenon is being measured, it can be done with different units of measurement.

Researchers have created categories or levels for the types of observations that can be made. The term **nominal scale** is used when the observed objects, persons or characteristics can only be placed into categories. The different categories in a nominal scale can differentiate elements, but lack the ability to distinguish the elements in terms of how much of a difference there is between them. For example, religious affiliation represents a nominal scale; we can label people by their religious membership and we can place them in categories. Contrast this level of classification with an interval scale such as height. An **interval scale** is a true numerical scale because the units that make up the scale are uniform. The height scale measures the length of an object in inches (or centimeters) and the units along the scale are the same length. When we measure the heights of people we can order the people from shortest to tallest, but we can also say how far apart they are from each other (e.g. 2 inches or 5 centimeters). For religious affiliation we can only say that this person or that persons belongs to a certain religious group.

In between the nominal and interval scale lies the **ordinal scale**. Like a nominal scale, an ordinal scale can identify and label elements. Like an interval scale, an ordinal scale can distinguish elements and place them in an order. However, there is not a precise numerical difference between the units of an ordinal scale. Thus, we can only rank elements in an ordinal scale (which is not possible with a nominal scale). In an ordinal scale we know A is bigger than B, but we do not know how much bigger it is. We'll use an example to better illustrate what this last statement means.

Frequently researchers construct ordinal scales. The scale consists of ordered categories and can be used to classify beliefs or attitudes. For instance we could have a scale regarding people's level of support on a public issue. People would be asked to indicate their level of support by picking from categories such as:

1) Strongly Agree
2) Agree
3) Indifferent
4) Disagree
5) Strongly Disagree

It's clear a "Strongly Agree" response represents more support than choosing "Agree". Observe also that those two choices are 1 unit apart on the scale (Strongly Agree is a 1 and Agree is a 2). It is also true the "Agree" response represents more support than an "Indifferent" answer. Again note that these two responses are also 1 unit apart on the scale (Agree is a 2 and Indifferent is a 3). In each comparison, the numerical difference in the scale is 1 unit. But in the mind of those answering the question, is the size of the difference between "Strongly Agree" and "Agree" the same size as the difference between "Agree" and "Indifferent"? Researchers are rarely in a position to say that the size of the differences is the same. The units of the scale are just assigned a value by the researcher. In reality the size of the difference between scale units is not known. In our example, they are 1 point apart on the scale, but the researcher only constructed the numerical scale - he or she made it up. The researcher assigned a 1 to Strongly Agree and a 2 to Agree, etc. The assignment of the 1, 2, 3, 4 and 5 was arbitrary. We say "arbitrary" since we could have used different numbers and people would still answer the same way since they are focused on the terms, not the numbers assigned to the terms.

In contrast, look at the interval scale of heights. The difference between 5 inches and 6 inches is the same as the difference between 6 inches and 7 inches. Both times the difference is 1 inch. The height scale has uniform differences between its levels - an ordinal scale does not. The size of the scale units in an ordinal scale is not uniform. In the interval scale we have the property of a constant amount of difference between adjacent scale points. Look at time - time for mice to complete a maze. If mouse A takes 24 seconds to run a maze and mouse B takes 48 seconds, we can use quantitative terms to distinguish the two mice. It took mouse B two times as long to run the maze. But if person A picks category 2 (i.e., Agree) in our attitude scale and person B picks category 4 (Disagree) we can not say that person A is two times more supportive than person B.

In an interval scale we can calculate statistics and they make sense. A mean height of 67.5 inches with a 5.1 standard deviation is valid and useful information. However, you can't do these statistics on a nominal or ordinal scale and get the same level of useful information. The categories of a nominal scale aren't numeric to begin with. With ordinal scales you could calculate a mean and standard deviation for the numbers assigned to the response categories. But unless we can assume the units between categories are equivalent, we may not be able to make much sense out of the statistics.

The three scales (nominal, ordinal and interval) can be thought of as in a hierarchy:

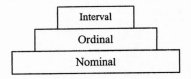

At the lower end is the nominal scale. At the top is the interval scale and in between the two is the ordinal scale. Based on this discussion of scales here are some examples - can you correctly identify them as nominal, ordinal or interval?

    A.   Coin flip
    B.   Job satisfaction rated as high, medium or low
    C.   Student ages
    D.   States in which students are born
    E.   Economic categorization as upper, middle and lower class
    F.   Football jersey numbers

The answers to questions are:

A. The coin toss represents a measurement that is nominal. A coin toss has no only two outcomes and those outcomes can not be ranked in terms of which one is better or higher or larger than the other one.

B. Question B on satisfaction levels represents an ordinal scale measurement. Satisfaction levels can be ranked, but the size of the units between categories is not a constant.

C. Age (question C) is a true quantitative measure with equivalent units between categories and meets the requirements of an interval scale.

D. The states make up a nominal scale measurement. We can place a student in a state category: Ohio, New York, Michigan, California, Illinois, etc., but we can't order the categories. Contrary to what many New Yorkers may believe, being born in New York doesn't allow a person to be ranked higher than one who was born in Michigan.

E. Question E on economic status allows us to rank economic levels, but the size of the difference between levels is not necessarily uniform so economic levels are an ordinal scale measurement.

F. The last item, football jersey numbers, is a nominal scale even though the variable is expressed as a number. The football jersey number is simply a convenient way to classify the players. You certainly can't calculate means and standard deviations on those numbers and end up with a meaningful result. This last example shows that the presence of a number doesn't guarantee that you have an interval scale measurement.

Sometimes what category a variable belongs to is not clear. Take letter grades - on the surface grades are in the ordinal measurement class. They clearly can be used to rank students, but you can't get an average till you convert the letter grade to a numerical equivalent. When we assign weights to grades (such as A = 4 points, B = 3 points etc.) and calculate a GPA we end up with a useful statistic. The underlying scale for a letter grade in many cases is based on a numeric scale of 90 or above = A, 80-89 = B. etc. and that numeric scale is similar to an interval scale. Consequently, we will see that sometimes a scale that appears to be in the ordinal class is treated as if it were an interval scale. Frequently the crudeness of a measuring devise obscures the underlying continuity of the units of measurement for a variable. The variable may in fact be at a higher scale level, but the way the data are presented places it in a lower category.

There are things you can do with an interval scale measurement that you can't do with an ordinal scale measurement so there is an incentive for researchers to want to claim that the variable they are using is an interval scale measurement. In many cases the decision to treat the variable as an interval scale measurement is justified. However, these types of judgments are best made by professionals in a given area of expertise. You need to have a deep understanding and a reasonable amount of experience with the variable when you give it a higher classification than it appears to merit on the surface.

Why learn about measurement scales? The measurement classification for a variable is important to statisticians since different statistical testing methods can be done on each scale. In general, the higher the scale the more powerful the statistical procedure. In later chapters we will review some of the most popular statistical methods used with nominal, ordinal and interval scaled measurements. We will see that for a nominal scaled variable a somewhat limited number of manipulations and statistical tests are appropriate. Nevertheless, because of the limited number of test procedures, the nominal methods available are most important. We will study the chi square test, a frequently used procedure for nominal variables. For variables with ordinal measurements, more options and procedures are available than those that rely on the nominal scale. We will learn about statistical procedures such as the Spearmen Rank Correlation method. Finally, the greatest flexibility lies with variables based on interval scale measurements. Most of our attention will focus on statistical approaches for interval scaled variables such as the analysis of variance.

You may read or hear researchers talk about non-parametric and parametric techniques. **Non-parametric** methods are appropriate for data measured by either the nominal or ordinal scale. **Parametric** methods are applied to data measured by the interval scale.

In addition to classifying measurements in terms of different types of scales, we can also identify certain properties that are associated with measurements. Different measurements will have different levels of reliability or validity. We will now turn our attention to a discussion of those properties.

Reliability

Reliability refers to the consistency of a measurement. It makes sense that we would want to use measurements that were reliable. In every day life, we are very dependent on the reliability of instruments - when we turn on the ignition switch we want the car to start, when we hit the brake pedal we want the car to stop, when we are told that the phone repair technician will be at our home at 9:00 A.M. we want that person to be on time.

A measurement is reliable or unreliable to the extent that a certain action produces the same response. If, when we first weigh a package we plan to mail, we get a certain weight, we would say our scale is reliable if a second weighing gave the same result. If the weight changed every time we weighed the package, we would judge the scale as unreliable.

Even though there is a single concept of reliability, it can be measured differently. Statisticians talk of stability, equivalence and internal consistency when referring to reliability.

Stability

In the package-weighing example, we were worried about reliability in terms of consistency over time. This dimension of reliability is referred to as **stability**. We examine the stability facet of reliability by doing repeated measurements over time. Of course, we need to keep extraneous factors constant when we do repeated measurements, for they can be the cause of any apparent changes.

Stability asks the question - over time, given no extraneous interference, do we get the same result? In our weight example an extraneous factor such as opening the package and removing part of the contents would obviously exonerate the weight scale as being unreliable. We can come up with many other examples - Does a person's intelligence remain the same over time? Yes it does or at least it should, but will the measuring devise, an IQ test, reflect that state of constancy? A board doesn't change size, but do I get the same result each time I measure the board? Unfortunately some of my wood working attempts suggest I don't quite measure enough times.

In these examples the measurement involves both the instrument and the operator. Although the scale may be just fine, the inconsistencies inherent in the operator become part of overall measurement accuracy. A reliable device in the hands of an inconsistent operator can result in an unreliable measurement.

Equivalence

Another idea associated with reliability is **equivalence**. In this case we are looking at consistency between different devices and operators. If two experts observe the same behavior will they come up with identical assessments? If the assessments are the same, there is high equivalence; to the extent they differ there is lower equivalence.

Think of an IQ test, an instrument or device to measure intelligence. There is more than one version of the IQ test - do you score the same on all the versions? It's not uncommon to find there are slight differences between various versions of IQ tests. So here we have a second aspect of reliability - the equivalence of measuring devices or techniques.

Internal Consistency

A third way to look at reliability has to do with **internal consistency**. Here the focus is on the multiple elements that make up the measuring system. In general we would not want to be tested for our knowledge of geometry based on a single question. The inclusion of many questions increases the likelihood that the phenomenon under study has been comprehensively measured. The place you usually run in to internal consistency is with tests. Do all the questions in a test relate to the same phenomenon? The phenomenon may be directed at how much a student knows about Greek history. The civil service exam is another good example - are all the questions pertinent to the job for which one is an applicant? All items on the test should relate to capability to do the job in question. The fewer number of irrelevant and immaterial items, the greater the internal consistency of a test.

Note that these three dimensions of reliability (stability, equivalence and internal consistency) may not be present in every measurement situation. The weight scale example involves observations by a single person using the same instrument. On the other hand, if a research project requires before and after observation of many people's behavior as they participate in a training program, multiple observers may be required. In the weight scale example we could test for stability but not equivalence. In the observation of behavior case, we could test for equivalence (are the behaviors reported by the different observers consistent?), but not stability (the intentional introduction of the training program was designed to change behavior).

The measurement may consist of a single element - taking a weight or many elements such as the questions that make up an intelligence test. An internal consistency assessment is possible for the intelligence test, but not the weight scale situation.

To obtain a stability assessment it is necessary that the same measurement be taken over time. To obtain an equivalence measurement it is necessary that there be alternative ways of making the assessments. To obtain an internal consistency measurement it's necessary that there be multiple elements, which make up the measuring device.

Usually the extent of reliability for stability, equivalence or internal consistency is determined by means of a correlation coefficient. We will examine correlation in a later chapter, but for the present suffice to say that we can not expect a measurement to be completely free from inconsistency. In addition, even though we may be able to calculate the degree of reliability present in a measurement, whether the value is too high, low or just fine remains a judgment call.

Validity

Validity is measuring what you say you're measuring. Now this last sentence may sound obvious, but in fact, we rarely directly measure the phenomenon we are truly interested in. When we give an IQ test we are trying to measure intelligence. Yet, there has been concern about whether IQ tests truly measure intelligence. Scholastic aptitude tests may be designed to measure aptitude to do college level work, but there is controversy about whether these tests truly and fairly measure this ability.

Let's take an example to see more closely what is going on when we concern ourselves with the validity concept. Assume that the goal is to create a drug that will cure a patient of his or her infection. The objective, cure the infection, sounds clear-cut but how sharp is the definition of a cured infection. In a research setting, a "cured infection" would be considered a **concept** (i.e. an abstract or general idea). That concept, to be useful for research purposes, needs to take on quantitative features.

Going from concept to quantification of the concept is referred to as **operationalizing** the concept. The term, operationalize, was chosen based on one of its definitions - the ability to make something function as in "getting the factory operational". What we do when we operationalize is to transform a concept into a surrogate notion that can be measured in a quantitative manner. We can not directly measure most concepts, but we can come up with an **operational definition** of the concept that can be measured. The degree to which the operational definition does in fact measure the underlying concept is what validity is all about.

There are many operational definitions for important concepts. Take the concept of wealth - what do we actually measure when we want to know a person's wealth? His or her income might be one way to measure wealth or how about a person's assets or better yet the assets minus the liabilities.

In our case of an infection, what's the operational definition of a cure? A cure implies absence of the infection so we would want to measure temperature since we know infections cause a fever. We also would want to look at white blood cell count (WBC) since WBCs are part of the way the body fights an infection. Or how about a direct count of the invading organism, assuming the method to do this exists?

This latter proposal implies we are getting at the true concept, curing the infection by measuring the amount of offending organism in the system. However, we actually only measure a sample of the organism and in some cases we do not even do that directly. For instance, in the case of AIDS the most common diagnostic test does not directly measure the Human Immunodeficiency Virus (HIV). Instead, we rely on a count of the blood cells that are designed to fight HIV to tell us how bad the HIV infection has become.

To sum up, in research projects we frequently use a surrogate measure(s) that we can quantify to evaluate the phenomenon we're truly interested in. We frequently end up with a number of measures since what the phenomenon represents may be obscure or it may be complex with numerous dimensions requiring different ways to get at it.

Notice that for different phenomenon, different levels of measurements may be used as well. To assess whether an increase in fares will affect train ridership we could survey riders on the basis of whether they will or will not ride with the higher fare imposed (a nominal scale measurement) or we could ask if they will ride the same, a little less, a lot less, etc. (an ordinal scale measurement). We can also do pilot tests in various geographical areas and obtain the percent change in ridership (an interval scale measurement).

Validity Principles

Just as reliability has different forms, there are also different ways, in a research setting, to think about validity. We'll refer to the two forms discussed below as validity principles.

INTERNAL VALIDITY. **Internal validity** refers to how tight or sound the conditions of an experiment are. Is the research conducted so that there is a clear cause and effect relationship between the dependent and independent variable? From chapter 1 we learned that the independent variable is the one that the researcher controls and presumably the one that causes a change in the dependent variable. In an ideally controlled experiment the response of the dependent variable is not due to extraneous forces and, thus, that kind of experiment would have high internal validity.

EXTERNAL VALIDITY. The term **external validity** refers to the usefulness of the study results for other conditions that extend beyond those used in an experiment. Most research uses a defined group of subjects and conduct the study under a relatively strict protocol or set of rules. As we will see later, there is an important statistical reason for maintaining tight controls. But by being very strict, we can limit our ability to generalize the results to other situations or environments.

For example, if we want to evaluate a new drug we might select a hospital setting for the trial. In a hospital, we could be sure the drug was properly administered and the right amount of drug would be given at the right time. We would also have the advantage of an aseptic environment, three healthy (not necessarily tasty) meals every day, 24-hour care and other features, to ensure that the study protocol is adhered to. Now what if the drug is found to be effective in this trial, how sure are we that it will work in a different setting? Will it work as well when used by patients in their home? Are the findings in the hospital trial generalizable to a home setting? Perhaps for the drug to work, the hospital environment is essential.

In this example the generalizability of the hospital trial may be jeopardized by the strict conditions imposed. We are not arguing against strict controls. But it's important to point out that the more we control the testing environment, the greater the likelihood of constraining the generalizability of the results.

As a second example think of a study done only on people older than 60. Would the results be applicable to those who are in their 50s? How about those in their 20s or 30s? The extent to which the conclusion of a study can be generalized to conditions, other than those used in the experiment, constitutes external validity.

Validity Measurements

Another dimension of validity has to do with the ways it can be measured. Below we present three types of validity measurements that are commonly used.

FACE VALIDITY. Often people talk about face validity. **Face validity** is essentially based on opinion - hopefully expert opinion. Based on our own knowledge, do the measurements look as if they are the right ones? On the surface, do we expect the measurements to give us sound information about the underlying phenomenon? If people answer "yes" to these questions, you have face validity.

CONTENT VALIDITY. There is also content validity, which is a type of validity concerned with how completely the measurements used encompass the underlying concept. A test of math skills, which does not include questions on how to handle fractions, would not be a comprehensive test of mathematical competence. To have **content validity**, a measurement device should assess a representative sample of the information felt to be inherent in the concept being studied.

CRITERION VALIDITY. **Criterion validity** is another type of validity and involves a comparison between the measurements being used and those traditionally used to measure the concept. If you came up with a new test to measure intelligence, you would be expected to show how closely your test agrees with the traditional tests being used. Be aware that if you found low criterion validity, it does not necessarily mean your alternative is a failure. Criterion validity establishes the old measurement as the presumed "right" answer and you have to see how consistent your test is compared to the established measurement. You may have a better test (i.e., it's a better measure of intelligence), but if it did not correlate well with the traditional tests you would have low criterion validity. Most likely, you would need to show by other means that your test was a better way to determine intelligence, before your new test would be accepted.

Obstacles to Valid Research

Researchers, of course, want their studies to be valid. Due to limited money, inadequate facilities, lack of experience or even just plain bad luck research projects frequently do not always reach this goal. Whether you are involved in doing a research project or reading about work done by others, it's important to be alert to the obstacles that can affect validity. The following factors need to be controlled in order to end up with a valid study.

In this discussion we will frequently refer to the independent variable as a **treatment**. In many types of research a researcher administers a treatment and then studies the affect of the treatment on the dependent variable.

Maturation

Living and non-living things change over time. Machines wear out. In animals there are spurts of positive development and periods of decline. Individuals grow and groups mature - an individual develops better coordination, a group replaces old members with new ones. Studies that examine behavior at two points in time are vulnerable to these maturation processes. The people you studied at the beginning of a study are not necessarily in the same state later in the study. Failure to recognize this aging effect may lead the researcher to believe the experimental treatment produced an effect while in reality the effect came as a result of the maturation in the people being studied. For example, what if the research project involves measuring athletic skills developed by teenagers included in a long term physical fitness program? Any improvement may well be influenced by the teenager's maturation and not just the athletic program provided.

Concurrent Events

During the life of a study, events are happening in the environment that can influence the dependent variable. Societal norms, attitudes and political beliefs change over time. Publicity about these factors during a study can affect the response of the dependent variable. Thus, studies that examine behavior at two points in time are susceptible to the influence of concurrent events. For instance, participants in a behavior modification program, which included assessing attitudes towards minorities, would be vulnerable to very positive or negative news stories dealing with minorities.

Subject Selection

In many experiments two study groups are formed - one to receive a new experimental method and the other to receive the established approach that has become the recognized standard. Typically the treatments are allocated to the subjects by a random process. (Participants in a research project are frequently referred to as **subjects**). The goal is to have the two groups of subjects comparable in terms of physical, mental, environmental, etc. factors at the start of the experiment. If comparability is not achieved, it may be impossible to know if more favorable response on the new technique are due to the new technique or an extraneous factor that is more prevalent in one of the groups. For example, if the group receiving the experimental method has a much higher percentage of males than the other group, and attributes related to gender can affect the results, then it is not clear if a difference in results is due to the new method or the lack of equivalence in male and female representation.

Testing

You can think of testing as a practice effect. Imagine a testing device in which the performance of the subjects improves as they get more experience with the device. Think of a situation where such a device is used as a baseline measurement, and also at the end of a treatment period to measure change. If, with practice a subject can do better, any change in performance is suspect. Did the scores change due to the treatment or was it because the subjects were more familiar with the device and did better because of that additional experience? Use of a treadmill to measure endurance is a good example of a testing effect. Some people could be a bit nervous and unsure the first time on a treadmill, but this uncertainty could evaporate on subsequent tests. Consequently if a group improved between an initial test, just prior to starting treatment, and a follow-up test at the end of treatment, it would be hard to know how much of the improvement was due to the treatment and what amount was a result of being more comfortable on the treadmill.

Instrumentation

Instrumentation refers to a change in, or differences between, measuring devices. It is popular to use mechanical devices in research studies in part because of the objectivity and reliability of those devices. In addition, multiple instruments may be required because of different research locations or the need to examine many patients at one time. In spite of their advantages, mechanical instruments can reduce internal validity since they are not all identical and their performance can decline. Any inconsistency between, or degradation in, the instruments used in a study can make a subtle change in measurements of the dependent variable. Concerns with reliability, especially stability and equivalence that were discussed above, are pertinent to instrumentation as well.

Experimental Mortality

Experimental mortality is a relatively descriptive term since it refers to people who drop out of a study before its completion. However, some people prefer to use the expression

attrition for this event. In either case, when subjects do not remain in an investigation there is cause for concern. People leave the area due to marriage or a job change. They may be unavailable due to illness or travel. However they may also discontinue because of a study related event. Without follow-up information, the researcher is faced with a problem when those who start a study are unable to finish it. How those cases are handled can have a strong influence on the results. Important data may be lacking and a complete analysis impossible. To ignore the data that was collected can lead to incorrect decisions. How do we know that the drop out was unrelated to the treatment? Perhaps those with the best results are dropping out ("I'm doing so well I don't need to stay in the study any more"). Or maybe it's the people with the worst result who chose to discontinue ("This is not doing me any good so I'll quit"). We need to be particularly concerned and curious about studies with high drop out rates. If the attrition occurs in one of our studies we need to determine the causes and, if necessary, make adjustments for the incomplete cases.

Statistical Regression

Internal validity can be threatened whenever selection is based on subjects having an extreme characteristic or attribute. In way of an example, a study may be designed to compare high versus low achiever groups and a single test is used to place the subjects in one group or another. We each have our "good" days and "bad" days - on the good days we are sharp and perhaps a little lucky as well. But unfortunately, that may not be our true self - we just had a good day. On another day we may not do nearly so well. As a matter of fact, people who score the very highest on a testing instrument will tend to have their scores drop on subsequent tests - they become more "average". Conversely people who did very poorly will tend to score better on later tests. This phenomenon of having the extreme scores on one test become more ordinary on succeeding tests is called statistical regression. The scores regress - those at either extreme move closer to the middle. The implication for research studies should be clear, the changes in scores might be misleading due to an exaggerated reading at the beginning.

Depending on the nature of the study some obstacles may be a bigger threat to internal validity than others. It probably isn't possible to control all the obstacles, and even if we tried we might never know for sure how successful we were. There isn't an explicit way to know the extent to which a factor was present nor how well it was controlled. However, this doesn't mean we can be indifferent to the potential problems. The researcher should do his or her very best to reduce potential problems related to internal validity. Furthermore readers should read research reports with an awareness of the potential problems that might have occurred in the trial.

Sensitivity

One last quality, sensitivity, is another important feature of a measurement. **Sensitivity** is important because it examines whether the measurement can detect an important difference. In research a comparison is frequently made. The results on the new treatment are compared to the results on the old treatment. The subject's performance before a training program was instituted is compared to the performance after the program is completed. Sensitivity scrutinizes how well the measuring device detects a change in the condition being studied. Crude measuring devices can miss important changes. If we relied on holding the back of the hand to a person's forehead rather than a thermometer to detect fluctuations in a person's fever, important changes might be missed. Thus, sensitivity refers to the ability of a measurement to find important differences.

Summary

In this chapter we learned that we could classify and evaluate measurements in different ways. For example, measurements can be differentiated based on the type of scale used. The three main scales, for our purposes, were defined as the nominal, ordinal and interval scale. However we can also classify measurements based on a functional definition. How reliable and valid is the measurement? When we examine the reliability of a measurement we check it for stability, equivalence and/or internal consistency. When we worry about validity we talk about internal validity (the soundness of the experimental setting and process) or external validity (the ability to apply the results to other settings than those used in the experiment). However, validity also has other dimensions and we discovered that researchers could examine measurements in terms of face validity, content validity and criterion validity.

We also saw that there are countless ways in which research studies can become flawed. Seven areas of concern were reviewed and can serve as a checklist to evaluate a research project. We found we need to be sensitive to the effect of maturation, concurrent events, subject selection, testing, instrumentation, attrition and statistical regression since individually, or collectively, these threats to validity can cause misleading conclusions.

We also realized that a research project usually begins with a concept that needs to be converted into a measurable variable through a process called operationalizing. Last, but not least, we also learned that we must be concerned about the sensitivity of a measurement - can the measurement detect an important difference - if not, our research effort may be in vain.

PROBLEMS

1. To which measurement scale (nominal, ordinal or interval) do the following variables belong?
   a. Military positions (e.g., Private, Corporal, etc.)
   b. License plate numbers
   c. Socioeconomic status (e.g., Upper class, Middle class, etc.)
   d. Miles per hour
   e. College major (e.g., English, History, etc.)

2. The following case is to be used to answer questions 2.1-2.4.
   The local high school wants to conduct a study in which they will introduce a confidence building program for all senior students. One goal of the program is to build self esteem. To assess this goal, Dennis Murphy, an advisor to the research team, recommends the use of three clinical psychologists to evaluate student self esteem. The psychologists would make one assessment, at the completion of the program, on the degree to which student self esteem had changed relative to the status when the program began. Due to the number of students involved, three psychologists who had worked together in the past, would be trained to carry out the assessments. One-third of the students would be assigned to each psychologist. In addition, students' self assessments would be added to a survey which would be administered in a student's home room the day before the confidence building program begins and the day after the program ends. To get at the student assessment of self esteem, Murphy suggests that the following question be used:
   Indicate how valuable you feel you are to your community. (Students answer the question by selecting either the "Valuable" or the "Not Valuable" option.)
   The following comments were made in the discussion that followed Murphy's presentation.
   Sam Adams felt that only asking about a person's value to the community could overlook other aspects of self esteem. He recommended that three self assessment questions should be asked. Besides asking about value to community, Adams felt questions on the value to family and the value to the school should be added.
   Mary Baker wondered if one assessment of self esteem at the beginning of the program was enough since students could be "so unpredictable". She stated that it might be best to get two assessments before the program began, with a week between the assessments.
   Jennifer Chu stated that there should be 6 options not just 2 for the self esteem question because almost all students would chose the option "Valuable". She recommended that options such as "Extremely Valuable", "Very Valuable" and "Somewhat Valuable" be added.

Juan Damingo asked what would happen if Mary Baker's suggestion were adopted and there was an inconsistency between the answers a student gives on the two pre-test surveys.

Joan Esposito challenged the inclusion of the "community" question in the set of proposed self esteem questions. She argued that students probably had a good idea of their value to the school and to their families, but she felt that they might not be in a position to know their value to the community.

Colleen Finegan asked whether the three clinical psychologists would use comparable questions in the evaluation of student self esteem.

Rami Hadad said he didn't like the whole idea because, based on his knowledge of self esteem, he didn't feel there was a way to measure it accurately among high school students.

George Inoyue also criticized the plan - he said that he was aware of a published 15 item scale for self esteem and unless he could be shown that the proposed questions do as well as the 15 item scale he could not support Murphy's plan.

Derek Jackson, a counselor at the school, joined in saying that people with low self esteem have feelings of guilt, reject negative feedback and are hesitant to try new things. He didn't see how those behaviors would be picked up in the proposal.

Betty Gila noted that her sister, who lives in another state, is a principal at a middle school and if the confidence building program were successful she would urge her sister to use it at her school.

2.1.   Match the comments by Sam Adams, Mary Baker and Jennifer Chu with the following measurement properties (each individual should be placed in only one category).
   a.   Reliability?
   b.   Sensitivity?
   c.   Validity?
2.2a.   What type of reliability is Juan Damingo concerned with?
   b.   What type of reliability is Joan Esposito concerned with?
   c.   What type of reliability is Colleen Finegan concerned with?
2.3a.   What type of validity is Rami Hadad referring to?
   b.   What type of validity is George Inoyue referring to?
   c.   What type of validity is Derek Jackson referring to?
2.4.   What validity principle should Betty Gila be concerned with before she makes any recommendation about the program to her sister?

3.   The following case is to be used to answer questions 3.1 – 3.3.
     A study was conducted comparing treatment group A to group B. Group A received an appetite suppression pill, 1 hour of guided exercise (every day except Saturday and Sunday) and a recommended diet. Group B received a placebo pill plus the

identical exercise and recommended diet. Only women between the ages of 30 and 50 participated in the study which was conducted at two medical centers in a major city and lasted 6 weeks. The following assessments were made:

a. Weight just before they began treatment and at the end of each week during the 6 week trial.

b. A psychiatrist's evaluation of a subject's sense of well being (rated as Very High, High, Average, Low or Very Low) was obtained at the completion of the trial (i.e. week 6).

c. A self completed check list of side effects with daily recording for the following signs and symptoms: Nervousness, Dizziness, Dry mouth, Nausea and Irritability. Only women who showed none of these symptoms at the pre treatment visit were allowed to be randomized to one of the treatment groups.

At a pre-treatment visit, the baseline measurements were taken and the women were randomly assigned to a treatment group. The results were published in a leading journal and showed that group A, the group taking the active medication, experienced a greater weight loss than the group taking the placebo pill. Both groups received above average ratings for "sense of well being" but the ratings were better for the active treatment group. The placebo group had fewer complaints of nervousness and irritability, but for the other items the groups results were quite comparable. The following events occurred.

Event	Description
1	A company with an appetite suppression product on the market issued a report that noted that the active medication used in the study had an increase in all the side effects listed on the weekly check list. They emphasized that the incidence of these problems was much lower with their product.
2	The number of check lists not filled out was much higher in the placebo group. It was assumed in the analysis that these subjects had no side effects.
3	More women in the group on active medication had jobs and felt they were having successful careers.
4	In the middle of the trial, the FDA issued a recommendation that because of side effects, the dosage should be cut in half of the newest over-the-counter appetite suppression pill on the market.
5	It was discovered that more women in the placebo group had completed a similar trial the month before and had been evaluated on the same set of measurements as used in this trial.
6	One medical center had the same exercise instructor for all classes, but the second center used 3 or 4 different instructors depending on their availability.
7	It was found more women in the medicated group were pregnant before the trial began or became pregnant during the 6 week trial.

3.1. Match each event to one of the 7 threats to validity so each threat has one match.
3.2. Identify the assessment variable(s) (if any) that would be affected by an event and explain why it would be affected.
3.3. Identify what group would be advantaged or disadvantaged by the event - if you can't tell if any bias would occur, explain why.

The format and approach you should use in answering these questions is shown below:

Event	3.1 Threat	3.2 Variable(s) Affected	3.3 Group Bias
1 Competitor Report	Subject Selection	Self Esteem variable because	The placebo group's values would be biased in a favorable way because ....

4. Answer the following questions:
   a. In question 2, what would be the value of a control group (i.e. including a second group of students who would not receive the confidence building program)?
   b. A critic claimed that the side effect check list used in question 3 was inadequate because it lacked content validity. What did she mean?
   c. In the study described in question 3, a letter to the editor was received following publication stating that the report failed to show that the well being assessment possessed criterion validity. What did the writer mean?

5. Six hundred employees have left your organization. The table below gives the primary reason each employee left (note that each employee is placed in only one of the termination categories). Forty percent of those who left were males. Assume there is independence between sex, type of position and termination reason.

Termination Reason	Type of Position		
	Line	Staff	Total
Excessive Absenteeism	12	21	33
Insubordination	14	7	21
Medical Disability	6	12	18
Normal Retirement	81	90	171
Voluntary Resignation	147	178	325
Other	12	20	32
Total	272	328	600

Use the probability rules learned in chapter 3 to answer the following questions:
   a. What is the probability that the termination was due to insubordination or excessive absenteeism?
   b. What is the probability that a person in a line position voluntarily resigned?

    c. What is the probability that a staff person left for any reason?

    d. What is the probability that a person who left was either male or the termination was due to a medical disability?

    e. What is the probability that the person who left was a female who chose either normal retirement or a voluntary resignation?

6. The grant administrator of your organization submitted 3 applications to the federal government. Based on past experience the administrator is pretty certain that there is a .30 probability of each grant being funded. The funding of one grant has no effect on the funding of the other grants.

    a.   What is the probability of funding just 1 of the 3 grants?

    b.   What is the probability of funding exactly 2 of the grants?

    c.   What is the probability of funding none of the grants?

    d.   What is the probability of funding at most 2 of the grants?

    f.   What is the probability of funding at least 2 of the grants?

7. At the race track you decide to bet on the daily double (i.e., picking the winning horse in the first two races). Since you are unfamiliar with the horses you decide to bet on the favorites in both races. Assume the following probabilities represent the chances the favorite will win the first and second race respectively: .33 and .25.

    a.   What is your chance of winning the daily double?

    b.   What is your chance of picking exactly one winner is the first two races?

8. Think of a horse race with 7 horses who have the same chance of winning. You know nothing about the horses so your bet is just a wild guess.

    a.   What are the chances of your winning the trifecta (i.e., picking the exact order of finish for the first 3 horses)?

    b.   Are you more likely to pick the winners of the first three races in which there are 6 horses in each race and you have to make a wild guess for each winner?

9. You took a civil service exam and scored a 56. The mean of all who took the test was 52.1 with a standard deviation of 6.5.

    a.   What proportion of those who took the exam did better than you?

    b.   To score in the upper 10 per cent what score would you need?

There are no instructions in Excel for this chapter.

Chapter 7

Sampling, Standard Error and Sample Size

Introduction

Each academic discipline has a core set of principles that serve as the building blocks for the specialty. Sampling is a fundamental principle upon which much of statistical reasoning is based. In this chapter, we describe how a sample differs from a population and review ways we can choose a sample. We will also examine the statistics associated with samples and introduce the standard error and central limit theorem. We will use these features to do Z tests and find out how unusual a sample mean is. We will also explore confidence interval estimates for means and learn how to determine the sample size for a research study. We begin by defining a few terms.

Populations

A population is the collection of people or objects that we are interested in studying. We frequently want to learn something about a population - their average age, the proportion who are married, the number who have had a traffic accident, etc. The population could be all residents of New Jersey, all people who have AIDS, all third graders, or all boxes of corn flakes made during the month of June.

Ideally, to find the average age of the people living in New Jersey we would contact each resident and record his or her age as of a given point in time. But is that practical? The cost would be exorbitant and the chance we could locate every resident is unrealistic. So what do we do? We identify a smaller number of residents that we can access and use their ages to infer what the ages are like for all residents in the state.

Samples

A sample is a subset of the population of interest. The process used to take a sample is called the **sampling plan**. The number of people or objects selected is called the **sample size**. We frequently use a sample to tell us something about a population. It's obvious that the use of the smaller sample saves us a lot of time, effort and cost. However, there is no guarantee that the data from the sample will give us a correct population value. In fact it is almost certain that the statistics calculated from the sample will not be identical to that of the population. Therefore, it is important to find out how different your sample information is from the population figures.

There is a branch of statistics called **inferential statistics**, which studies the process of using a sample to give us information about a population. The term "inferential" is most appropriate since what we are doing is making an inference about a characteristic in a population based on our sample. For example, it is popular to periodically poll the voting public prior to an election to find out which candidate is likely to win the most votes. When we use the data from the sample to estimate the proportion of all voters (i.e., the population) who favor a candidate we are making an inference. However, we have learned that with each poll there is a "margin of error". A zone of uncertainty is placed around our inference. Much of what we will do in later chapters is learn how to ascertain the probability that an inference is incorrect or how large the zone of uncertainty should be.

The option of using a sample rather than trying to identify an entire population is very appealing. Yet, one of the problems with a population - trying to identify all of its elements - is not avoided when we take a sample. To take the sample, we still need access to the population - and if our sample is to be representative, we would like to have access to the whole population.

Assume we want to try a new remedy for obesity. Our population of interest is all obese individuals. Even if we came up with an acceptable definition of "obese" can you imagine how difficult it would be to identify all obese people? There are just too many and they're all over - we could never identify them all. Or say we want to test a promising prophylactic treatment for a disease, which is based on a genetic flaw. The population of interest is all people with the genetic flaw. But many of those people have not even been tested for the genetic defect so, for all intent and purposes, they are unrecognizable at this point in time.

In the cases cited above we must settle for something less than the complete population. Let's try to make our problem more manageable - it's not unreasonable to find that a registry is maintained on all people with the genetic flaw. We will not have the untested people, but we will have all those who tested positive for the genetic defect in the registry. What we would now do is to take a sample from the registry. The term "frame" is used for the registry. A **frame** is therefore the proportion of the population that is accessible.

What if you are asked by the local government to survey all property owners regarding their views on a reduction in the hours of the municipal clerk's office. The population in this case is all property owners in the municipality. You find that once a year a roster of property owners is generated which gives names, addresses and phone numbers. The list was last generated 6 months ago. Although the list does not contain the latest changes in property owners, it would probably be considered good enough to serve as your frame. From that list of property owners you could draw your sample. In all sampling operations, whether one specifies it or not, there is a frame.

Think of a sampling plan where one stands on the corner of Main St. and First Ave. and asks every third person who walks by if they think the President of the United States is doing a good job. In that circumstance, the people who walk by that corner on the days the interviewer is there, constitute the frame. From that frame you take your sample. Your sampling plan is to take every third person in your frame. If there were 600 people passing the corner and you succeeded in interviewing one-third of them, then your sample size is 200.

You can probably think of some frames, but two of the most popular ones are voter registration lists and phone book listings. Other frequently used frames are the membership list of organizations.

Advantages of Sampling

We have already seen that two major advantages of sampling, versus examining the whole population, are a savings in time and a savings in cost. To that list we can add other reasons why sampling may be better under special circumstances. These are:

1. The Destructive Nature of the Test. If we wanted to make sure the bullets sold to the US Army would work, we need to fire the bullets and keep track of how many exploded properly and how many misfired. After a moments thought, we can see that testing the entire population of bullets would have dire consequences for our military preparedness.

2. The Infinite Universe. The population in some cases approaches infinity. Imagine a project where we want to measure the amount of an enzyme in the black ant. Not only is the population of gigantic size, but we also must be aware of the constant increase in the black ant population due to the birth of new ants.

3. Accuracy. On the surface the claim of accuracy may seem absurd. After all, if the goal of sampling is to estimate a population characteristic how could the sample, which by definition is incomplete compared to the population, be more accurate than the population? The answer comes not from theory, but from the realities of life. Assume the survey involves interviews by people who must have a unique profession. Furthermore, assume that the interviewers must be given meticulous training before they can go out into the field. Also assume there is a shortage of people with the professional specialization. Under these circumstances it will be

much easier to hire a limited number of interviewers, train them and then conduct a sample. The massive numbers needed to do the entire population will, very likely, result in inadequately prepared, poorly trained and unmotivated people. Furthermore, it will probably be a lot easier to keep up interest and maintain enthusiasm with the smaller staff. The net result is that the sample approach allows a far better performance by the staff and, as a result, an increase in accuracy.

A chapter on sampling can hardly be written without reference to the Literary Digest case. In the 1936 presidential election the Literary Digest, a major magazine at the time, published the results of a poll that they had taken which showed that Alf Landon, the Republican candidate, would win 60% of the vote and easily beat the Democratic nominee, Franklin Delano Roosevelt. The actual winner won with 62% of the vote - unfortunately for the Literary Digest the winner was Roosevelt, not Landon. The question is frequently asked; how could they be so wrong? Any ideas?

Many times it's suggested that they had too few people in their survey - a reasonable guess, but the sample size was in the thousands - they had a far greater number than most political surveys taken in current times. The answer lies in the frames they used to obtain their numbers. They reportedly used telephone lists, auto registrations and subscription lists to their magazine. Can you now see what went wrong?

Remember the time is the 30s. Without casting any aspersions on the many good Democrats living at that time - would Democrats or Republicans more likely have a phone in the 30s? Would Democrats or Republicans more likely own a car in the 30s? Would Democrats or Republicans more likely subscribe to a magazine with "literary" in the title? The Literary Digest case clearly shows that quality is far more important than quantity when it comes to sampling. Large sample size can not compensate for a poor choice of the sampling frame.

Random Sampling

If you use a process where every member of the frame has an equal chance of being included in the sample you are doing random sampling. The steps in selecting the members for a random sample are critical. Let's say that our frame has only 20 objects in it and we want a sample size of 10. We could assign a distinct name or number to the 20 objects and then write those identifications on slips of paper, place them in a hat and draw 10 slips. That seems easy enough. But what if the frame has 20,000 items in it and we want to have a sample size of 500. In that situation forget the slips and hat process - we need something more efficient.

What's usually done is that 20,000 items would be given a number beginning at 1 and ending at 20,000. It's quite legitimate to assign the first item on the list a 1, the second item a 2, etc. till the last items is assigned the number 20,000. At the next step we'd rely on the computer to produce say 550 random digits between 1 and 20,000. The number of digits is slightly higher than the number required because there can be duplicate number produced by the computer so to get 500 unique numbers we must produce a slight excess. The first 500 unique numbers on the computer list would be used as the sample.

It's possible to be too enamored with random sampling and fail to see its flaw. A random sample does not guarantee a fair or representative sample. Granted we use random sampling because we want that feature - we want the sample to be typical and representative of the population, but the sampling process does not insure that this will be the outcome. In the majority of cases our random sample will be a very close approximation to the population, but a given sample can be skewed.

We can illustrate this with a small example. Imagine a frame containing just 100 people, half are males and half are females. We'll draw a sample of just 6 people from this frame. Most of the samples we could draw will have a nice mix of males and females, but there will be one here and there that will be "odd". There is even a sample possible, which contains only males. That's an unlikely event. How unlikely is it you may ask? We have learned how to determine that probability - can you remember how it's done and determine the probability of only selecting 6 males?

You could use the General Multiplication Rule (see chapter 3) since we have an AND probability situation in which the events are not independent - once you draw a person, he or she is removed from the frame and, thus, each selection changes the odds of the subsequent selection. Since we begin with 100 people and 50 of them are male our equation becomes:

$$P = \frac{50}{100} * \frac{49}{99} * \frac{48}{98} * \frac{47}{97} * \frac{46}{96} * \frac{45}{95}$$

Solving the equation gives us a probability of .0133. A most unlikely event, but not an impossible one. If you realize there are also other unbalanced outcomes, such as all females, you can see that the chance of a "bad" sample is not all that remote. We would not want to say such distorted samples are representative of the population, but keep in mind that they are legitimate random samples.

The situation is not all that bad since as the sample size increases the chances of a serious distortion falls. Although there is no magic number at which you have protected yourself from an odd result, many researchers try to have samples of size 30 or larger.

Statisticians often talk about **sampling error**, which refers to the inaccuracy of a sample relative to the population from which it is taken. Any sample can have too many individuals of one type and too few of another type. This disproportionate result is not controllable and it's not intentional - it just happens when a sample is drawn. In some respects the term error may be misleading since sampling errors are the result of the chance fluctuations inherent whenever a sample is taken. Hence, all samples will have some degree of sampling error. However, we can compound the problem when we fail to take random samples. In the case of non-random samples we know in advance that the sample will have distortions relative to the population. The inaccuracy of the sample, compared to the population, is assured and therefore considered intentional. As a result one needs very strong arguments to justify using non-random samples.

Sampling Plans

A sample is always based on a sampling plan - how are we going to select our objects from the frame? Sampling strategies can be dichotomized into two general types - probability samples and non-probability samples. The distinctive characteristic of the **probability sample** class is that the selection is based on chance. Conversely, a **non-probability sample** makes its selection without invoking chance. The selection process for the non-probability class can be described as subjective or arbitrary. We will present four types of probability samples that all use random sampling: simple random, systematic, stratified and cluster. We will then discuss two non-probability types of sampling plans - judgmental and convenience.

Simple Random Sampling

As the name implies simple random sampling utilizes random sampling without any complications. In simple random sampling all elements in a frame are included and there are no other additional procedures required before making the selections. In some forms of random sampling there are preliminary steps that take place before a sample is drawn (e.g. see stratified sampling below).

Systematic Sampling

A systematic sampling plan is pretty straight forward. You pick every nth item from your frame. The n in "nth item" is a value you pick based on the sample size you want. The formula is:

$$n = \frac{\text{Total No. Available}}{\text{Desired Sample Size}}$$

If you have a frame of 600 items (the "Total No. Available") and want 30 (the "Desired Sample Size") in your sample your n will be 20.

$$n = \frac{600}{30} = 20$$

Thus, you pick every 20th item or numbers 20, 40, 60, etc. till you get to the last number in your sample - number 600. This selection process will give you a total of 30 elements - the number you wanted. As noted above, the n is arrived at by dividing the total number in the frame by the sample size, but if your answer is not a whole number, you must also round it down to the nearest whole number. For example had we started with 585 rather than 600 items in the frame our n would be:

$$n = \frac{585}{30} = 19.5$$

Your frame is based on whole numbers so your n must also be in whole numbers - you have a 19th and 20th member in your frame but there is no 19.5th member. You round down since if you round up you will not select an adequate number for you sample. Rounding up to 20 would mean you end up with 29 rather than 30 items selected. Rounding down means you pick every 19th item or numbers 19, 38, 57, etc. till you get to the last number in your sample - number 570. At the end of the process you will have 30 selected times.

Stratified Sampling

With a stratified sampling plan you first divide the elements in a frame into groups or strata. You then take a random sample from each group. Stratified sampling is appealing when the population you are using contains characteristics for which you have a special interest or concern. For instance, you may set out to do a survey among your employees concerning their satisfaction with the fringe benefit package they have. You are confident you'll have a sufficient number of people who are between 30 and 50, but are concerned that a simple random sampling plan will give you too few people who are in their 20s or over 50. To insure an adequate number in all age groups, you establish strata for employees who are in their 20s, 30s, 40s, 50s, and 60s. You then need to set the sample size desired for each age strata. From personnel records we could identify the number of employees in each age group and then calculate the proportions who needed to be selected. We might then draw random samples from each age strata consistent with the proportion in the population. Since the sample sizes will represent different proportions within each age group this approach is called a **disproportionate stratified sample**.

As you might guess there is also a plan called a **proportionate stratified sample**. In this version, employees who belonged to each age group are identified. Since there are 5 age strata we would then draw 20% of our sample from each strata. Note that a proportionate stratified sample ends up with a sample that is not representative of the population. Some strata in the sample are over represented and other strata are under represented compared to the population. Proportional stratified sampling would be an efficient method if our goal were to find differences between the five age groups rather than estimate overall population characteristics.

Cluster Sampling

In research projects, which involve face to face contact with individuals who are included in the sample and who live in a large geographic area, cluster sampling is often preferred. An appropriate example would be obtaining information by means of a personal interview with the head of households for residents of Alaska. A simple random sample could easily end up selecting households in very remote parts of the state. In this event, most of the day would be spent traveling from town to town. In a **cluster sample** you first select the locations from which you will draw your sample. The locations, selected randomly, may have subdivisions before you reach the final level of selection - the household.

How would cluster sampling work in our example of interviewing households in Alaska? We could first select, at random, a limited number of counties within the state e.g. 4). We then could select, for each county, some townships (e.g., 3). Again that selection would be a random choice. Within the township we could use street maps to randomly select blocks (e.g. 2 blocks per township). We would have a total of 24 blocks in our sample (4 [counties] * 3 [townships] * 2 [blocks] = 24). We could then interview all households for the 24 blocks selected. Alternatively, we could select a limited number of households on a block. The households selected would be based on a random process.

Here is another example. If we wanted a sample of voters for a national poll, we could first take a random sample of states and, from the states selected, take a random sample of legislative districts. From the districts you can then take a random sample of citizens to interview. As illustrated above, large studies often do sampling in stages and the term **multi-stage random sampling** is used for this approach.

We will now look at two non-random sampling plans.

Judgmental Sampling

The term "judgmental sample" is a euphemism for a personal choice selection process. In a judgmental sampling plan the investigator looks through the frame and selects the elements to be included in the sample. Since the probability that any element will be chosen is unknown (inclusion or exclusion is at the discretion of the investigator), judgment sampling is considered a non-probability sampling method.

Judgment sampling is often done because the person doing the selection wants to keep the rejection rate low - i.e., he or she wants to minimize the number of people who, after being selected as part of the sample, refuse to participate. The investigator therefore, selects elements (e.g. people who are more likely to participate such as friends or people who are especially interested in the topic being investigated). Sometimes judgmental sampling is done because the investigator wants to be sure certain elements are included in the sample and doesn't want to leave that choice to chance.

The danger of judgment sampling is that bias may creep into the sample and no one is aware of it. A good illustration of this is provided in the following example. Two researchers (Dr. Able and Dr. Baker) are studying a new extract that is believed capable of stimulating a sexual response and the product is being tested in young rabbits. The study is organized with an experimental group (rabbits chosen by the investigator to receive the active extract) and a control group (rabbits chosen by the investigator to receive an inert substance). The first set of experiments showed the extract was highly effective in producing the desired response. Dr. Able selected the animals and did all the injections. It was decided to repeat the study before making the findings public. This time Dr. Baker picked the animals and gave the injections. Lo and behold, when the results were examined, there was no difference between the active and inert preparations. Do you have any idea of what happened?   That's a very hard question, but we can give one hint - the problem is a result of the method of selecting the rabbits for the two study groups and it is related to the maturity of the animals.

In this case the investigators used a judgmental sampling plan - they chose which rabbit would receive the active extract and which ones would get the inert substance. As it turned out Dr. Able tended to pick older more mature rabbits and Dr. Baker picked young more immature rabbits. The older rabbits selected by Dr. Able were more developed and their organs were able to respond to the extract. The younger rabbits that Dr. Baker selected were too underdeveloped and consequently the extract had no effect. There was nothing deceitful or devious in the actions of the investigators, but they may be faulted for being a bit careless when they decided to use a judgmental sample. A good rule of thumb is to do probability sampling simply to avoid unintentional problems. The cost is usually minimal and the protection you get is well worth it.

Convenience Sampling

The easiest way to describe convenience sampling is to say that the researcher takes what's available. The Literary Digest example is a case of convenience sampling. It was easy for them to use their subscription rolls. It takes less time to pass out a questionnaire to those who are nearby, it takes less effort to ask your friends to be interviewed, etc. However, by taking these short cuts you may jeopardize the usefulness of your research.

It should be clear that judgmental and convenience sampling are dangerous ways to select a sample. Bias is highly likely and there is an increased risk of coming to incorrect conclusions. You may be able to get away with judgmental and convenience sampling when you are in the early stages of a research project - trying to define and develop your methodology, but when you set about doing the definitive work it's important to protect yourself and rely on probability sampling techniques if at all possible.

Sample Statistics

In chapter 1 we learned how to calculate measures of central tendency (mean, median and mode) and measures of dispersion (standard deviation and variance). Those calculations were appropriate for statistics coming from a population. The same statistical names are used for a sample, however, there a few differences as will be pointed out below.

Whether you are dealing with a population or a sample, the formula for the mean is unchanged but remember that there is a change in the symbol used for the sample mean. The μ symbol (the Greek letter mu) is reserved for a population mean; for a sample mean the $\bar{X}$ symbol (an X with a bar over it) is used.

For the variance and standard deviation both the formula and the symbol changes when you work with a sample. Let's first look at the variance. The formula for a **population variance** is:

$$\sigma^2 = \frac{\Sigma(X_i - \mu)^2}{N}$$

The formula for a **sample variance** is:

$$S^2 = \frac{\Sigma(X_i - \bar{X})^2}{N-1}$$

There are 3 differences - can you find them?

The differences are:
1.   The symbol for the variance has changed. The symbol S^2 is used for a sample variance while a σ^2 was used for a population variance. Actually σ is the Greek letter for s so the two symbols are related.

2. In the sample variance formula there is a change in the numerator. The $\overline{X}$ symbol replaces the μ notation, but both symbols still refer to a mean.
3. There is also a difference in the denominator. The population variance has an N in the denominator and the sample variance has an N-1.

In respect to changes in the standard deviation, they will parallel those mentioned for the variance since the standard deviation is the square root of the variance.

A **population standard deviation formula** is:

$$\sigma = \sqrt{\frac{\Sigma(X_i - \mu)^2}{N}}$$

and a **sample standard deviation formula** is:

$$S = \sqrt{\frac{\Sigma(X_i - \overline{X})^2}{N-1}}$$

For a sample standard deviation, the changes are identical to those for the variance - σ changes to S, the mean symbol in the numerator changes from a μ to a $\overline{X}$, and in the denominator the N changes to an N-1.

We can summarize the change in symbols between a population and a sample statistics in the following table:

Table 7-A Symbols for Population and Sample Statistics

Statistic	Population Symbol	Sample Symbol
Mean	μ	$\overline{X}$
Variance	σ^2	S^2
Standard Deviation	σ	S

Note that the symbols for the population statistics are Greek letters and the sample statistics use conventional letters. The mean, variance and other descriptive statistics are often referred to as **parameters** when they refer to a population. These same elements are called **statistics** when the reference is to a sample. However, there is more than one meaning for the term statistics (e.g. statistics could be simple counts or a field of study). Consequently, when the term "statistics" is used it does not always refer to the same concept.

We'll now explain how the population variance and standard deviation differ from their counterparts, the sample variance and standard deviation. You will see that the degree of the difference is a function of the sample size. If N is large the difference between a population parameter (e.g. a population standard deviation) and a sample statistic (e.g. the sample standard deviation) is insignificant. But if N is relatively small then there is a large difference. The math involved in showing the relationship is straight-forward, but a tad tedious. If your instructor doesn't mind you might want to skip it. However, if you choose that course of action, you still must remember that the variance and standard deviation for a sample and a population are different and that the amount of difference depends on sample size. For small sample sizes the differences are considerable, but they become insignificant as the sample size increases.

We'll work with a hypothetical example. In our example, assume the population variance is 10.0 and we have a sample size of 100. A sample size of 100 would be considered large and therefore we would not expect much difference between the population variance and the sample variance. In fact, this data set, treated as a sample rather than a population, would result in a variance of 10.101. Compared to the population variance of 10.0, there is a 1% difference - an insignificant difference.

You convert a population variance (σ^2) to a sample variance (S^2) by the following formula:

$$S^2 = \sigma^2 * \frac{N}{N-1}$$

In our case σ^2 is 10.0 and N is 100 so the formula becomes:

$$S^2 = 10 * \frac{100}{99} = 10.101$$

and the value we get for S^2, as noted above, is 10.101.

On the other hand what if the population variance were still 10, but the sample size was only 6? The conversion of σ^2 to S^2 involves the following calculation:

$$S^2 = 10 * \frac{6}{5} = 12$$

This data set, when treated as a sample instead of a population, produces a variance of 12. The difference between 12 and 10 (the variance when the data set was treated as a population) is 20 percent and that is a sizeable difference.

Technically a given data set matches a population or a sample - it can't be both. The above illustrations, where we used the same data set as a population and a sample, was introduced only to show the effect of sample size on population and sample variances. It may be worthwhile to recall that when you collect data on <u>all</u> the items you are interested in then you have a **population data set**. If you took measurements on only a subset of the population, then you'd end up with a **sample data set**.

To show the amount of difference between a population and sample standard deviation we repeat what we did above using the standard deviation formula in place of the variance formulae. Again we will see that if N is large the difference is insignificant.

We'll use the same hypothetical data set and determine the population standard deviation. Since the population variance was 10 the corresponding standard deviation is 3.16. (Remember the standard deviation is just the square root of the variance). The sample size remains at 100. You convert a population standard deviation (σ) to a sample standard deviation (S) by the following formula:

$$S = \sigma * \sqrt{\frac{N}{N-1}}$$

In our case σ is 3.16 ($\sqrt{10} = 3.16$) and N remains 100 so the formula becomes:

$$S = 3.16 * \sqrt{\frac{100}{99}} = 3.18$$

The value we get for S is 3.18. The difference between 3.18 and 3.16 is insignificant.

On the other hand, what if the population standard deviation remained at 3.16, but the sample size was only 6? The conversion of σ to S, for this set of data, involves the following calculation:

$$S = 3.16 * \sqrt{\frac{6}{5}} = 3.46$$

The sample standard deviation is now elevated to 3.46. The difference between 3.16 (the population standard deviation) and 3.46 (the sample standard deviation) is not trivial. Again we see that sample size plays a major role when comparing population and sample standard deviations.

For reasons stated earlier, most statistical work involves samples rather than populations. It is also true that the statistic of interest for a sample is often the mean. This makes sense since the mean is an efficient representative statistic for a data set.

The One Sample Z Formula

Back in chapter 5 we learned how to use the Z formula. Applying the formula could tell us a lot about a result for a given individual relative to other members of the same data set. The Z formula we used was:

$$Z = \frac{X - \mu}{\sigma}$$

When we applied the Z formula we used the X in the formula to represent an individual's score or result, and we could tell how unusual that individual result was compared to other individuals in the data set. It would be nice to be able to obtain the same kind of information for a group of people as we did for a single person.

A typical question might be: How unusual is the mean based on a sample of people, who just completed a training program. We would want to compare that sample mean to the mean for all those who previously completed the program. Note that in this problem, we want to know something about a mean based on a single sample rather than obtain information about a single observation. Therefore, we need to work with a slightly modified Z formula. The version of the Z formula we want is called a "One Sample Z Formula". We will use that formula to compare the sample mean to a population mean (i.e. in our example, the mean for all those who previously completed the program). The **One Sample Z Formula** is:

$$Z = \frac{\overline{X} - \mu}{SE}$$

Note that the $\overline{X}$ refers to the sample mean for those just completing the program and μ refers to the population mean for all the people who previously completed the program. In order to define the newest element in this Z formula, the SE, we need to do a bit of explaining. It will be a while before we can get back to applying the Z formula so you'll have to be patient.

Standard Error

We are familiar with most of the components that are included in the One Sample Z Formula. $\overline{X}$ replaces X because of our interest in a mean rather than an individual value. The population mean, μ, is common to both Z formulae. What is new is the SE in the denominator. SE stands for standard error and replaces σ in the original Z formula. Not surprisingly SE and σ have similar functions relative to μ. As you probably recall, σ is a standard deviation - it is the standard deviation for a set of individual values. (The X in the Z formula represents the individual value of interest). The **standard error** is also a standard deviation, but it is the standard deviation for a set of means. Sometimes the words "of the mean" are added to the term "standard error". The full expression "standard error of the mean" is appealing since it clearly indicates we are looking at a characteristic for a distribution of means. However, parsimony dictates the shortened version.

To better appreciate SE, imagine taking samples from a population data set (See Table 7-B below). Assume the mean of that data set is 58 with a standard deviation of 20.4.

Table 7-B Population Data Set – Individual Values (Partial Listing

39	96	51	70	42	29
40	93	84	34	40	50
54	57	68	80	76	87
48	41	60	62	58	etc.

where etc. means there are many more individual values in the data set.

All the samples we take are of the same size (for illustrative purposes we will assume the sample size is 16). We can easily calculate a mean ($\overline{X}$) for each sample taken. We could continue taking samples so we'd end up with a data set containing many mean values. It could look something like this (see Table 7-C below).

Table 7-C Sample Means

58

60

54

67

52

etc.

where etc. means continue taking samples and recording the mean value of the sample.

For the set of numbers (i.e. means) in Table 7-C, we could easily calculate the overall mean and the standard deviation. Assume we do that, and obtain a mean of 58.2 and a standard deviation of 5.2. The resulting standard deviation is called the **standard error** and is represented by the notation SE. SE is clearly a standard deviation - in this case a standard deviation for means. In contrast, the standard deviation (σ in the original data set is a standard deviation for individual values. Note that σ measures the spread of individual results (i.e. Xs) around μ and SE measures the spread of sample means (i.e. $\overline{X}$s) around μ.

You now should have a good understanding of what SE is - it is simply a standard deviation of means. In the real world we would not want to obtain our standard errors by the approach we used above - it would be time consuming and costly. You'd also have to find populations from which you would extract many samples - not a very practical approach.

We can find the value of a standard error mathematically. To determine a standard error all you have to do is to divide the standard deviation of the underlying data set by the square root of the sample size. The **Standard Error Formula** is:

$$SE = \frac{\sigma}{\sqrt{N}}$$

In words the SE formula is

$$\text{Standard Error} = \frac{\text{Population Standard Deviation}}{\text{Square Root of the Sample Size}}$$

Note that we used the standard deviation of the underlying distribution, the population standard deviation, to find SE. The population standard deviation is not always available, but for the purposes of this chapter we have assumed that it is known and has a value of 20.4. We also said that the means we calculated were based on a sample size (i.e. N) of 16. We can now calculate our SE.

$$SE = \frac{\sigma}{\sqrt{N}} = \frac{20.4}{\sqrt{16}} = 5.1$$

Central Limit Theorem

Remember what we just did - we began with a Population Data Set. We randomly drew out a fixed number of elements from the Population Data Set. We did this many times. In other words we created samples from the Population Data Set. For each of the samples that we created, we calculated a mean. We formed the means into elements of a new data set. We'll call this new derived data set a Sample Means Data Set.

The elements in our Sample Means Data Set will form a pattern - a distribution. Look back at the numbers shown above in Table 7-C (i.e., 58, 60, etc.). We can easily create a distribution with those numbers. A valuable feature of such distributions is that they tend to form a normal curve even if the underlying Population Data Set is not normally distributed. The underlying data set, the Population Data Set, could be skewed, rectangular or have some other irregular shape. Nonetheless, the means of the samples extracted from that data set will take on the shape of a more normal distribution. The larger the sample size of the means that go into the Sample Means Data Set, the more likely we will get a distribution that is normal.

The tendency for means to distribute themselves normally as the sample size increases, even if the underlying distribution from which the elements are taken is not normal, is referred to as the **Central Limit Theorem**. The affinity for means to form a normal distribution is very useful. We remarked, when we began the discussion of the standard error, that it would be useful to be able to use the Z formula for means. But you may recall that when we introduced Z values in chapter 5 we said that to use Z we had to assume that the data set we were working with was normally distributed. Consequently, it would be easy to conclude that if the underlying distribution was not normal then we couldn't use the Z formula for a mean extracted from the non-normal data set. However, the central limit theorem says that if our mean is based on a sufficiently large sample size we can assume normality. Thus, we can use the Z test for means even if the underlying distribution is not normal.

What is sufficiently large is not easily answered since it depends on how non-normal the Population Data Set you're using is. There are obviously all sorts of ugly looking distributions and it's impossible to generalize on what to do given all the possibilities. However, you can look at the distribution you are working with and see if it looks something like a normal curve. If it does you are probably quite safe if the sample size is around 30. However, if in doubt, it is wisest to seek the advice of a statistician who can do more formal inspections of your data and advise you accordingly.

Let's get back to applying the Z formula for means. Frequently people in academia like to use hours of study time as a way to gauge how motivated students are. What if we calculated the mean study time per week for full time freshmen students. We might be interested to see if the mean study time for this year's new students is different from that of students who preceded them. Assume that gathering data on study times has been done for many years and the resultant data base represents the population of freshmen students who have entered the school. We take a sample of 36 current freshman and find their average study time is 25.8 hours. Our data base on all previous students has a mean of 27.9 and a standard deviation of 7.8. Before we can run the Z test, we need to use the sample size (36) and the population standard deviation (7.8) to derive the standard error. Using the formula given above for the standard error we get:

$$SE = \frac{7.8}{\sqrt{36}} = \frac{7.8}{6} = 1.3$$

The standard error of 1.3 can now be inserted in the One Sample Z formula along with the means for freshman ($\overline{X}$ =25.8) and all previous students (μ=27.9):

$$Z = \frac{\overline{X} - \mu}{SE}$$

$$Z = \frac{25.8 - 27.9}{1.3} = \frac{-2.1}{1.3} = -1.62$$

The Z value of -1.62 is used to find the area under the normal curve and using the AUNC table in the Appendix translates into an area of .0526. Figure 7-A illustrates the situation.

Figure 7-A

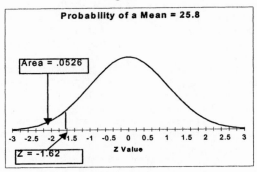

What can we conclude? It appears that the motivation of the new class is suspect since their average study time is quite unusual - only about 5 per cent of the means of previous students would be this small or smaller.

Finding a Mean that Delineates a Given Proportion of a Distribution

When we studied the Z test in chapter 5 we found we could use it to answer other questions about an individual's result. We can answer other questions about the mean for a group in a similar fashion. For instance we could find out what mean would be required so that the class average would be at the 25th percentile (i.e., 25% of the classes did worse and 75% did better). The area under the curve that we're interested in is the area on the left-hand side that cuts off .25 of the total area. That area corresponds to a Z of -0.67. The formula we use to solve this problem is:

$$\overline{X} = (Z*SE) + \mu$$

Note that this formula has the same elements as the Z formula only the elements are transposed. Hence it is really the Z formula in a different version.

We next substitute the values for Z, μ, and SE into the revised Z formula:

$$\overline{X} = (-0.67*1.3) + 27.9 = -0.9 + 27.9 = 27.0$$

Our answer is that a mean of 27.0 hours is required for the freshmen to be at the 25th percentile level.

We can also use this technique to cut off a central or middle part of a distribution. Suppose we had the means for all freshmen classes and wanted to know what scores identified the middle 50% of that distribution. We will still have a mean of 27.9 and a standard error of 1.3 for this example. However in this latter problem we are interested in the central part of a curve, not a tail.

Looking for the middle 50% means we will have an upper and lower limit so we'll need to find a Z value for both locations. The middle 50% is identified by the 25th and 75th percentiles. We therefore need a Z that cuts off .25 of the area in the lower tail and a Z that cuts off .25 area in the upper tail. The Z for the lower tail is -0.67 and for the upper tail it is +0.67. We now need two formulae for our interval - one for the lower limit and one for the upper limit. The formula for the lower limit is:

$$LL = (Z_l * SE) + \mu$$

where LL stands for lower limit and Z_l refers to the Z value corresponding to the lower tail. The formula for the upper limit is:

$$UL = (Z_u * SE) + \mu$$

where UL stands for Upper Limit and Z_u refers to Z value corresponding to the upper tail. Note that the Z in the above formula is the same absolute value, 0.67. Only the sign is different. Z_u is positive in the upper limit formula and Z_l is a negative value in the lower limit formula. We complete our work by solving the two equations. The answers are:

$$LL = (-0.67 * 1.3) + 27.9 = 27.0$$

and

$$UL = (0.67 * 1.3) + 27.9 = 28.8$$

We conclude that the mean scores for freshmen classes that cuts off the middle 50% of the distribution of all freshmen means is 27.0 to 28.8. Many times textbooks will condense these two formulae into one formula and add a ± sign. In this case the formula becomes:

$$\text{Limits} = \pm(Z * SE) + \mu$$

which requires the same calculations as our two formulae approach.

Sample Size

When a researcher wants to conduct a survey, the question always arises about how large a sample he or she should collect. The calculation to make the determination varies depending on the class of variable you are dealing with. We'll confine our discussion to the two most likely cases - when the principal variable of interest is either a continuous or a binomial variable.

The N for a Continuous Variable

Let's assume we want to survey recent graduates of a masters degree program to find out how much they are earning one year after they finished their training. Based on past surveys the standard deviation, σ, is known. For our example σ is $2,500. We, for time and cost reasons, decide to sample the recent graduates rather than conduct a full canvassing. We want to use a sample to tell us something about a population. To determine the sample size we need to specify two things in advance. We realize that we'll have to tolerate some inaccuracy in the sample estimate since we've decided against doing the entire population. So one thing we need to do is to specify how close we want the sample mean to be, compared to the earnings mean we would get if we did all of our current graduates (i.e., the population mean). Are we willing to be off by $100, a $1,000 or even more? We'll use the expression, E, to indicate the amount of <u>error</u> we can tolerate.

The second thing we need to do is to state how certain we want to be that we will not exceed E. For example if we chose $1,000 as the amount of error we can live with, then do we want to be 75 per cent certain our sample is within $1,000 of the true mean? The per cent value we select becomes our assurance level.

Once the researcher decides these two issues the sample size can be determined. The **Sample Size Formula** is:

$$N = \left(\frac{Z * \sigma}{E} \right)^2$$

What if we go with a $500 error tolerance and an 80 per cent assurance level - what will our required sample size be? As we examine the sample size formula we see that we need to specify a Z. The value of Z is based on what we want as our assurance level. When we say we want to have 80 per cent assurance it means we are willing to have 10 per cent of the samples above our tolerance limit and 10 per cent below our tolerance limit. Visualize a normal curve. What we are saying is that we want our sample mean to be in the central portion of the curve - we want to be in the middle 80 per cent. See Figure 7-B.

Figure 7-B

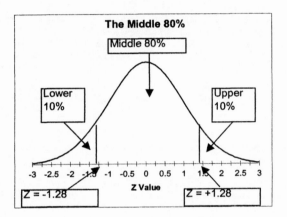

To be in the middle 80 per cent area means 10 per cent is cut off at each end of the curve. There is a Z value that corresponds to these locations. We want the Z that cuts off a .10 proportion on each end. We know how to find the Z for the upper .10 cut off point - we've done it before. We look up in the Z table under the column headed "Area in Right Tail" for the closest value to .10. The Z value that comes closest to cutting off .10 in the upper portion is 1.28. Due to the symmetry of the normal distribution the Z for the lower .10 cut off is also 1.28. However, since we are working in the lower portion of the curve there is a minus sign affixed to that Z. Nonetheless, because of the way the sample size formula is constructed, only one Z value is required. That Z value needed is the positive value, or in our case, 1.28.

To find the sample size we now perform the following calculation:

$$N = \left(\frac{1.28 * 2500}{500}\right)^2$$

$$N = (6.4)^2$$

$$N = 40.96$$

The number 40.96 must be rounded to a whole number and you always round up so the necessary sample size is 41. With 41 survey responders we can have 80 per cent assurance that the sample mean will be within $500 dollars of the population mean.

Frequently, when we set out to do a survey we are too idealistic. We often set our tolerance error too low and find we need a sample size well beyond what we can afford to do. What if we had chosen $100 instead of $500 as our acceptable amount of error? After all it is a lot better to be off by only $100 compared to $500. Can you now calculate a sample size for this situation?

You should have arrived at a sample size of 1,024 based on the following calculation:

$$N = \left(\frac{1.28 * 2500}{100} \right)^2$$
$$N = (32)^2 = 1,024$$

Our sample size has zoomed to 1,024. We know we can increase E to lower this sample size requirement. Is there any thing else we can do? Take a look at the formula for N. In theory σ is fixed (because it is a population parameter based on past experience) so we can't change that value to lower the sample size number. However, we can change Z, our assurance requirement. If we are willing to have less assurance that the sample mean will be close to the population mean, then we will decrease the N required. Instead of an assurance of 80 per cent, let's use 70 per cent. Substitute a Z of 1.04 (the Z corresponding to .15 of the area in each tail) into the formula gives us a sample size of 676. This number is still quite large, but clearly smaller than the 1,024 we had when the assurance value was 80 per cent.

A common quandary that one must face when trying to come up with a sample size is determining the population standard deviation value since it is rarely known. In some cases σ can be estimated from past data. One may also come up with a good approximation by working with the likely range and expected distribution of the variable. If the distribution can be expected to be normal and you think you know where the range likely will fall you can assume σ is equal to the Range divided by 6. The formula is:

$$\sigma \cong \frac{\text{Range}}{6}$$

The $\cong$ symbol means approximately equal to and is based on the fact that almost all the observations in a typical distribution are contained in an interval that is 6 σ long (i.e. from -3σ to +3σ). If you have reason to believe the range for a variable is 50 units than your estimated σ would be 8.33 (i.e. 50/6=8.33)

The N for a Binomial Variable

There may be a situation in which the variable of interest is a binomial variable. For example it is not unusual to see a survey where the question asked is whether the person expects to vote for a Democrat or a Republican. Political party preference, when the choice is confined to Democrat or Republican, is a binomial variable. The results of the polling will give us the proportion who intend to vote for a Republican and the proportion who intend to vote for a Democrat. We will need to do some calculations involving proportions so we can designate either the proportion favoring a Republican or a Democrat as p. If we elect to represent the proportion favoring a Republican as p, then the proportion favoring a Democrat will be 1-p. Once you make your choice as to what group is p (and therefore what group is represented by 1-p) you must consistently observe those designations.

The formula for the sample size of a binomial variable is the same as that for a continuous one:

$$N = \left(\frac{Z^* \sigma}{E} \right)^2$$

The Z and E have the same interpretation as that given above, but the formula requires us to know σ for a proportion. We have not yet learned to calculate σ for a proportion, but it's easy to do. The formula is:

$$\sigma = \sqrt{p^*(1-p)}$$

If p were .3 then σ would be:

$$\sigma = \sqrt{.3^*(1-.3)} = \sqrt{.3^*.7} = \sqrt{.21} = .46$$

Note that σ means we need the population standard deviation. This means we need to know what p is. In practice, we usually rely on the p from a previous poll with a large N and have reason to believe that the poll is equivalent to the one we are planning. We use the p from that poll to calculate our σ.

The formula for the mean and standard deviation do not look like the original formula we developed in chapter 1. Nevertheless, a proportion for a binomial variable consists of only two responses. Suppose the proportion is based on a question such as "Do you support public policy X". An affirmative response indicates a respondent fully agrees with the question and a negative response indicates the respondent totally disagree with the question. We can therefore assign a value of 1 to those who answered affirmatively and a value of 0 to those who answered negatively. If there were 30 responses - 9 affirmative and 21 negative we would have a data set that looked like his:

$$
\begin{array}{cccccccccc}
1 & 1 & 1 & 1 & 1 & 1 & 1 & 1 & 1 & 0 \\
0 & 0 & 0 & 0 & 0 & 0 & 0 & 0 & 0 & 0 \\
0 & 0 & 0 & 0 & 0 & 0 & 0 & 0 & 0 & 0
\end{array}
$$

If we applied the formulae we learned in chapter 1 for the mean and standard deviation to this data set what value would we find for the mean and the standard deviation?

You should have obtained a mean of .3 and a standard deviation of .46. The calculations would be:

$$\overline{X} = \frac{\Sigma X_i}{N} = \frac{9}{30} = .3$$

and

$$\sigma = \sqrt{\frac{\Sigma(X_i - \mu)^2}{N}} = \sqrt{\frac{6.3}{30}} = \sqrt{.21} = .46$$

Thus, even though the mean and standard deviation formulae for a proportion look different than the formulae presented in chapter 1, they give the same result.

Sometimes statisitcs appears confusing because different formulae are used for the same type of calculation. However, the discipline is not as confusing as it may appear. Frequently, as in this case, alternative formulae are introduced because they are easier and quicker to apply, but the same basic calculation is being performed, even if that is not apparent by looking at the formulae.

To show how we obtain a sample size for a proportion, suppose we want to poll the citizens in our town to determine if they plan to vote for a Republican and Democrat for Congress in the next election. Previous polls, similar to ours, had a p of .20 and a standard deviation of .40 ($\sqrt{.2*.8} = .40$). We'll set the error tolerance value at .01.

Remember we are dealing with proportions so if the vote splits 50/50 it means each party wins .50 proportion of the vote. An error tolerance of .01, when converted to a percentage becomes 1 per cent. Consequently, what we are saying is that we don't want our sample mean, relative to the population mean, to be off by more than 1 per cent.

For the assurance level we'll use a figure of .90. We want to have 90 per cent assurance that our error tolerance is not off by more than 1 per cent. To convert the .90 assurance value to a Z we need to isolate out the top and bottom .05 proportions of the normal curve and find the Z that matches the .05 cut off point. Can you find the correct Z?

You should have obtained a Z of 1.64 or 1.65. Actually the .05 proportion is right in between a Z of 1.64 and 1.65. We'll be conservative and use the larger value since that will give us a larger and therefore "safer" sample size.

Substituting the values for Z, σ and E in the formula give us an answer of 4,356 (see below).

$$N = \left(\frac{1.65 * .40}{.01}\right)^2 = 66.0^2 = 4,356$$

Unless you are well funded the sample size may be beyond reach. You therefore need to revise Z or E or both. You can accept a shorter assurance zone or be willing to tolerate a larger error. You might even need to adjust both criteria. Let's use an assurance value of .80 (a good choice since we know from the preceding example that the Z is 1.28) and accept an error rate of 4 per cent or .04 when expressed as a proportion. Now use the Z formula and see what the N for the survey should be.

You should have gotten an answer of 164. This is a much more reasonable number. If you failed to get the right answer, here is the equation set-up you should have used.

$$N = \left(\frac{1.28 * .40}{.04}\right)^2 = 163.8 \cong 164$$

One interesting feature, when determining sample size for a binomial variable, is that you don't need to know the population standard deviation. There is a <u>maximum</u> standard deviation value for binomial variables, which is not true for continuous variables. By using the maximum standard deviation you can use the sample size formula even if you don't know σ. However, you pay a price since you will end up with a larger sample size than if you knew σ. The maximum standard deviation value for a binomial variable is .50. Thus, if in the last example, we didn't know the standard deviation we could substitute .50 for σ in the formula. Now what is the required sample size?

Your answer should be 256. If that is not what you found, check your problem set up. All you need to do is substitute .50 for the .40 is the formula shown above.

In the work we did above, we calculated a sample size for one variable. Yet, in the real world, we rarely do a survey for a single variable. It is usually necessary to have a series of questions in a survey. In that case what should your N be? There is nothing mysterious in arriving at an answer. We have to use a common sense approach. A conservative approach is to determine a sample size for all the questions and then chose the largest one. However, you may find that the N required is very large. Another alternative is to select the N that satisfies the sample size necessary for a majority, but not all the questions. A third option is to base the sample size on the most important question in the set.

Many times we approach a survey with a sample size limit in mind - we can only survey so many people or units. What can happen is that for the Z and E we'd like, the sample size is too large. Reducing Z and E will give smaller sample sizes and is frequently necessary. However, realize that what you're doing is trading off assurance levels and tolerance limits. If you find that in order to have a sample size that fits your budget, you will have an unacceptable level of assurance or tolerance, you need to reconsider conducting the survey. Research is a costly enterprise and why waste time and money on a study that you know in advance will produce an unsatisfactory answer?

A final word about sample size. In this chapter we used a formula to determine sample size, but there are important qualitative guidelines we should be aware of before we start applying mathematical formulae. Normally the larger the sample, the safer the result. For instance, taking a sample of size 30 instead of 15 will help a lot to reduce sampling error. On the other hand, if your sample size is 500 an increase to 515 will have little effect. Furthermore, you can have satisfactory results with a sample size of 25 providing the variation is small.

The relationship between sample size and variability is clearer when we examine the sample size formula introduced in this chapter. The standard deviation appears in the numerator of the sample size formula (i.e. the upper portion of the equation). Consequently, small standard deviations reduce the sample size and large standard deviations increase the sample size. A good rule of thumb to follow, when you are working with proportions rather than continuous variables, is that if the proportion you expect to observe in your study is very large or very small, then you need relatively large sample sizes. For continuous variables the larger the standard deviation, the larger the sample size. However, remember that even a huge sample size can not overcome the intentional bias inherent in non-probability samples. The Literary Digest sample was extremely large, but its prejudicial result could not be overcome.

Summary

We saw in this chapter that a sample is simply a subset of a population. We learned that probability sampling (such as a simple random sample or a stratified sample) are superior methods of choosing a sample compared to non-probability methods such as judgmental or convenience sampling. We showed that the standard deviation and variance formulae are somewhat different for samples compared to populations, but the difference is insignificant when N is large. The standard deviation of a distribution of means is called the standard error and we saw how the standard error and a principle called the central limit theorem allowed us to use the One Sample Z formula and the AUNC tables to tell us how rare a sample mean is. A decision tree to help us select the correct formula when we are trying to determine the probability of an individual value versus the probability of a group mean is given on the next page.

We also studied a way to determine the percent of sample means that fall within a certain distance from a population mean. The formula to calculate sample size was introduced and we learned that to use the formula we need to know the population standard deviation for the variable of interest, the amount of error we can accept and the certainty we must have that that amount of error will not be exceeded.

Decision Tree for Selecting the Correct Z Formula

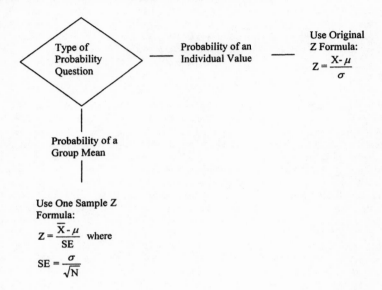

Type of Probability Question

Probability of an Individual Value

Use Original Z Formula:

$$Z = \frac{X - \mu}{\sigma}$$

Probability of a Group Mean

Use One Sample Z Formula:

$$Z = \frac{\overline{X} - \mu}{SE} \quad \text{where}$$

$$SE = \frac{\sigma}{\sqrt{N}}$$

PROBLEMS

1. The IRS wants to audit individual tax returns. Four possible ways to select the return to be audit are suggested. What is the name of the sampling plan for each suggestion?
    a. Identify recipients by five income levels: (under $10,000; 10,000-29,999; 30,000-49,999, 50,000-79,999 and 80,000 or more) and take a random sample of tax filers from each income group.
    b. Select 5 states at random, and do a random sampling of tax filers from the 5 states.
    c. Take a random sample of all tax filers.
    d. Organize tax filers by social security number and pick every 10,000th tax filer from the list.

2. A computer program you used calculated the population mean and standard deviation for a set of data as 35.6 and 6.2 respectively. If the data set is a sample rather than a population, what kind of qualitative change will there be in the mean and standard deviation if the number of items in the data set is:
    a. 1000
    b. 10
    c. Under what circumstance is there a major difference in the sample and population standard deviation?

3. Age information is obtained from a sample of 652 union members. The age variable is normally distributed with a mean of 35.8 and a standard deviation of 12.8.
    a. Where do the central 50% of the individual ages fall?
    b. What is the age standard error?
    c. If many samples were taken, where would the central 50% of age means fall?

4. Expense accounts are monitored by your agency. The auditor reports that based on a sample of 44 meal charges, the mean cost for dinners during the first quarter of the year is $16.65. In preceding years, after adjusting for the cost of living, the cost was $15.82 with a standard deviation of 4.23. How unusually is this year's mean dinner costs?

5. You are asked to survey your alumni to see what they will give this year. You want to have 80% assurance that your sample estimate is not off by more than $200. Contributions to colleges such as yours have a standard deviation of $800. How large of a sample do you need?

6. The Department of Health wants to know what proportion of high school students in the county smoke. Statewide surveys in prior years found that the standard deviation for such surveys was .40. If you want to have an error of no more than 3 per cent and 90 per cent assurance that you do not exceed this level, how many useable responses from the survey will you need?

7. Fishing is an important leisure activity in you region and, for the latest season, a sample of those fishing in the area indicated that the catch for the region was 234,689 pounds. Records have been kept on recreational fishing in your region for years and show that the seasonal mean has been 278,215 pounds with a standard error of 30,575 pounds. How unusual is this year's catch?

8. A random sample of revenue for community based concerts this year indicates that there has been a profit of $3,672 per concert. Records for previous years, adjusted for inflation, show an annual profit of $2,689. The standard error, using the population standard deviation and sample size, is $485. What quantitative comment can you make regarding this year's profit from the concerts relative to the average in previous years?

9. You are asked to survey the teachers at the largest state university to determine what proportion favor forming a union. You are told to have no more than a sampling error of 5 per cent and that you should have 85 per cent assurance that that tolerance limit is not exceeded. You find that there are no relevant past surveys to help you plan your research. How many teacher responses will you need?

10. Your boss asks you to let him know what your department needs to average so it ranks in the top five per cent of the departments in the organization in respect to average employee contribution to the United Fund campaign. Historically the average employee contribution has been $352 with a standard error, derived from σ and the sample size, of $49. What do you tell her?

11. You have a grant to sample 500 people for you research project. Your boss tells you that she learned that the variable you plan to use in your research project is not normally distributed. She asks you why your are proceeding anyway. What do you tell her?

12. There are different types of data in the database we have been using for the computer exercises. Assign these variables to the nominal, ordinal or interval scale.

13. A new friend of yours announces that she has psychic powers. To test her claim you ask her to call the outcome of a coin toss before the coin hits the ground. The coin is tossed 4 times. On the assumption she has no psychic power, what are the chances:
    a. She will call each toss correctly?
    b. She calls 2 out of the 4 tosses correctly?
    c. She calls only 1 toss correctly?
    d. If you pay her $4 for each correct guess and take away $5 for an incorrect answer, what is the expected value of the game?

14. Assume you have the following shirts: 8 White, 4 striped and 4 blue. In addition you have 5 pairs of shoes - 2 are brown and 3 are black. If you select a shirt and pair of shoes at random, what is the probability of the following selections:
    a. A blue or white shirt?
    b. A striped shirt with black shoes.
    c. A blue shirt or stripped shirt with black or brown shoes

15. Assume a variable has a normal distribution, a mean of 190 and a standard deviation of 25. How likely are the following results?
    a. 170 or less
    b. 280 or more
    c. 165 or less
    d. 120 or more

16. What is the difference between internal and external validity?

Excel Instruction

In chapter 7 we introduced the following sample statistics: mean, standard deviation and variance. The command we learned in chapter 1 to find a mean applies to samples as well as populations. Thus all you need is to recall that the Excel format for a mean is: "=AVERAGE(First cell:Last cell)". In the original database, ages are in cells C4 to C42 so if you wanted the mean age you would find an empty cell, type "=AVERAGE(C4:C42)" and Excel will return the correct value, 38.5. In chapter 1 we also obtained the population standard deviation by using the Excel function STDEVP. You can obtain the sample standard deviation by using the Excel function STDEV. Note that the difference between the population and sample standard deviation is the presence or absence of a P at the end of the function name. This makes sense because the P at the end stands for population. Excel uses a similar approach to distinguish between a population and sample variance. The function for a sample variance is VAR and it is VARP for a population variance.

We'll work an example where we determine the sample standard deviation, sample variance, population standard deviation and population variance for the AGE variable. Copy the AGE data to a new worksheet where the heading AGE is placed in cell C3 and the age data begins in cell C4. We'll select 4 cells to enter the results of our statistics. We will enter the name of the statistics we will calculate in column A. In cell A44 type "Sample SD", in A45 type "Sample Var", in A46 type "Population SD" and in A47 type "Population Var".

We will use column C to record the respective statistics. In C44 enter "=STDEV(C4:C42)" and Excel will return the sample standard deviation value of 14.0. We enter "=VAR(C4:C42)" in cell C45 and Excel returns 196.3 as the sample variance. In cell C46 we want the population standard deviation so we enter "=STDEVP(C4:C42)" and the answers is shown as 13.8. Finally in cell C47 we type "=VARP(C4:C42)" to obtain the population variance and the value Excel returns is 191.3. Note that the population and sample statistics are not very different because the sample size (39) is relatively large.

The standard error must be calculated manually, but it is a straight-forward calculation which only requires the sample standard deviation to be divided by the square root of N. We have already learned how to do that type of calculation, but the steps will be repeated here in case you have forgotten how it would be done. Assume we want the standard error for age in cell C48. We'll type in "Stand Error" in cell A48. To find the standard error we will use the sample standard deviation which we already determined and placed in cell C44. We next need to divide by the square root of N. Therefore, type the following "=C44/sqrt(39)" in cell C48. The standard error for the test scores, 2.24, will appear in cell C48.

We also calculated sample sizes in this chapter. There is no built-in function for this statistical function, but it can be calculated manually without much trouble. You need to know E, σ and the assurance level. Let us assume those values are 50, 200 and 90% respectively. On our current worksheet enter the following terms in columns E1 through E3: "Error (E)" in E1, "Sigma" in E2, and "Assurance" in E3. In cells F1 through F3 type in the data, i.e. 50 in F1, 200 in F2 and .90 in F3. We need to convert the assurance level to an area under the curve in a form that Excel can use to give us our Z value. We'll call our new entry "Area" and enter that term in cell E4. The area we want to designate for our Z value can be determined by the following calculation:

$$\frac{1-\text{AssuranceLevel}}{2} + \text{AssuranceLevel}$$

In our case the assurance level is .90 so the area we want is:

$$\frac{1-.90}{2} + .90 = \frac{.10}{2} + .90 = .5 + .90 = .95$$

Consequently we type the following expression "=(1-F3)/2+F3" in cell F4 and Excel returns the value .95. We will also need a place to record the Z value corresponding to the area value we just obtained. In cell E5 type "Z". To obtain the Z value enter "=NORMSINV(F4)" in cell F5 and Excel returns the Z value we need, 1.64. In cell E6 type "N" and move to cell F6. To find N, the required sample size, refer to the mathematical formula:

$$N = \left(\frac{Z*\sigma}{E}\right)^2$$

In Excel, this formula becomes "=(F5*F2/F1)^2". Entering that expression in cell F6 returns the number 43.29 which must be rounded up to 44.

As noted in Chapter 7, random numbers are frequently used to select a random sample from a population. Random number generation is easily done in Excel. Assume we want to obtain random numbers for a study in which two different treatments are to be allocated to 50 subjects. We will arbitrarily assign one of the treatments the code number 1 and the other treatment code number 2. To generate the allocation schedule, open a new page in your workbook and enter "Subject No." in cell A1 and "Treatment" in cell B1. In cells A2 through A51 we will want the subject numbers 1 to 50 to appear. As described in Chapter 1 you can enter the numbers efficiently by using the "Edit" / "Fill" / "Series" commands. Begin by highlighting cells A2 through A51. Then use the sequence of commands described above (i.e. "Edit" etc.). In the dialog box that appears after you select "Series" be sure that the "Step Value" box has a "1" in it. When you hit OK the numbers 1 through 50 should appear in cells A2 to A51 respectively.

To generate the random treatment schedule to be given to the subjects, position the pointer in cell B2. Then go to the Standard toolbar and select the Function icon. Select "All" for the "Function Category" and browse through the listing that appears until you reach "RANDBETWEEN". Some versions of Excel do not contain the RANDBETWEEN command and in that case you will need an alternative method to create random treatment schedules which will follow the RANDBETWEEN explanation.

Click on "RANDBETWEEN" and hit "OK". You will now have the dialog box for "RANDBETWEEN" on your screen (see below).

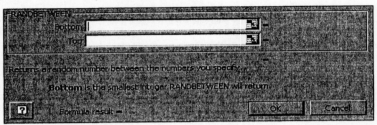

In the "Bottom" box enter 1 and in the "Top" box enter 2 (1 and 2 represent. the two treatment codes we have chosen). Hit Finish and cell B2 should have either a "1" or a "2" in it. If there is a "1" in cell B2 it means subject 1 should be given the treatment coded as 1 and if there is a "2" it means subject 1 should be given treatment 2. The next step is to copy the instruction in B2 to the other cells for which a treatment code is needed (i.e. cell B3 through B51). When you have finished copying, you should have a full treatment allocation schedule for all 50 subjects. If you want to retain this schedule it's best to copy it to a new location. Any activity within the set of numbers (e.g., a font change) will cause a new calculation which replaces your original set.

We will locate the permanent allocation schedule in columns C and D. Copy all the cells (i.e. cells A1 to B51) and then place your pointer in cell C1. Go to the Menu Bar and Select "Edit". From the dropdown menu select "Paste Special" You will then have the "Paste Special" dialog box to complete. Complete the items in the dialog box so that there are dots in the "Value" circle of the "Paste" section and the "None" circle of the "Operation" section. Your screen should match the screen shown below.

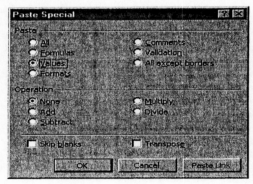

Hit the OK button and your allocation schedule will appear in cells C1 through D51. To avoid confusion delete the information in columns A and B.

If you cannot find a RANDBETWEEN command, then you can use a random number generator built into Excel. To access the generator, select Tools from the main menu followed by "Data Analysis". From the options available, search down the list until you come to "Random Number Generator" and select it. The following dialog box appears:

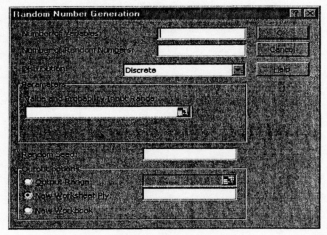

The number of random numbers is determined by the first two entries of the dialog box. We need a single column of random numbers so type 1 in the "Number of Variables" box. The "Number of Random Numbers" is now self evident – if you want 10 random numbers type in 10, type in 25 if you want 25 numbers, etc. Click on "Uniform" for the distribution type. By selecting "Uniform" the Parameters section will change (see the figure on the next page). For the new "Parameters" area enter .5 in the first box and for the last box type in the number of treatments you plan to use plus .5. If you have two treatments you type in 2.5, if there are four treatments type in 4.5, etc.. Leave the Random Seed box blank and then identify where you want the result to appear in the Output Option section. If we wanted a 50 patient randomization schedule using this method, our completed Random Number Generation screen would match that shown on the next page.

The output in our case should consist of numbers between .5000 and 2.5000. Our goal is to have our output represented in whole numbers rather than decimal numbers so we need to round the numbers so they are either a 1 or a 2. It is best to have the whole numbers adjacent to the decimal numbers so go to the cell that is next to the first random number in your listing.

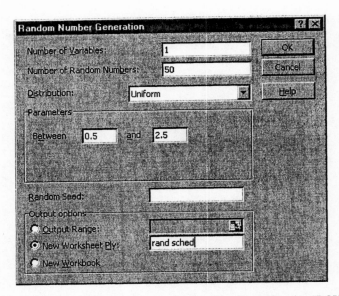

We still need to round the decimal numbers and this is done by the function "ROUND", one of the Excel functions. When you select ROUND the following dialog box appears.

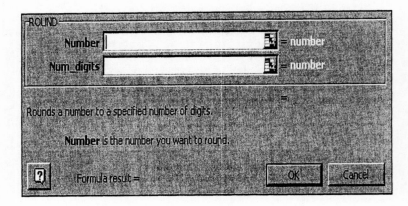

For the first entry, "Number", enter the cell location of your first random number (e.g. B3, or G5, etc). For "Num digits" enter 0. The 0 means we want no decimals. Remember the random numbers are now in digital form and we need whole numbers (a 1 or a 2). Copy this last command so that there is a rounded number for all of your 50 decimal numbers. After hitting OK you should have only 1s and 2s in your random number schedule.

Excel also can select random samples from a listing that is already on a worksheet. Let us assume we want to pick a random sample of 10 employees from our database. We can identify the 10 people by their ID NO. which is given in column A of the database. From the Menu Bar select "Tools" and from the dropdown menu select "Data Analysis". On the listing that appears on the next screen, select "Sampling" and then hit OK. The "Sampling" dialog box appears. The entries for the three parts of the dialog box are given below.

Input Range. Enter "A4:A42" because that is the range of the ID NO.s.

Sampling Method. If necessary click on the box for "Random" so there is a dot in that box. The entry for the "Number of Samples" box should be "12". Technically we only want to select 10 people, but since some ID No.s may be repeated it's best to ask for a few more and then overlook the ones at the end of the list that you don't need.

Output Options. On the assumption that we want the randomly selected Id No.s to appear on a new worksheet, be sure that the "New Worksheet Ply" box has a dot in it. The adjacent box requires a name of the worksheet. We'll use "Sample" so type that term in the adjacent box. At this point your screen should look like this.

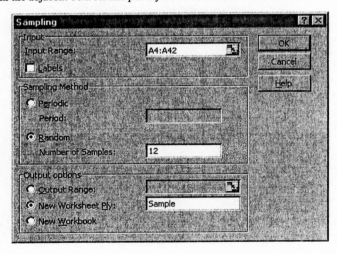

Hit OK and on the worksheet called "Sample" you should have 12 randomly selected ID NO.s. You would use the first 10 unique numbers from that listing for your sample.

Some exercises you might want to do to test your knowledge of these instructions are given below: The correct answers for some problems are shown in brackets.

1. Calculate the variance and standard deviation for the test scores assuming they are from a sample. [Standard Deviation = 12.8 and Sample Variance = 162.9]

2. Calculate the standard error for the pay variable assuming that the data are a sample. [1.79]

3. Refer to problem 5, 6 and 9. Use Excel to obtain the answers to those problems (due to rounding there may be a slight discrepancy between the answers.)

4. Create a treatment allocation schedule for 25 subjects who are to be allocated to 2 different treatments using one of the methods described in this chapter.

5. Select at random 15 of the people in your database for a focus group session.

Chapter 8

Statistical Inference: Estimation and Hypothesis Testing

Introduction

In this chapter we will examine an important statistical practice called statistical inference. We mentioned inferential statistics in the last chapter and noted that a **statistical inference** uses sample results to draw conclusion about population characteristics. The two methods of statistical inference that will be discussed in this chapter are estimation and hypothesis testing. The first method we will explore, estimation, uses statistics from a sample to deduce what are the most likely population parameters. The second topic is hypothesis testing. The name, hypothesis testing, implies that we will set up tentative assumptions (i.e. <u>hypotheses</u>) about the population and then use the results of a sample to <u>test</u> to see if the assumptions are true.

Estimation

There are two major types of estimation - they are called point estimates and confidence intervals. A **point estimate** uses a single sample statistic to estimate a value of the population. For example we are aware that we can use the sample mean to estimate the true value of the population mean. If we found a mean of 24 for a sample we would say that 24 is our estimate (a point estimate) of the population mean. The term "point" means there is a single value or point that is being used to estimate the population mean. However, although the sample mean may be our best guess as to what the population mean is, it's probably wrong. From sampling theory (chapter 7) we would expect our sample mean to be close, but not equal to the population mean. It, therefore, would be helpful to provide a more comprehensive assessment of the population mean than simply providing a point estimate. This is done by developing confidence interval estimates of the population mean.

Recall that in chapter 7 we learned how to identify scores that cut off a certain percent of the elements in a data set. For example, we found the range of scores that cut off the middle 50% of the freshmen means - the range was 27.0 to 28.8. The formulae we employed were:

$$LL = (Z_l * SE) + \mu \text{ and } UL = (Z_u * SE) + \mu$$

In that example we used the characteristics of a population (e.g. mean and standard error) and determined certain characteristics for samples that would come from that population (e.g. an interval that contained a given proportion of the elements in a distribution). In developing confidence intervals we will reverse the process and take the results of a sample to reach conclusion about the population. The confidence interval formulae will be very similar to the above formulae, but since we start with a sample mean, the μ in the formulae must be replaced with $\overline{X}$.

Confidence Interval

To begin our discussion of a confidence interval it will be useful to work a problem using the methods we learned in chapter 7 to derive an interval. Assume we have a population with a mean of 25. We'll also assume that the standard error is 5.

We need to find the scores that cut off the middle 90% of that distribution. Looking for the middle 90% means we need to identify an upper and lower score for the interval. There is a Z value for both locations. The middle 90% is circumscribed by the 5th and 95th percentiles (see Figure 8-A for a graphical representation)

Figure 8-A

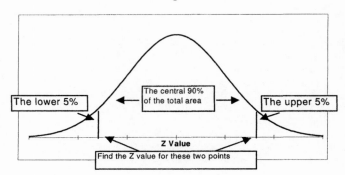

We next need to find a Z that cuts off .05 of the area in the lower tail and a Z that cuts off .05 area in the upper tail. By using the AUNC table we determine that the Z for the lower tail is -1.65 and for the upper tail it is +1.65. The formulae for the lower and upper limits are repeated below.

$$LL = (Z_l * SE) + \mu$$

and
$$UL = (Z_u * SE) + \mu$$

After substituting our values for the Zs (-1.65 and +1.65), SE (5) and μ (25) we get the following result:

$$LL = (-1.65 * 5) + 25 = 16.75$$

and
$$UL = (1.65 * 5) + 25 = 33.25$$

We conclude that the scores that cut off the middle 90% of the distribution are 16.75 and 33.25. We will come back to this result because it is linked to our understanding of a confidence interval.

Derivation of the Confidence Interval

The logic that allows us to derive the confidence interval is difficult to understand. If you don't grasp the concepts, there is probably no need for alarm. At the end of the chapter we will discuss how confidence intervals should be used, as well as interpreted, and learning how to do that will be sufficient for most practical purposes.

Let us again work with the above population statistics - a known standard error of 5 and a population mean of 25. From this population, let us draw a sample and calculate the sample mean, $\overline{X}$. Now assume we <u>aren't told</u> what μ is. We know the standard error but μ is unknown. What if we wanted to identify the interval that marked off the central 90% of the distribution. We could <u>not</u> determine this interval because we <u>don't know</u> the population mean, μ, and the formulae requires that information. However what if we substitute the sample mean, $\overline{X}$, for μ? The formulae would look like this:

$$LL = (Z_l * SE) + \overline{X} \text{ and } UL = (Z_u * SE) + \overline{X}$$

We use these formulae to obtain a **confidence interval**. For example, if we assume that the sample we drew has a mean ($\overline{X}$) of 23, then our 90 per cent confidence interval would be: 14.75-31.25.

$$LL = (-1.65*5) + 23 = 14.75$$

and

$$UL = (1.65*5) + 23 = 31.25$$

Note that the population mean, 25, is included in the interval. Essentially we are adding and subtracting the value of (Z*SE) from the mean of 23. The value of (Z*SE) is 8.25 and we can refer to that value as the half-interval.

Suppose we take another sample from the population that is the same size as the first one and again calculate the sample mean. This time, let us find that the mean from sample 2 is 29. Our confidence interval, when $\overline{X}$ = 29, is: 20.75-37.25. The half-interval, 8.25, is again added and subtracted from the sample mean which in this case is 29. Once again the true population mean of 25 is included in the interval.

We could go on taking samples from the population and calculating the sample means. Most sample means will be around the population mean, but some will be much smaller or larger. For example we could obtain a mean from one of the samples that was 15. The 90 per cent interval for a mean of 15 would be: 6.75 - 23.25. This confidence interval would fail to include the population mean which is 25. The three intervals we calculated are shown in Figure 8-B below.

Figure 8-B

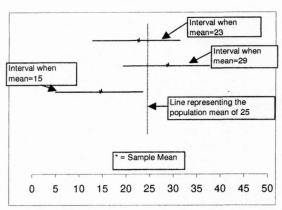

It is clear that for some samples our interval will include the population mean, but that is not true for all the intervals. As a matter of fact, since we are using an interval that is designed to include the central 90% of a distribution, we should find that after many samples are taken that approximately 90% of the intervals would contain the population mean and 10% would not.

Next let us determine the range of sample means that would contain the population mean. What we want to find is the sample mean that has an upper limit of 25 and a second sample mean that has a lower limit of 25. All sample means between these two means will have ranges that include 25 (the population mean). To find the sample mean that has an upper limit of 25 we simply subtract the half interval, 8.25, from 25. The subtraction (25-8.25) gives us a value of 16.75. A sample mean of 16.75 would have an interval of 8.50-25.0. Since the upper limit of this interval is 25 we know that any sample mean less than 16.75 would not include the population mean.

In a similar fashion we can identify the sample mean that has a lower limit of 25. To find the sample mean that has a lower limit of 25 we simply add the half interval, 8.25, to 25. The addition (25+8.25) gives us a value of 33.25. A sample mean of 33.25 would have an interval of 25.0-41.50. In this case the lower limit of the interval is 25 and we know that any sample mean greater than 33.25 would not include the population mean. We can conclude that the intervals for sample means between 16.75 and 33.25 include the population mean and those outside that range do not.

We now can look back at the interval we constructed for the mean scores that cut off the middle 90% of the distribution when we <u>knew</u> what the population mean was. That interval was also 16.75-33.25. What has happened is that we developed a process (the use of sample means) that can gives us the interval that includes the true population mean 90% of the time. We can therefore say that a **confidence interval** is a range of values within which there is a specific degree of certainty that the population value occurs. So much for the rationale behind confidence intervals - let's see how a confidence interval is used.

Using Confidence Intervals

In the real world when we want a confidence interval for a population mean, we collect only one sample - it would be very costly and, as shown above, unnecessary to take many samples from the population. However, we never know for sure if the interval we construct from that single sample contains or doesn't contain the true population mean. However, we do know that we have used a process in which there is a specified percentage of the time (e.g. 90% in our example) that the population mean will be included in the interval. We can therefore say that we have <u>90% confidence</u> that the interval we developed includes the population mean. Of course, the interval either contains the population mean or it doesn't. As a result we shouldn't say that there is a <u>90% probability</u> that the interval contains the population mean because the interval either contains it (and the probability = 1) or it doesn't (and the probability = 0).

We have a choice of how tight or loose the confidence interval will be - it depends on how sure we want to be that the population mean is included in our range. In our example, we used a 90% confidence interval, but we might wish a higher degree of assurance that the population mean is included. If we wanted greater assurance we could create a 95% confidence interval - all we'd need to do is to find the Z values that cut off the central 95% of a distribution. For a 95% confidence interval the Z values would be ±1.96 and including those values in the formulae would result in a wider interval than the one we would produce for a 90% confidence interval. Using our formulae, the 95% confidence interval is 15.2-34.8.

$$LL = (-1.96 * 5) + 25 = 15.2$$

and

$$UL = (1.96 * 5) + 25 = 34.8$$

In practice, to determine a confidence interval, the experimenter first determines how confident he or she wants to be that the interval calculated will contain the population mean. If one wants to be "real sure" then a relatively broad range will be produced. However, if you're willing to take a chance that your interval may miss including the population mean, then you can have a shorter interval. In other words, if you need to have a lot of confidence that the confidence interval includes the population mean you'll most likely end up with a relatively large interval. Conversely, if you can be less confident (assured) that the interval contains the population mean, you will end up with a shorter interval. In actuality the population mean is a fixed value - we don't know where it is, but we can reason that a wider zone is more likely to contain that mean then a shorter zone.

The confidence level is stated as a per cent - 80 per cent confidence, 90 per cent confidence, etc. To have 80 per cent confidence means we will create an interval that covers the middle 80 per cent of the distribution. The curve will be centered at the sample mean value and the area will be shaped by the required number of standard errors or Z values to the left and right of the center.

It's time to work an example. Let us assume that we took a sample so that we could estimate the population mean. There were 25 people in the sample, the sample mean was 35 and the known standard deviation was 10. At this point we need to work with known standard deviations. We need the standard error so we must calculate it. The standard error is:

$$SE = \frac{\sigma}{\sqrt{N}} = \frac{10}{\sqrt{25}} = \frac{10}{5} = 2$$

What will our confidence interval be if we want to have 80% assurance of including the population mean in the interval? In this case the Z values that exclude 10% of the area in each tail are ± 1.28. Excluding 10% in each tail leaves 80% in the center which is what we want (see Figure 8-C on the next page)

Figure 8-C

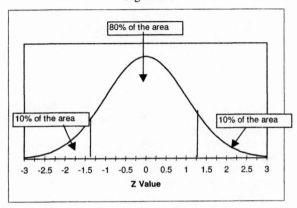

We can now use our confidence interval formulae. After inserting the statistical values we have calculated for this problem we get:

$$LL = (-1.28 * 2) + 35 = -2.56 + 35 = 32.44$$

and

$$UL = (1.28 * 2) + 35 = 2.56 + 35 = 37.56$$

Our answer is that the 80 per cent confidence interval is 32.44 - 37.56. It is tempting to say that there is an 80 per cent chance that the interval 32.44 - 37.56 contains the population mean. As noted above, technically that statement is not quite correct and statisticians will object to its use. It's more accurate to state that one is 80% confident that the population mean is between 32.44 and 37.56.

If we wanted to be more certain that we had a range that contained the population mean we would simply increase our confidence level above 80%. For example, often a 95% confidence interval is used in statistical work. For that case our Zs become ±1.96 - those are the values that cut off 2.5% of the area in each tail leaving a central zone of 95%. The calculation of the 95% confidence interval would result in the following limits:

$$LL = (-1.96 * 2) + 35 = -3.92 + 35 = 31.08$$

and

$$UL = (1.96*2) + 35 = 3.92 + 35 = 38.92$$

See what you'd get if the confidence level for our example is set at 90%.

You should have gotten an interval of 31.7 - 38.3. The Z corresponding to a 90% confidence interval is +1.65 for the upper limit (1.64 could also be used, but it's safer to use the larger value) and -1.65 for the lower limit.

It was noted that a confidence interval becomes wider when greater confidence is desired. However, it should be noted that larger sample sizes produce smaller intervals. This is so because the standard error gets smaller as the sample size (N) is increased. If you inspect the standard error formula, you will see that N is in the denominator (i.e. the bottom portion) so as N gets larger the standard error becomes smaller As a result, the researcher may add greater confidence without expanding the interval by increasing sample size.

It should be clear that a confidence interval has definite advantages over a point estimate. Let us assume you use a sample size of 100 in a study and another researcher investigating the same problem uses a sample size of 25. You will both be able to calculate a sample mean from your respective data sets, which can be used as the point estimate for the population mean. If you both had the same mean, there would be no way to differentiate the two studies. However, if you each reported a confidence interval, the study with the larger sample size will have a shorter interval and be the more appealing result.

It should be emphasized that the confidence interval gives researchers more important information than a simple point estimate. It gives a range of values rather than a single value. The confidence interval also reflects the degree of assurance the researcher has that the population mean lies within the interval.

Confidence Interval for a Proportion

Before we leave the topic of confidence intervals we need to show that we can also calculate a confidence interval for a proportion. In fact, we are more likely familiar with this use because when survey results are reported they almost always refer to a "margin or error" of ± X per cent. The " ± X per cent" is the width of the confidence interval for a proportion.

A proportion can be thought of as a binomial variable (i.e. the variable is present in only two states), in which there is a proportion of the population with the characteristic of interest (p) and a proportion without that characteristic (1-p). To use our confidence interval formulae we need to define the mean (μ) and standard deviation (σ). The formulae we used in the past are repeated below. Remember the symbol p is used for the population **mean for a proportion.**

$$p = \frac{\text{Number with the charactistic of interest}}{\text{Total number}}$$

The **standard deviation of a proportion** (σ) is:

$$\sigma = \sqrt{p * (1-p)}$$

The **standard error of a proportion** (SE) is:

$$SE = \frac{\sigma}{\sqrt{N}}$$

If our population consists of exactly 100 people and we know that there are 60 Republicans and 40 Democrats, then these are the corresponding statistics for the proportion of Republicans:

$$p = \mu = .60 \ (60/100)$$

$$\sigma = .49 \ (\sqrt{.60 * .40}$$

$$SE = .049 \ (.49 / \sqrt{100})$$

In working a proportion problem, the formula we use requires us to know the population standard deviation. We'll work an example. Assume the result of a poll of 100 voters is that 36% favor a proposition and 64% oppose it. The p or mean in this case is .36. Assume we know the population standard deviation and it is .4. The standard error is therefore $.4 / \sqrt{100}$ or .04. Assume we want 95 per cent certainty that we have an interval that includes the population portion favoring the proposition. The area in each tail is .025 and the Z is +1.96 for the upper limit and -1.96 for the lower limit. Substituting the required values in our confidence interval formulae gives us

$$LL = (-1.96 * .04) + .36 = -.08 + .36 = .28$$

and

$$UL = (1.96 * .04) + .36 = .08 + .36 = .44$$

We can now say with 95% confidence that the population proportion favoring the proposition is .28 to .44. We could also rephrase the statement and say that the population that supports the proposition is .36 ± .08. The width of the interval (i.e. ±.08) is called the **"margin or error"** and is usually stated as "plus or minus 8 per cent".

If in the above example we didn't know the population standard deviation, we would use a worst case scenario - a standard deviation of .5 to calculate the standard error. We say "worse case" because no matter what p really is, the maximum value for the standard deviation of a proportion is .5. Our SE would be $.5/\sqrt{100}$ or .05. Substituting this value in our equation gives the following results:

$$LL = (-1.96 * .05) + .36 = -.10 + .36 = .26$$

and

$$UL = (1.96 * .05) + .36 = .10 + .36 = .46$$

We now have a margin of error of ±.10. Note that had we used a different confidence level we would get a different answer for the "margin or error". For this problem calculate a 80% confidence interval using the same n of 100 and assume you know the population standard deviation is .4.

Your answer should be .31 - .41. When polls are reported there is no mention of the confidence level used - we assume it is 95%, but it is rarely if ever stated. I suspect the omission of the confidence level is intentional since lay people would not quite grasp what it means.

In these exercises we assumed the standard deviation was known. More than likely that will not be the case so we need to be able to obtain confidence intervals using a sample standard deviation. How this can be done is shown in the next chapter.

For-Your-Information Statistics

Note that in estimation, we use the sample mean to give us information about the population mean. We are free to use this knowledge about the population mean whatever way we wish. In chapter 7, we compared the sample mean to the population mean and, using the Z score, we found the probability associated with the observed value. The probability we obtained was also for our information - we could use it as we thought appropriate. For example, we could have decided the sample mean was not the least bit unusual (i.e. there was a high probability of occurrence) and therefore called for no action on our part. On the other hand we could have decided that the sample mean was so unusual (i.e. a very small probability of occurrence) that we should take some action. In these cases we are reacting to a statistical finding <u>after</u> it's calculated. Alternatively, we could specify, <u>in advance</u>, an action to be taken or a decision to be made. For example, we could say that if this result has a probability of .05 or less, then we will always take a certain action. However, if the probability is greater than .05, then we will always take a different action. What we are doing in this case is making an a priori commitment to act in a certain way based on a statistical result. We set a criterion in advance - a probability this small or smaller will prompt one action on our part. If the probability is greater than the criterion value, then we will take a different action.

We could also use the statistical result to make a decision rather than to take action. We could decide in advance that, if a favorable mean difference between a new and standard treatment occurs only five per cent of time or less, then we will accept the new treatment as better than the standard treatment. The use of our probability findings in the above situations falls into a category of statistics called hypothesis testing.

Hypothesis Testing

Suppose we draw a random sample from the entering class at a college and compare men and women on the basis of their SAT scores. The result might show that the female average is higher than the male average. The difference may be a reflection of a true difference in the male and female populations from which the samples are taken. On the other hand, the difference may be due to errors inherent in random sampling, which we defined earlier as sampling errors. It is possible that the population means for males and females are the same and, by chance, we drew too many high SAT scores in the female sample and too many low SAT scores in the male sample. Consequently we found a difference in the SAT scores for the samples that was due to the errors that can occur in random sampling. We need a process to see if our sample difference can be assumed to be a true population difference or is it just a case of sampling error. The process used is called hypothesis testing.

Hypothesis testing is a formal process used in research studies to determine if one treatment or experimental condition is different from another treatment or experimental condition. The process begins by establishing two hypotheses - a null hypothesis and an alternative hypothesis. We will use the notation H_0 for the null hypothesis and H_1 for the alternative hypothesis.

A **null hypothesis** is a statement of no difference - treatment A is no different than treatment B. The **alternative hypothesis** is a hypothesis that says there is a difference between treatment A and treatment B. In most cases, it is not likely that experimenters really believe that two treatments are the same, usually they are doing the experiment in hope that one of the treatments is better than the other. In that sense the null hypothesis is a strawman - a proposition set up with the intention of disproving it. If the null hypothesis is proven wrong, then we can accept the alternative hypothesis - one treatment is better than the other. In fact, we could say that our research objective is to disprove the null hypothesis.

We can display our research situation in the form of a decision table. Under the hypothesis testing procedure, when we do a study only two conclusions are possible - accept the null hypothesis or accept the alternative hypothesis. The "conclusion" we reach is therefore one dimension of a 2 by 2 decision making table that we can create. The other dimension is based on what is the actual case. In actuality, the two treatments are truly alike or they are not. Hence the second dimension of the table could be titled the "True Situation". The column and row headings of our 2 by 2 table look like this:

Conclusion:	True Situation	
	H_0 is True	H_1 is True
Accept H_0		
Accept H_1		

We can now fill in the four interior cells of our table. In the cells we are simple going to indicate whether our actual conclusions are good or bad decisions - are they correct or incorrect? Look at the cell in the upper left hand corner - the one at the intersection of the "Accept H_0" and "H_0 is True" categories. If we accept the null hypothesis, and if in fact the null hypothesis is true, then that is a good decision. The conclusion we reach is the correct conclusion so we can write "Correct" in that cell. On the other hand the cell on the right is a mistake (i.e. the cell where we accept H_0, the null hypothesis, but the alternative hypothesis, H_1, is true). We are concluding the treatments are the same (accept the null hypothesis) when they are different (the alternative hypothesis is true). That would be a mistake so we can write "Mistake" into that cell. Our table at this point looks like this:

Conclusion:	True Situation	
	H_0 is True	H_1 is True
Accept H_0	Correct	Mistake
Accept H_1		

In the second row, we find the first cell refers to a situation in which we accept the alternative hypothesis (H_1) when the null hypothesis (H_0) is true. That also is a mistake so we can again write "Mistake" in that cell. Finally, the last cell in the lower right hand corner calls for us to accept the alternative hypothesis (H_1) when in fact the alternative hypothesis is true so that is another good choice and we can write "Correct" in that cell. Our table now looks like this:

Conclusion:	True Situation	
	H_0 is True	H_1 is True
Accept H_0	Correct	Mistake
Accept H_1	Mistake	Correct

Our correct decisions occur when what we conclude matches what is true. Our bad decisions or mistakes occur when what we conclude is inconsistent with the true state of affairs.

Type 1 and Type 2 Error

The cells in which a mistake is made are given names. The expression "**Type 1 Error**" is used when we conclude that the treatments are different (i.e. we accept H_1) when in fact they are the same. That is the case for the cell in the lower left-hand corner of the table. The expression "**Type 2 Error**" is used when we conclude that the treatments are the same (i.e. we accept H_0) when in fact they are different. That is the case for the cell in the upper right hand corner of the table. We can also state the definitions in a slightly different way. A Type 1 Error occurs if we accept the alternative hypothesis when it is false. A Type 2 Error occurs if we accept the null hypothesis when it is false. We'll add the Type 1 and Type 2 Error terms to our table.

Conclusion:	True Situation		
	H_0 is True	H_1 is True	
Accept H_0	Correct	Mistake ◄———	Type 2 Error
Accept H_1	Mistake	Correct	

Type 1 Error ——⌐↑

The Jury Analogy

It is sometimes valuable to relate this situation to a jury decision - a jury must decide if a party is guilty or innocent. In truth, of course, the party is in fact either guilty or innocent, but the jury does not know which condition prevails. Hence, they hear the evidence and then make a decision - either guilty or innocent. They may be right or wrong in their judgment. Furthermore, there is the concept of presumed innocence and this belief can be viewed as our null hypothesis. Therefore, innocence and the assumption of no difference (i.e. accept H_o) can be viewed as equivalent situations. Similarly a guilty verdict is the other choice and can be equivalent to accepting H_1. A researcher is faced with the same type of situation as the jury - test two treatments and based on the evidence decide if the treatments are the same or different. Our jury situation is displayed below:

Jury	True Situation		
Decision	Party is Innocent	Party is Guilty	
Innocent	Good Decision	Bad Decision	◄——— Type 2 Error
Guilty	Bad Decision	Good Decision	

Type 1 Error ——⬏

In the jury situation, the jury does not want to make a wrong decision. Furthermore, in a jury trial the worse decision is to convict an innocent person i.e., to commit a Type 1 Error. There must be strong evidence to find the party guilty. From a statistical standpoint, we can say this means that the jury wants to avoid a Type 1 Error. However we never know the truth, for sure, so the best that can be done is to make the risk of a Type 1 Error small.

The same logic dominates the experimental testing situation. We do not know with certainty what treatment is better, but we want to make sure the risk of a Type 1 Error is small. Consequently to say one treatment is better than another, when they are not truly different, also needs strong evidence and a low risk of making that kind of mistake (a Type 1 Error).

Let's work with an example. We'll compare an old treatment, which is the established way to deal with a condition, against a new treatment that promises to be a better way. Our null hypothesis is that the new and old treatments are equally effective. Our alternative treatment is that the new treatment is better than the old treatment.

We can now present a decision-making table for this specific example. Since accepting the alternative hypothesis is the same as rejecting the null hypothesis we will modify our table to use the latter expression. Furthermore, it may be better to reword our explanation of truth as well. "Ho is True" means the treatments being compared are equal and "H_1 is True" means the new treatment is better, so we will substitute that wording in our table.

Conclusion:	True Situation		
	Treatments are Equal	New Treatment is Better	
Accept H_0	Correct	Mistake	◄——— Type 2 Error
Reject H_0	Mistake	Correct	

Type 1 Error ——⬏

Alpha and Beta Risks

There's a risk of making each type of error and there are terms used to refer to each kind of risk. The terms and their definitions are 1] the **Alpha Risk,** which is the chance of rejecting the null hypothesis when the treatments produce equivalent results; and 2] the **Beta Risk**, which is the chance of accepting the null hypothesis when the treatments produce different results. The Alpha and Beta Risks have been added to the decision table below.

Conclusion:	True Situation		
	Treatments are Equal	New Treatment is Better	
Accept H_0	Correct	Mistake ⟵	Beta Risk
Reject H_0	Mistake	Correct	

Alpha Risk ⤴

Note that an Alpha Risk is the risk of making a Type 1 Error and that the Beta Risk is the risk of making a Type 2 Error. In some respects Alpha Risk and Type 1 Error appear to be synonymous. However, note that one is an a priori risk during the planning stages of a study (the Alpha Risk) and the other term refers to an incorrect decision at the end of the study (the Type 1 Error). By the same reasoning, the Beta Risk is an a priori consideration and Type 2 Error is a mistake that occurs when a conclusion is drawn at the end of a study.

Conducting a Hypothesis Test Using the Critical Value Method

We can use our comparison of a new treatment versus the established treatment to demonstrate how we could go about doing a hypothesis test. We need to set our probability of rejecting the null hypothesis in advance. If we followed custom, the Alpha Risk would be set at a probability of .05. We next need to know what kind of result will cause us to reject the null hypothesis. The formula we will use to compare the treatments is the One Sample Z formula described in chapter 7. We applied that formula when we compared the mean of the new treatment, $\overline{X}$, to an established treatment, μ. When the One Sample Z formula is applied to a hypothesis testing situation, it is frequently referred to as the **One Sample Z Test**. Hence, the formula for the One Sample Z Test is:

$$Z = \frac{\overline{X} - \mu}{SE}$$

To test the null hypothesis, we first need to find the value of Z that will cause us to reject the null hypothesis. The critical Z is the value that corresponds to our Alpha Risk. We set the Alpha Risk at .05 so the critical Z is the Z value that cuts off the upper 5% of the normal distribution. We require the upper tail because $\overline{X}$ must be greater (i.e. better) than μ and, therefore, Z has to be a positive number. From Table 1 we see that the Z value that cuts off .05 of the area in the upper tail is Z = 1.65. Hence, our critical Z is 1.65.

Rejection and Non-Rejection Regions

We can divide all possible Z values into two categories – 1] Z values that will cause us to reject the null hypothesis and 2] Z values that will lead us to accept the null hypothesis. In general, the Z value that begins to define the rejection region is considered the **critical value**. In our example the critical Z value is 1.65. The area to the left of Z = 1.65 is the **non-rejection region**. The area that begins with, and is to the right of Z = 1.65, is the **rejection region**. If the observed Z value falls in the non-rejection region we do not reject the null hypothesis and, therefore, we would conclude that the evidence isn't strong enough to claim a difference between the treatments. If the observed Z falls in the rejection region, we reject the null hypothesis and conclude that there is a difference between the treatments. Since, in our example, the critical Z value is 1.65, the rejection and non-rejection regions are separated at that Z value (see Figure 8-D below).

Figure 8-D

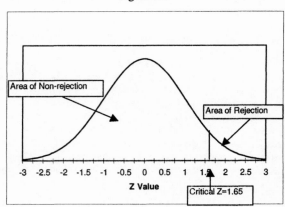

Now that the critical value is defined, and the rejection and non-rejections regions identified, we can continue working our example. Recall that the established treatment has a mean (μ) of 50 and a standard error of 10. We will assume that the new treatment has a mean ($\bar{X}$) of 60. We will first calculate the Z value. If the Z we calculate is in the rejection zone (i.e. it is equal to or greater than the critical Z of 1.65) we will reject the null hypothesis. If the Z we calculate falls in the non-rejection zone (i.e. it is less than the critical value) we will accept the null hypothesis. Often when the critical value is equaled or exceeded, researchers say that "There is a statistically significant difference between the treatments". Conversely when the calculated Z value is less than the critical value, researchers often say "There is no statistically significant difference between the treatments." Frequently the clause "at the α level of significance" is added to the end of each statement. Actually, the numerical value for the Alpha Risk (e.g. .05) would be used in place of "α" in the statement. The calculated Z value for our example is:

$$Z = \frac{60-50}{10} = 1.0$$

As shown in Figure 8-E below, a Z of +1.0 falls in the non-rejection area so we would accept the null hypothesis and conclude that there is not a statistically significant difference between the established and new treatments at the .05 level of significance.

Figure 8-E

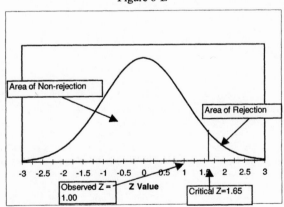

Note that even a mean for the new treatment of 65 would not cause us to reject the null hypothesis (The Z for a mean of 65 is 1.5). A mean of 65 may seem quite different than the established treatment mean of 50, but according to the rules of hypothesis testing you must conclude that the evidence isn't strong enough to say there is a difference between the treatments. The expression "the evidence isn't strong enough" is a reasonable way to report a test result that fails to reject the null hypothesis. Using this clause means you are not concluding there is absolutely no difference. Who knows, there may be a real difference that simply was not detected.

Statistical Significance

The probability level we set in advance (.05 in our example) in addition to being referred to as the Alpha Risk is frequently called the **"level of significance"**. In a report an experimenter might write that "The statistical tests were performed at the .05 level of statistical significance" or "the tests were done using a .05 Alpha Risk". The expression **statistical significance** means that the observed result is less probable than the probability level set for the Alpha Risk. When researchers declare a result is "statistically significant" they are rejecting the null hypothesis in favor of the alternative hypothesis. They are concluding that there is sufficient evidence to believe that there is a difference between the treatments being compared.

By convention the Alpha Risk for most experimental work has been set at .05. If there is one chance in 20 (.05), that the new treatment came from the sampling distribution of the established treatment, then we will say that the two treatments are different. It is frequently argued that the Alpha Risk should not be set at an arbitrary .05 level and the critics are correct. The seriousness of making an error would be a better standard to use. In fact, a researcher may chose to use a higher standard such as a .01 Alpha Risk and no one would object, but using a looser standard (e.g., .10) may not be received well by the scientific community.

There is an important advantage in always requiring a fixed Alpha Risk of .05 or smaller. Remember that in hypothesis testing, we want our decision making to be based on rules that are set in advance of doing the study. Requiring a maximum value of .05 for the Alpha Risk avoids the possibility of a researcher revising the Alpha Risk based on the study results. For example, assume a researcher sets up a study using an Alpha Risk of .05. However, the study ends up with a Z that translates into a probability of .08. Hence, there is no statistical significance. However the researcher desires a positive result and, therefore, revises the Alpha Risk so it is .10 and reports that there is a statistical significant difference at the .10 level of significance. Requiring all testing at a .05 Alpha Risk avoids this temptation to change the rules of the game after the fact.

The Beta Risk has been generally ignored and for good reason. The **Beta Risk** is the chance you will conclude that the treatments are the same when they are, in fact, different. Although there is a relationship between the Alpha and Beta Risk, the mathematics of that relationship are complex. Nevertheless, it's important to note that as you improve an Alpha Risk (i.e., use a smaller probability) the Beta Risk becomes larger (i.e., there is a higher risk of concluding the treatments are the same when they are really different). To pre-set Beta requires a lot of information which is rarely available, but the Beta Risk can be calculated fairly accurately after the fact. For that reason experimenters frequently set the Alpha Risk and worry about the Beta Risk later.

A statistical feature that is related to the Beta Risk is called power. The **power** of a statistical test is the probability of correctly rejecting the null hypothesis because it is false. It is the likelihood you will conclude there is a difference when in fact that is the case. Algebraically it is $1 - \beta$ (where β represents the Beta Risk). Thus, power is a probability value that lies between 0 and 1. Some researchers will report the power of their study. Clearly, since power is a measure of making a correct conclusion, researchers want power to be high in their studies. If the power of a study were reported to be .50 it would mean there is a 50-50 chance that the study would have found a difference if it existed. It is important to remember that simply because we conclude that there is no difference between two treatments, it does not mean conclusively that there is, in fact, no difference. Maybe there is a difference, but the power of the study is just too low to detect it.

One and Two Tail Tests

In the example we just worked, our null hypothesis was that the new and established treatments were the same. Our alternative hypothesis was that the new treatment mean, $\overline{X}$, was better than the established treatment, μ. We calculated a critical value and found the appropriate rejection and non-rejection areas using the right hand side of the standardized normal curve. We were on the right hand side of the curve because $\overline{X}$ must be larger than μ if we are testing to see if the new treatment is better than the established treatment. We are assuming that the highest mean is associated with the best response.

We could have a situation in which the alternative hypothesis is that "the new treatment is worse than the established treatment". In this case our critical value would be a Z value in the left-hand portion of the standardized normal curve. It is on the left-hand side because the ($\overline{X} - \mu$) portion would be a negative number since worse means $\overline{X}$ is less than μ.

We also have another possibility. We could have a situation in which the alternative hypothesis is that the new and established treatments are different without specifying how they are different. They could be different in that the new treatment was better than the established treatment or they could be different if the new treatment was worse than the established treatment. In the case of "different", without specifying the direction of the difference, we need critical Z values for both the right and left-hand sides of the standardized normal curve. We also would have a critical value in the right tail and one in the left tail. We would be doing a test in which critical values are needed for two tails. Thus, the expression a **"Two Tail Test"** would be appropriate.

When we were only looking for the possibility that the new treatment was better than the established treatment, we were concerned with only one tail (refer to Figure 8-D). We therefore could say that we were doing a **"One Tail Test"**. Similarity if we wanted to know if the new treatment was worse than the established treatment we would be interested in just the left hand tail and again we would do a One Tail Test.

In a One Tail Test the Alpha Risk (i.e., the probability of concluding that the alternative hypothesis is true when it isn't) is confined to the tail of interest. If we want the Alpha Risk to be .05, we need to find the critical value for Z that cuts off 5 per cent of the curve. When our alternative hypothesis was that the new treatment was better than the established treatment the upper 5% area was used to identify Z. If our alternative hypothesis is that the new treatment was worse than the established treatment, the bottom 5% area is used. In both cases all our risk is represented in one tail. The Z that cuts off 5% of a tail is ±1.65. The critical value for the "new treatment is better than the established treatment" proposition is at Z = +1.65. The "new treatment is worse than the established treatment" situation requires a critical Z of -1.65.

When we do a Two Tail Test, we must split our risk - half goes to one tail and the other half goes to the other tail. For example if we want our Alpha Risk to be 5% in a Two Tail Test we must split it in half - 2.5% of the risk must go to the upper tail and 2.5% to the lower tail. In this way our <u>total</u> risk is 5%. The Z values that cut off 2.5% of each tail are ±1.96. Observe that for a Two Tail Test there are two critical values: +1.96 and -1.96. If your calculated Z is +1.96 (or greater) or if it is -1.96 (or less), then you can claim a statistically significant result. In either case, you reject the null hypothesis and conclude that the two treatments are not the same. You accept the alternative hypothesis that the treatments are different.

Let's work an example. Suppose we want a 5% Alpha Risk and we're interested in knowing if a new treatment (A) is <u>different</u> than the current treatment (B). Furthermore, since treatment B is an established treatment we'll assume we already know the mean for treatment B. The treatment A mean will be based on a sample. By "different" we mean we're interested in whether treatment A is <u>better or worse</u> than treatment B. Hence we have a two tail situation. The 5% risk must be spread between each tail of the standardized normal curve (2.5% in the upper tail and 2.5% in the lower tail). The critical Z value for the upper tail is +1.96 and for the lower tail it is -1.96.

We'll assume we know the standard deviation and the corresponding standard error, which is 8. If treatment A has mean of 80 and treatment B mean is 100, we would calculate a Z value and see if it falls in the rejection or non-rejection zone. Our calculated Z is:

$$Z = \frac{80 - 100}{8} = -2.50$$

The calculated Z of -2.50 is in the rejection region (i.e. it is more extreme than Z=-1.96 (see Figure 8-F below)

Figure 8-F

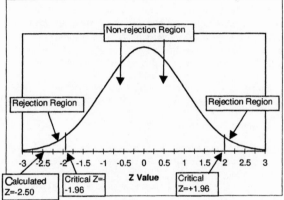

Consequently, we reject the null hypothesis and conclude that treatment A is worse than treatment B at a statistically significant level.

Why Do Two Tail Tests

It is appropriate to ask, why do Two Tail Tests since the statistical standard to declare a difference seems more difficult to meet. For the same Alpha Risk, .05, the critical Z required for a One Tail Test is 1.65, and although we have two critical Zs in a Two Tail Test, they are both more extreme (i.e. ±1.96). However, note that in the Two Tail Test you also are in a position to detect whether a treatment is 1] better than the other treatment and if it's 2] worse than the other treatment. When you perform a One Tail Test you only test in one direction - you can test only one possibility. You can see whether treatment A is better than treatment B in a given test. To determine if treatment A is worse than treatment B would require a separate test. In a Two Tail Test you get to consider both situations in a single test.

The general rule of thumb is to use One Tail Tests when you are absolutely sure that you have no interest in the outcome you're excluding - e.g., that the new treatment is worse than the established treatment. If that result would be of no value to you, then a One Tail Test would be appropriate. In most, but not all scientific studies, Two Tail

Testing is done. It's a little more conservative, but it represents a more robust kind of testing situation.

One danger with a One Tail Test is that you can miss finding the unexpected result. If you expect the new treatment to be better than the established treatment, you may say I can do a One Tail Test. But what if the established treatment outperforms the new treatment, your design does not allow you to know if that difference is statistically significant. You might wonder why you just don't re-analyse the data using the Two Tail Test - unfortunately, that alternative is considered inappropriate. As noted in the beginning of this discussion on hypothesis testing - you set up rules in advance. Under those rules you agree to let probabilities guide your decision making. If you change the rules because of the outcome (i.e., change from a One Tail to a Two Tail Test) you distort the probabilities and your testing is flawed. Thus, it is generally a good practice to do Two Tail Testing unless you can say I'm absolutely uninterested in one of the outcomes.

Conducting a Hypothesis Test Using the Probability Value Method

Earlier we saw how we conduct a hypothesis test using the Critical Value Method. There is a second method that can be used which will result in the identical conclusion. It is called the "Probability Value Method" and differs from the Critical Value Method only in terms of what statistic you use to make your decision regarding statistical significance.

In the Critical Value Method we decided on our Alpha Risk, based on that risk we created critical value(s) and then we found the observed or calculated Z value. We took note of whether the calculated Z value was in the rejection region or non-rejection region. We then came to a conclusion regarding the presence (calculated Z is in the rejection region) or absence (calculated Z is in the non-rejection region) of a statistically significant difference.

In the **Probability Value Method**, we note what our Alpha Risk is (e.g. .05), calculate the Z value and then look up the Z value in terms of it probability of occurrence. If the probability is less than the Alpha Risk, it means your observed result is even more rare than the risk you were willing to take to declare statistical significance. Therefore, if your Alpha Risk is .05 and your calculated Z value is associated with a probability of 0.05 or less, you have statistical significance. On the other hand, if your calculated Z is associated with a probability value greater than the Alpha Risk, it means your observed result can occur with greater likelihood that the standard you set for the Alpha Risk. In that instance you can not reject the null hypothesis and must conclude that the evidence isn't strong enough to declare statistical significance.

In the example we worked for the Critical Value Method, the established mean (μ) was 50, the new treatment mean was 60 ($\bar{X}$) and the standard error was 10. The 10 point difference in means gave us a Z of +1.00.

In the Probability Value Method we now find the probability of Z = +1.00 from Table 1 in the Appendix. The probability of a Z = +1.00 or more is found to be .16. This result means that there is a .16 probability that the two means could be different by 10 points or more by chance. Since that probability is greater than the Alpha Risk (.05) we were willing to take, we conclude that the evidence isn't strong enough to declare statistical significance. The abbreviation **p-value** is often used for the probability value reported for a test statistic such as Z. Statistical significance is present for p-values that are equal to or smaller than the probability specified in the Alpha Risk and absent for p-values that are greater than the probability specified in the Alpha Risk. A graphical representation of the Probability Method for the example we just worked is shown below.

Figure 8-G

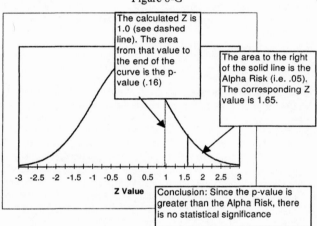

Let us summarize this section on hypothesis testing using the Probability Value Method. The test of a null hypothesis gives us the probability that the null hypothesis is true. The probability is usually represented by lower case p and thus p = .16 means that there is a .16 probability that you could have obtained the observed difference by chance. If the study produced a p-value of .02, then there is a .02 probability that the observed results were due to chance. Obviously the second case (i.e., p = .02) is more convincing evidence that the null hypothesis is not true. When the probability gets small, the researcher concludes that the null hypothesis is not true and declares that there is a difference between the groups. A probability of .05 has become the customary level to reject the no difference hypothesis and accept the alternative hypothesis that there is a difference. Many researchers prefer to report results using the p-value because it is a simple statistic that summarizes the results of a hypothesis test.

There is always some degree of probability that the null hypothesis is true. Selecting .05 means that we will be wrong when we reject the null hypothesis 5 times in 100. In a sense we are taking a calculated risk that we might be making an error. This risk of an error is our Alpha Risk. It is the risk of rejecting the null hypothesis when the hypothesis is true. As noted in the section on "Statistical Significance" there are different ways we can express the fact we rejected the null hypothesis, A common way this is done is to state that "the differences are statistically significant at the .05 level.".

When you determine your p-value, you may find that it is unexpectedly small, for instance .01. When p is .01 researchers are even more sure that the sample differences reflect true population differences. Hence they may note that the null hypothesis was rejected at the .01 level because that is a stronger statement than rejecting the null hypothesis at the .05 level.

Summary

In this chapter we learned we can use samples to give us information about a population. The term estimation is used for this process and consists of finding a point estimate or a confidence interval. A point estimate is a single value whereas a confidence interval is a range of values. The size of the range is a function of the amount of confidence a researcher wants to have that the interval includes the population mean. A confidence interval not only provides a range of feasible values for the population mean, it also is a way of informing others just how much confidence you have in that range.

The chapter also introduced the hypothesis testing procedure which relies on a decision matrix where one dimension is the conclusion a researcher will reach when comparing treatments and the other dimension is whether the response to the treatments are in fact the same or different. Two incorrect conclusions can be made by a researcher – a Type 1 Error (claiming a difference exists when there is no difference) or a Type 2 Error (claiming a difference doesn't exist when it does). Furthermore we learned that an Alpha Risk is the chance one will make a Type 1 Error and the Beta Risk is the chance of making a Type 2 Error.

A researcher also has the choice of using a One or a Two Tail Test. In a Two Tail Test, the researcher can test two hypotheses in a single study. For example, the researcher can see if 1] treatment A is better than treatment B and 2] whether treatment B is better than treatment A. In a One Tail Test only one of the above hypotheses is tested.

A claim of statistical significance is made when the observed result is more rare than the probability specified as the Alpha Risk. Two methods were described for the declaration of statistical significance – the Critical Value Method and the Probability Value Method. It doesn't matter what method is used since the two methods always reach the same conclusion.

PROBLEMS

1. Determine the confidence limits for the mean under the following situations: (In addition to calculating the interval, mark off the Z values corresponding to the confidence interval area on a hand drawn standardized normal curve).
   a. Confidence level = 80%; Mean = 65; SE = 7
   b. Confidence level = 99%; Mean = 26; SE = 1.2
   c. Confidence level = 90%; Mean = 125; SE = 18

2. Complete the following problems:
   a. A survey of 200 people shows that 72% think candidate Jones is doing a good job. If the population standard deviation is .45 what is the 95% confidence interval?
   b. A poll reports that 400 people were surveyed and 32% plan on voting in the forthcoming election. The σ is known to be .47. If a 98% confidence limit is used, what margin of error will be reported?
   c. On a questionnaire, it was found that 72 out of 1055 people who responded, admitted to binge eating. With a population standard deviation of .25, what is the 75% confidence interval for the reported prevalence?

3. In a research report it is stated that after talking to 37 representative families, the mean amount of money spent on "junk foods" is $585 per year. The population standard deviation for these costs is $108. What is the 90% confidence interval?

4. An organization's current method of operation is capable of processing 24.4 forms a day with a standard deviation of 6.4. A new technique is available and a trial is designed to test it. Assume the values given above for the current method are population values. The trial for the new machine involved testing it on 16 different days.
   a. If the goal is to determine whether the new method is better than the current method, using an Alpha Risk of .05, what is the critical value of Z?.
   b. If the goal is to test whether the new method is worse than the current method, using an Alpha Risk of .10, what is the critical value of Z?
   c. If the goal is to test whether the new method is different than the current method, using an Alpha Risk of .01, what are critical values of Z?
   d. Refer to part a. If the new machine has a mean of 27.5 forms per day is there a statistically significant difference between the machines?
   e. Refer to part c. If the new machine has a mean of 20.0 forms per day, is there a statistically significant difference between the machines?

5. If 27 members of a department are asked if they want to hold a holiday party and 33% say "yes", what is the 70% confidence interval when $\sigma = .40$?

6. A standardized test has a mean of 88 and a standard deviation of 12. The test is given to 100 students who commute.
    a. What are the critical values, for a two tail test with an Alpha Risk of .02
    b. If the hypothesis is that commuters did better than the population average of 88, what is the critical value for an Alpha Risk of .04
    c. If the sample mean is 90 for commuters, would you conclude that there was a statistically significant difference for the situation described in part a?
    d. If the sample mean is 90 for commuters, would you conclude that there was a statistical significant difference for the situation described in part b?

7. What is the difference between the following?
    a. Alpha Risk and Type 1 Error
    b. Beta Risk and Type 2 Error
    c. Estimation and hypothesis testing

8. What is meant by the following terms/concepts?
    a. Statistical significance
    b. Margin of error
    c. 90% Confidence Interval
    d. A 5% Alpha Risk

9   The maker of a health food product claims that people will lose weight after three days use. The known standard deviation is .96 pounds. A randomly selected group of 36 people were contacted and it was found they had a mean loss of .35 pounds after using the product for three days. It was decided to test whether the observed mean was associated with a statistically significant weight loss. Assume that an Alpha Risk of .05 would be appropriate for this situation.
    a. What is the null hypothesis and what change in weight would corresponds to that statement?
    b. What is the standard error?
    c. What are the critical and calculated Z values?
    d   Use the Critical Value Method to determine whether there is a statistically significant weight loss.
    e. Use the Probability Value Method to determine whether there is a statistically significant weight loss.
    f. Based on the analysis, how would you state your conclusion?

10. Assume your club has 18 people without a college degree, 45 with a bachelor's degree and 6 with a graduate degree. What is the probability of picking, at random, the following individuals?
    a. A person who does not have a college degree or a person who has a graduate degree.
    b. Three different students and each one possess a graduate degree.
    c. At the first pick, a person with a graduate degree and at the second pick, a person with any college degree.

11. In a box there are 38 white balloons and 5 black balloons. For seven draws in a row, a white balloon has been chosen and replaced. What's the probability on the eight draw to pick:
    a. A white balloon.
    b. A black balloon.
    c. A white balloon which floats away and then a black balloon.
    d. A black balloon which you return to the urn and then you pick another black balloon.

12. Suppose you have 4 balls that have a distinct letter on them.
    a. We want to select all four balls and use the letters formed by the balls as secret code names. How many 4 letter code names can be produced?
    b. We want to select only 2 balls and use them to form 2 letter code names. How many 2 letter code names are possible?

13. You need to select three employees to appear on a float at the town's centennial observation. The first person chosen will be in the most prominent position, and the second choice will have a somewhat precarious location. The third person will be in a relatively insignificant but safe place on the float. If there are 29 people in the department how many ways are there to select the three people for the assigned locations on the float?

14. Refer to problem 11. You will draw 6 balloons, one at a time from that box and replace each one after it is drawn. After the 6 draws, you note that 4 selections resulted in white balloons and 2 produced black balloons - what is the probability of that outcome?

15. A number of citizens are pushing for a new library. The town council wants to obtain a better idea on community sentiment and is willing to have a survey taken. However, they want to be 90% sure that the estimated percentage is not off by more than 5%. How many citizens need to complete the survey?

16. Over a long period of time it has been found that the mean number of rejected applications at your place of work is 47. The standard error for this situation is 7. What number of rejected applications corresponds to:
    a. The 98th percentile (Remember at the 98th percentile there 98% of the items below and 2% above the item of interest).
    b. The 25th percentile.
    c. The 12th percentile.

17. A survey of 64 people reveals that the average cost for lunch in your area is $7.60. The population mean and standard deviation for the state are $8.20 and $3.36 respectively. What is the probability of obtaining the $7.60 mean or less?
    a. Before you do the calculations, and based on an intuitive guess, will the probability be relatively large (<.30), small (<.10) or somewhere in between?
    b. What is the actual probability of getting this result?

Excel Instructions

In chapter 8 we developed confidence intervals and there is a function in Excel that allows us to do this. Go to the Standard toolbar and select the Function icon. Select "All" for the "Function Category" and browse through the listing that appears until you reach "CONFIDENCE". Click on "CONFIDENCE" and hit "OK". You will now have the dialog box for "CONFIDENCE" on your screen.

Descriptions for the three items in the dialog box are given below.

Alpha. The "Alpha" entry is related to the confidence level in the following manner:

Alpha = 1-Level of Confidence

Therefore, if we assume the desired confidence level is .95, the Alpha value is .05 (i.e., 1-.95).

Standard dev. The "Standard dev" entry is the population standard deviation and is therefore not a new concept. Let us assume the standard deviation is 10.

Size. Finally "size" refers to the sample size. We'll assume the sample size is 35.

At this point our CONFIDENCE dialog box looks like this.

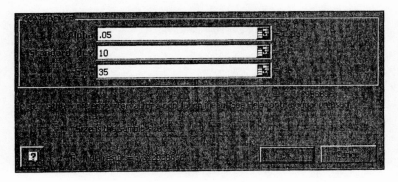

After you enter the appropriate information, you complete the last step in the CONFIDENCE function by clicking on OK. The number returned by Excel, which is shown in the at the bottom of the above screen as "Formula result=3.312938896, is the half-width of the confidence interval. It may seem strange that what Excel returns is half the width rather than the whole width. However it's more convenient to have the half-width. To create a full confidence interval we add and subtract a half-width to the mean. Therefore its easier to work with the half-width Excel provides. In fact, you can write a small equation to automatically calculate the upper and lower values by incorporating the mean with the CONFIDENCE function. How this is done is illustrated in the following example.

Suppose the mean for the confidence function we calculated was 50. Subtracting our half-width (3.31 after rounding) from the mean of 50 we obtain a lower limit of 46.69 and adding 3.31 to the mean gives an upper limit of 53.31, Hence our 95% confidence interval is 46.69 to 53.31.

Note that the entry in the cell containing the value 3.31 has the following format: CONFIDENCE(0.05,10,35) in it. That is the way Excel translated the information we supplied when we used the CONFIDENCE function. Once we are familiar with the elements of a function we can bypass completing the dialog boxes and directly enter the information required for a function.

We could calculate the upper and lower limits directly, (i.e. in one step) by combining the mean value (which is 50 in our example) with the confidence calculation (see below):

$$=50+CONFIDENCE(.05,10,35)$$

If you type that information in a cell Excel will return 53.31, the upper limit. For the lower limit you would only need to write a new equation by changing the plus sign to a minus sign in the above formula. If you wrote:

$$=50-CONFIDENCE(.05,10,35)$$

Excel would return 46.49, the lower confidence limit.

In chapter 8 we also calculated critical Z values so we could do hypothesis testing. To find the critical Z value we need to work with the Alpha Risk. Although there is no command in Excel to do the calculations directly, what we have already learned will allow us to obtain a critical Z value quite easily.

First we need to determine what part of a standardized normal curve we are interested in. The different possibilities are based on whether the problem involves one or two tails. If it is a one tail situation, we need to decide if our interest is in the upper or lower tail. The area we need to use for each possibility is given below:

Lower tail only: The Alpha Risk
Upper tail only: 1- The Alpha Risk
Two tails: Lower limit = The Alpha Risk/2
         Upper limit = 1-(The Alpha Risk/2)

If the Alpha risk is .10 our areas would be:

Lower tail only: .10
Upper tail only: 1-.10 = .90
Two tails: Lower limit = .10/2 = .05
         Upper limit = .1-(.10/2) = 1-.05 = .95

Now that the needed areas are identified we can use the NORMSINV command, which we learned about in chapter 7, to find the appropriate Z value. We will illustrate how this is done for a two tail test with the .10 Alpha Risk given above. For the critical Z needed in the lower tail, we would write "=NORMSINV (.05)" and Excel would return -1.64. and for the upper tail the expression is "NORNSINV(.95) giving a critical Z of 1.64.

In Chapter 9 we will also learn how to do a confidence interval when the standard deviation is unknown using a test statistic called t in place of Z. In that situation you can not use Excel's CONFIDENCE function. The CONFIDENCE function may only be used when the standard deviation is known and the Z procedure is appropriate.

The other calculations performed in chapter 8 also involve using Excel skills we already learned. For example we performed a Z test to obtain a calculated Z value which we compared to a critical Z value to see if there was a statistically significant difference. The formula for the Z test is:

$$Z = \frac{\overline{X} - \mu}{SE}$$

so if we had a sample mean ($\overline{X}$) of 40, a population mean (μ) of 30, a standard deviation of 20 and an N of 16 we would begin by calculating the standard error (SE). We use the expression "=20/sqrt(16)" to find SE. That value is 5 so we would next write "=(40-30)/5" and Excel would return a Z value of 2. Alternatively we could save ourselves a little work by using the STANDARDIZE function we learned in chapter 7. By using the expression "=STANDARDIZE(40,30,5)" Excel would also return the Z value of 2. We would compare that Z value to the critical Z value to see if there was or was not a statistically significant difference.

Here are some problems you can practice on. The correct answers for some problems are shown in brackets.

1. Use the Excel CONFIDENCE function to check the answer for problems 2 and 3.
2. Use the Excel NORMSINV function to provide the critical value for the following problems 4a, 4b; 4c, and 6b.
3. Use Excel to obtain answers to problem 4d and 4e.
4. Assume the test scores in your database represent a sample and that the population standard deviation is 12. Use Excel to find the 95% confidence interval for the test score mean. [72 - 80]
5. Assume that in a sample of 500 organizations the average cost of training programs was $6000 and the distribution is normally distributed. If the population standard deviation is $1200, what is the 95% confidence interval for the population mean? [5895 - 6105]

Chapter 9

Differences Between Sample Means: The t and Z Tests

Introduction

In the last chapter we learned to use the Z test to determine if there was a statistically significant difference between a sample mean and a population mean. In this chapter we will expand our analysis capability by learning how to make comparisons between two sample means using the t Test.

We will begin with a quick review of Z values. In chapter 8 we used the One Sample Z Test to compare a single sample mean, $\overline{X}$, against an established or population mean, μ. The One Sample Z Test formulae is:

$$Z = \frac{\overline{X} - \mu}{SE} \text{ where } SE = \frac{\sigma}{\sqrt{N}}$$

In the formula, $\overline{X}$ is the sample mean, μ the population mean and SE the standard error. Note that the SE is based on the population standard deviation (σ).

We interpret the Z value we obtain as a count of the number of standard errors that exist between the sample mean, $\overline{X}$, and the population mean, μ. As a result, a Z value of 1 indicates that the there is one standard error separating the means. A Z value of 1 also indicates that the area of the normal curve to the right of the Z value is .1587. We can also interpret this area as a probability. There is a .1587 probability of getting the observed $\overline{X}$. If the observed probability (i.e. .1587) is less than the Alpha Risk we conclude that it is sufficiently rare and claim that there is a statistically significant standard deviation is known. When the standard deviation is unknown we need to use a new procedure called the One Sample t Test. It is worthwhile to point out that lower case t is always used when referring to this test.

One Sample t Test

In the One Sample Z Test, as well as the One Sample t Test, we compare a mean from a single sample to an established or population value. However, for the Z Test, we have a known population standard deviation, but in the t Test, the population standard deviation is unknown and we rely on the sample standard deviation.

The One Sample t formula is essentially the same as the One Sample Z formula, but t replaces Z and the SE is based on S (the sample standard deviation) rather than σ (the population standard deviation). The formula for the **One Sample t Test** is:

$$t = \frac{\overline{X} - \mu}{SE} \text{ where } SE = \frac{S}{\sqrt{N}}$$

Note that the One Sample t Test formula looks much like the One Sample Z Test formula, but we solve for t not Z. For a Z Test we determined how unusual our $\overline{X}$ was, using the AUNC table. We need to use a different table to find the probability associated with a t value. For a t Test we use a table called the t Table. The AUNC table uses a single set of Z values because there is only one standardized normal curve. On the other hand, we rely on different sets of t values in a t Table. This is because there is a different t distribution for each sample size (N) associated with the sample mean.

Compared to the normal curve, the t distribution for small Ns is flatter and has longer tails. However as N increases the t distribution becomes more and more like the normal distribution. The table on the next page (Table 9-A) is a modification of what t Tables look like. See Table 2 in the Appendix for a more complete listing of t values.

The first column heading in Table 9-A is labeled "Df" whereas all the other column entries (.10, .05 etc.) refer to Alpha Risks. The Df notation stands for "degrees of freedom" and is directly related to the sample size (i.e. N). Sometimes degrees of freedom is also abbreviated as df. For a One Sample t Test the df = N-1. For a Two Sample t Test, which we will discuss later, the df usually equals $(N_1-1) + (N_2-1)$ where N_1 is the sample size for one group and N_2 is the sample size for the second group.

The notion underlying the concept, **degrees of freedom,** is based on the observation that when we calculate a test statistic, for example the mean, there are N values that go into that calculation. Assuming that the sum of those values is fixed, then all the numbers in the set can vary, except one. Since one number is not allowed to vary, we can say that N-1 numbers are free to vary. Consequently, if N is 10, we say there are 9 degrees of freedom and if N is 20 we would say there are 19 degrees of freedom.

Table 9-A Critical t Values for Various Alpha Risks

			One-Tail Test			
Df	.10	.05	.025	.01	.005	.0005
			Two-Tail Test			
Df	.20	.10	.05	.02	.01	.001
1	3.078	6.314	12.71	31.82	63.66	636.6
2	1.886	2.920	4.303	6.965	9.925	31.60
3	1.638	2.353	3.182	4.541	5.841	12.94
4	1.533	2.132	2.776	3.747	4.604	8.610
5	1.476	2 015	2.571	3.365	4.032	6.859
6	1.440	1.943	2.447	3.143	3.707	5.959
7	1.415	1.895	2.365	2.998	3.499	5.405
8	1.397	1.860	**2.306**	2.896	3.355	5.041

One-Tail Alpha Risk

Two-Tail Alpha Risk

Due to the mass of values that would be required if t Tables tried to replicate the same amount of information as a AUNC table, t Tables are usually set up so they list the critical values of t for only the most common Alpha Risks. In other words, the table is usually restricted to only a limited number of Alpha Risks. The t Table is also organized so both one and two tail critical t values appear in a single column. Note that in our abbreviated table above, there are only six Alpha Risks for one tailed tests (.10, .05, .025, .01, .005, and .0005). For the two tailed test there are also only six Alpha Risks listed (.20, .10, .05, .02, .01, .001). Also note that the same column of critical t values is used for one tail tests and two tail tests. For example, for a one tail test at an Alpha Risk of .10, the critical t values are the same as those for a two tail test with an Alpha Risk of .20.

If you are doing a two tail test with an Alpha Risk of .05 and had 9 observations in your sample, you would go to a t Table and locate the .05 value in the "Two-Tail Test" row of the t Table. (The Alpha Risks for a two tail test are shown in the fourth row of the above table). At the intersection of the column for a .05 Alpha Risk and the row for df = 8 (9-1), you find the number 2.306 - that number (it has been placed in bold font in the above table) is your critical value for t.

If we were doing a Z test, rather than a t Test you might remember that the critical Z value (for a two tail test with an Alpha Risk of .05), is 1.96. That value, 1.96, is obviously less than the 2.306 we found in the t Table for a two tail test with an Alpha Risk of .05. It is clear that the One Sample t Test, with a df of 8, requires a larger critical value than a One Sample Z test. The larger critical value for the t test means we will require relatively larger difference between the means before we can declare statistical significance. Some people find the imposition of a higher standard reasonable because in a t Test the population standard deviation is unknown, whereas it is a known value for the Z Test.

The chance we know the population standard deviation is unlikely for most variables. A known population standard deviation is not required for the t Test and we can simply use the sample standard deviation in place of the known standard deviation. In this sense, the t Test has a much broader application than the Z test. The t Test can be used when the population standard deviation is unknown, which is, by far, the more likely case.

In the t Table (See Table 2 of the Appendix) note that as df becomes larger the t values come closer and closer to the Z values. For example if the df were 30, the critical t value for two tail tests at the .05 Alpha Risk level is 2.04. If the df were 60 then the t value is 2.00, almost the same as the critical Z value of 1.96.

Assume we want to compare the cost of graduate textbooks to undergraduate textbooks. We are told by the bookstore that the average undergraduate book costs $32. We take a sample of 36 graduate textbooks and find that the mean cost per book is $34 with a standard deviation of $9. Does this result mean that graduate textbooks cost more than undergraduate textbooks? In other words is there a statically significant difference between the textbook costs based on our sample result? We first need to calculate the SE for this case which is calculated using the sample standard deviation (S).

$$SE = \frac{S}{\sqrt{N}}$$

$$SE = \frac{9}{\sqrt{36}} = \frac{9}{6} = 1.5$$

and then we calculate t where $\overline{X}$ is the sample mean for graduate textbooks (34) and μ is the known mean for the undergraduate textbooks (32).

$$t = \frac{\overline{X} - \mu}{SE}$$

$$t = \frac{34 - 32}{1.5} = \frac{2}{1.5} = 1.33$$

We next need to look up the t value in a t Table to determine if there was statistical significance. As shown in Table 2 of the Appendix, for a two tail test with an Alpha Risk of .05 and a df of 35, our critical t is 2.03. We do not have a statistically significant difference since our calculated t is 1.33. As you can see, the Critical Value Method we used with the Z Test also works when we do a t Test.

In this age, computers can do most of the tedious mathematical calculations and report the actual probability of the result we obtain. Furthermore, computers can do the work more accurately. There are so many computer programs available it is not possible to demonstrate how each one reports a t Test result. As shown below, the One Sample t Test performed by a computer program may request summarized data such as the sample size, sample mean, calculated standard deviation and the hypothesized (i.e. comparison) mean.

One Sample T-Test on Summarized Data
Test Direction:
U [Upper] L [Lower] T [Two tailed] ➡ [   ]
Data
Sample Size
Mean
Std Deviation
Hypoth. Mean

The Test Direction section gives us three options:
    Upper or right-hand tail
    Lower or left-hand tail
    Two tailed

We also need to specify the sample size. Note that we have two means in our example, but only one was based on a sample (the graduate book cost). The other mean (undergraduate book cost) is a mean we decided to use for comparison purposes and can be considered a population mean. The "Hypoth. Mean" refers to the mean that we want our sample mean compared to. In our case it is the undergraduate mean which is 32. We would input the information required so the input table would look like this:

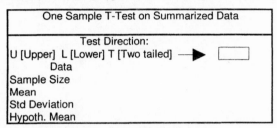

One Sample T-Test on Summarized	
Test	
U [Upper] L [Lower] T [Two] ➡ [T]	
Data	
Sample Size	36
Mean	34
Std Deviation	9
Hypoth. Mean	32

The computer output for this example is shown on the next page. We are familiar with all of the output information. The most important item is the last item, "Prob.", which stands for probability. The probability value of 0.19 is the likelihood that we could have obtained a mean of 34 if, in fact, we were sampling from a population in which the mean was 32. If we were doing hypothesis testing and set our Alpha Risk at .05, we would conclude that there is no statistically significant difference because the probability value is greater than .05. We employed the Probability Value Method and decided that the evidence isn't strong enough to conclude that the cost of graduate and undergraduate books are different.

One sample T-Test Result		
Count	36	
Mean	34	
Std Deviation	9	
Std Error	1.5	
95% C.L. of Mean	30.95493	37.04507
Hypothesized Mean	32	
Difference	2	
T Value	1.333333	
Prob.	0.191	

Now that we understand how to do a One Sample t Test, we can look at a somewhat more complex situation. It is very likely that we may conduct a study where we have <u>two sample</u> means. For instance, a study in which we test a promising experimental drug against a placebo to see if the drug is effective. In this case, we have a sample of people on the drug and a sample of people on the placebo. We, therefore, end up with sample means for both groups. What we obviously want to do is to compare the two sample means to see if there is a statistically significant difference.

Two Sample t Test for Independent Samples

There is more than one t formula that is appropriate for the comparison of two sample means. The first set of tests we will examine are samples that are independent, which means each of the sample means is based on a unique group of individuals. Later we will see that you can also use the same group of individuals as your two samples. There are two versions of the Two Sample t Test for independent samples depending on whether we assume equal or unequal standard deviations for the two samples. The formulae presented on the following pages will be useful to those who plan to do their statistics without the help of a computer program. However, for those who use a computer program, the information presented will help you appreciate the testing process. Your ability to interpret computerized statistical results should thereby be enhanced. In working through this section also realize that all the t formulae take on the same form which is:

$$t = \frac{\text{Difference between Group Means}}{\text{S tan dard Error}}$$

The numerator never changes for any version of the t test and you will see that the differences are confined to different formulae used to calculate the denominator, i.e. the standard error.

t Test When the Standard Deviations are Unequal

The formula to test means when you assume that the standard deviations are unequal is:

$$t = \frac{(\overline{X}_1 - \overline{X}_2)}{SEu}$$

where $\overline{X}_1$ is the mean of the first sample and $\overline{X}_2$ is the mean of the second sample. The formula for SE_u is:

$$SE_u = \sqrt{(SE_1)^2 + (SE_2)^2}$$

where SE_1 and SE_2 are derived from the respective sample standard deviations, S_1 and S_2. Those formulae are:

$$SE_1 = S_1 / \sqrt{N_1} \text{ and } SE_2 = S_2 / \sqrt{N_2}$$

The subscript u for SE is used to indicate that we are dealing with unequal standard deviations. We'll work a problem in which the necessary statistics for the two samples we want to compare are given in Table 9-B below:

Table 9-B

Statistic	Sample 1	Sample 2
Mean	65	60
Standard Deviation	9.25	6.38
Sample Size	36	25

The calculation of the numerator, $\overline{X}_1 - \overline{X}_2$, is straight forward (65-60=5). We obtain the SE for each sample using the formulae presented above. Those formulae produce the following results:

$$SE_1 = \frac{9.25}{\sqrt{36}} = 1.54$$

$$SE_2 = \frac{6.38}{\sqrt{25}} = 1.276$$

$$SE_u = \sqrt{(1.54)^2 + (1.276)^2} = \sqrt{4.001} = 2.00$$

We can now proceed to applying the t Test for a hypothesis testing situation. If we want a two tail test and an Alpha Risk of .05, our critical t values are found in Table 2 of the Appendix for a two tailed test at the .05 level and 59 degrees of freedom. The calculation of df was described earlier, but for this version of the t Test the df is not always equal to the $(N_1-1)+(N_2-1)$ formula. In this example the df conforms to the formula, but that is not always the case. There is no entry in Appendix Table 2 for 59 df, but there is one for 60 which we will use since it is so close to 59. The value we find is 2.00 and that becomes our critical t value. In other words, our rejection region is marked by t values greater than or equal to +2.00 and t values less than or equal to -2.00. If our calculated t is equal to or over +2.00, then there is sufficient evidence to claim that Sample 1 is better than Sample 2. There is an assumption here that a higher mean is associated with a better result. However if t is less than or equal to -2.00 than we would conclude that Sample 2 is better than Sample 1. Any calculated t value that falls in the non-rejection region, (i.e. t values greater than -2.00 but less than + 2.00) results in a conclusion that the difference between the samples is not statistically significant.

We substitute the information we calculated above in the t formula, and get:

$$t = \frac{(65 - 60)}{2.00} = \frac{5}{2.00} = 2.50$$

The calculated t falls into the rejection region so we have a statistically significant difference in favor of Sample 1.

In solving this problem we needed to do some calculations, but computer programs can do that chore more quickly and with more precision. Many statistical programs allow you to input your raw data and the machine will do the rest of the work.

A typical computer program will require you to identify the location of your raw data on the computer file, the Alpha Risk and the mean difference expected (under the null hypothesis that difference is 0). Once you have provided the necessary input, your output will look similar to that shown on the next page (the output shown is for the data given in Table 9-B and assumes the Alpha Risk is .05 and the hypothetical difference is 0).

For each sample we are given the Mean, Variance and number of Observations (i.e., the N for each sample). As you may recall the square root of the variance is a standard deviation so we can easily also know the value of the respective standard deviations. The "Hypothesized

Mean Difference" is the $\overline{X}_1 - \overline{X}_2$ difference expected under the null hypothesis. In most all cases that difference is 0. For this unequal variance version of the t Test, the "df" is calculated by the computer program and turns out to be 59. The "t Stat" notation refers to the calculated t value which is 2.50. Below the t value we are given the probability corresponding to a one tail test. The notation for this entry is P(T<=t) and the resulting value is 0.008. The notation means that .008 is the probability (P) that the calculated t (T) is less than or equal to (<=) to critical t (t). The next entry, "t Critical one-tail", is self-explanatory – it is the critical t value for a one tail test (1.67). The last two rows give the two tail p-value (0.015) and critical t value (2.00).

t-Test: Two-Sample Assuming Unequal Variances

	SAMPLE 1	SAMPLE 2
Mean	65.0	60.0
Variance	85.6	40.7
Observations	36	25
Hypothesized Mean Difference	0	
df	59	
t Stat	2.50	
P(T<=t) one-tail	0.008	
t Critical one-tail	1.67	
P(T<=t) two-tail	0.015	
t Critical two-tail	2.00	

The "t Stat" value given in the above table is the calculated t and is either positive or negative depending on whether the difference for ($\overline{X}_1 - \overline{X}_2$) is positive or negative. In our case ($\overline{X}_1 - \overline{X}_2$) is (65-60) so the t value is shown as a positive number. Be aware that in a one tail test the sign of the critical t (1.67) is determined by the researcher. He or she examines the difference between the means to see if the result is compatible with the alternative hypothesis. In this example, the alternative hypothesis is that the mean for Sample 1 is greater than the mean for Sample 2. Since Sample 1 has the higher mean, the result is compatible with the alternative hypothesis. Therefore, the critical t value is given the same sign as the t statistic (i.e. a plus sign which result in a critical t = +1.67). To determine whether there is statistical significance we compare the calculated t value of 2.50 (the t Stat value in the above table) to the critical t value (1.67). Since the calculated t value is greater than the critical t value, we have statistical significance.

The alternative hypothesis required Sample 1 to have a higher mean than Sample 2. If the reverse result occurred, the $\overline{X}_1 - \overline{X}_2$ difference would be a negative value and inconsistent with the alternative hypothesis. In our example, we have a statistically significant difference for the one tail test only when the alternative hypothesis is that the Sample 1 mean is greater than the Sample 2 mean. If we found that the Sample 2 mean was greater than the Sample 1 mean, there could not be statistical significance. The rejection region would still be in the upper tail, but the observed t would be in the opposite tail and, therefore, it would clearly fall into the non-rejection region.

What conclusion would we come to if we were doing a two tail test? We can use the Critical Value Method or the Probability Value Method to see if there is statistical significance between Sample 1 and Sample 2. If we use the Critical Value Method we observe that the calculated t (+2.50) is greater than the critical t (+2.00) and we have a statistically significant difference. Alternatively we could use the Probability Value Method. The p-value for the two tail test is given as 0.015 and this value is less than the .05 Alpha Risk that was specified. Consequently we know that the observed result occurs about 1.5 percent of the time. When we set the Alpha Risk at .05 we indicated that we were willing to take a 5 per cent risk of concluding that there was a difference when no difference exits. The observed probability is even more unlikely than the Alpha Risk so the result we obtained is sufficiently unique to declare statistical significance at the .05 level.

t Test When the Standard Deviations are Equal

You should note that the output table is labeled "t Test: Two Sample Assuming Unequal Variances". Although the term "Variances" is used the term "Standard Deviations" could have been substituted because, as we saw in chapter 1, the two statisitcs are based on the same information and the standard deviation is just the square root of the variance.

In practice, assuming equal variance is the more usual situation for researchers. The initial assumption for a two sample t Test is that the standard deviations/variances of each sample are equal. However, they may appear different due to **random variation** (i.e. unexplained variation that is not associated with treatment effects). Random variation can be considered variation that is due to chance and is inherent in the behavior of any variable.

Unless there is *a priori* evidence of unequal standard deviations/variances, most researchers initially use the t Test version which assumes equal standard deviations/variances. To test means based on interval scaled measurements, when you assume that the standard deviations are equal, the t formula is:

$$t = \frac{(\overline{X}_1 - \overline{X}_2)}{SEe}$$

The formula for SEe is:

$$SE_e = S_p * \sqrt{\left(\frac{1}{N_1} + \frac{1}{N_2}\right)}$$

where S_p is:

$$S_p = \sqrt{\frac{(n_1 - 1) * S_1^2 + (n_2 - 1) * S_2^2}{(n_1 - 1) + (n_2 - 1)}}$$

Observe that the e subscript in SE_e stands for "equal" and indicates that we are dealing with equal standard deviations/variances. The subscript p in the S_p formula stands for "pooled". If you study the S_p equation for a moment, you may be able to see that S_p is, in fact, a pooled standard deviation which is weighted by the sample size for samples 1 and 2. The rationale is that since the standard deviation is unknown, but assumed to be equal in Sample 1 and Sample 2, then both sample standard deviations are valid estimates of the true standard deviation. However, since the larger sample sizes tend to reflect better estimates than smaller sample sizes, they should get more weight in arriving at the population estimate. In fact, if the sample sizes are equal, the formula for S_p reduces to:

$$S_p = \sqrt{\frac{S_1^2 + S_2^2}{2}}$$

We'll re-work the problem given in Table 9-B, but this time we will assume the two standard deviations are equal rather than unequal. Technically, a researcher decides if the standard deviations are equal or unequal and does only one test. However, there are situations when it is hard to tell what condition prevails and the researcher may do both tests to see if the results differ. If there is no difference in results, the researcher will usually be content with the equal standard deviation assumption. If a difference occurs, the researcher will usually do further testing of the standard deviations to decide which assumption is more likely.

The calculations required are:

$$S_p = \sqrt{\frac{(35 * 9.25^2) + (24 * 6.38^2)}{(35 + 24)}} = \sqrt{\frac{2994.7 + 976.9}{(59)}} = \sqrt{67.3} = 8.2$$

$$SE_e = 8.2 * \sqrt{\left(\frac{1}{36} + \frac{1}{25}\right)}$$

$$SE_e \ 8.2 * \sqrt{0.0677} = 2.13$$

$$t = \frac{(65 - 60)}{2.13} = 2.35$$

Our critical values for t remain at +2.00 and –2.00 and since the calculated t of 2.35 falls into the rejection region, we again have a statistically significant difference in favor of Sample 1.

The computerized result we would have obtained for the example we used above, had we made the assumption of equal variances is shown below.

t-Test: Two-Sample Assuming Equal Variances

	SAMPLE 1	SAMPLE 2
Mean	65.0	60.0
Variance	85.6	40.7
Observations	36	25
Pooled Variance	67.322	
Hypothesized Mean Difference	0	
df	59	
t Stat	2.341	
P(T<=t) one-tail	0.011	
t Critical one-tail	1.671	
P(T<=t) two-tail	0.023	
t Critical two-tail	2.001	

The type of information is almost identical to the output for the previous t Test, but note this version gives the "Pooled Variance". The "Polled Variance" is actually our pooled standard deviation, S_p, squared. If we square our S_p of 8.2 we obtain the value 67.24 which, except for rounding matches the value in the table (67.3). Also be aware that the difference in the t value calculated above (2.35) and the t value in the table (2.341) is due to rounding. For this version of the t Test, df will always equal $(N_1-1)+(N_2-1)$.

You can see that the statistical results for the t Test provided by the equal variances version are almost identical to the results under the assumption of unequal variances. Among the differences are 1] a small changes in t (a little smaller for the equal variance assumption) and 2] a small change in the p-value (a little larger for the equal variance assumption). There is clearly no reason to alter the previous conclusion of a statistically significant difference between the samples.

This leads us to ponder - when do we use which version of the t Test. Usually t Tests are performed using the equal variance assumption, unless it is known in advance that the two groups have unequal variances. In addition, if a t Test assuming equal variances is run, but the results show large difference between the group variances, the researcher needs to reconsider the equal variance assumption. There are statistical tests that compare variances to determine if it's reasonable to assume equality (See the "F-Test Two Samples for Variances" in the Excel section of the next chapter.

One important point also needs to be made regarding the choice between the equal variance and unequal variance versions of the t Test. If you use the t Test for Unequal Variances, you have already concluded there is some kind of difference between the groups - the standard deviations are different. Whether the means are the same or not may still be important, but it is clear that that the groups are different due to the unequal variances.

Two Sample Z Test

All of the Two Sample t Tests we just reviewed can also be done using a Z test. But to perform the Z Test remember that the population standard deviation must be known. Essentially all that is needed is to exchange Z for t and σ for S in the formulae. The formulae for interval scaled measurements are shown below:

Group Standard Deviations	Z Formula	SE Formula
Unequal	$Z = \dfrac{(\overline{X}_1 - \overline{X}_2)}{SEu}$	$SEu = \sqrt{(SE_1)^2 + (SE_2)^2}$ where $SE_1 = \sigma_1 / \sqrt{N_1}$ and $SE_2 = \sigma_2 / \sqrt{N_2}$
Equal	$Z = \dfrac{(\overline{X}_1 - \overline{X}_2)}{SEe}$	$SE_e = \sigma * \sqrt{\left(\dfrac{1}{N_1} + \dfrac{1}{N_2}\right)}$

You can also test proportions using the t and Z tests, but usually the Chi Square test, described in a later chapter, is used.

Paired t Test

Another t Test we can conduct involves using the same person or subject to receive both treatments or experimental conditions. In this case our experiment is considered a paired or correlated comparison since the two treatments are given to the same subject. Generally, whether a design is paired or unpaired depends on how the experimental conditions are allocated to the sampling units. Sampling units can be students, patients, or even plots of land. To illustrate, we could give drug and then placebo to the same person measuring the patient's response following each administration of drug or placebo. We could also give drug to one group, and to a new group of people, give the placebo. When both treatments are given to the same sampling unit (the same person in our case) we have a **paired design**. When the treatments are given to independent samples (separate groups in our example) we have an **independent group design.** An independent group design is also called an **unpaired** or **uncorrelated design**.

In general, the means that result from a paired design are less subject to random error than means obtained from an independent group design. The groups receiving treatment in the paired design consist of the same people whereas the independent group design has two groups of different people. If the make-up of individuals influences the outcome variables, then there is less variance in the paired design and less random error.

You can also match people so they have the same or very similar characteristics. For example, we can pick two people who are the same age, sex, and approximate weight. To one we can give the drug and to the other person the placebo. The matched patient approach also is considered a paired design. In fact, it fits the notion of "paired" better than the approach which has the same subjects receive both treatments.

In terms of an example to illustrate a Paired t Test, imagine a course designed to improve reading comprehension. The same six people are tested before and after the program and we use their test results to tell us if the program is useful. We should specify the type of test situation (one or two tail) and Alpha Risk in advance. We'll use a two tail test. Although we really expect that the training had a positive effect, we'd like to be able to report if the unexpected happened and the training led to lower scores. We'll also conduct our test with an Alpha Risk of .05.

Assume these are the results we obtain from our pre and post program testing:

Table 9-C

Individual	A	B	C	D	E	F	Mean
Pre Score	65	78	60	88	73	80	74.0
Post Score	72	76	65	91	79	82	77.5

For computational purposes the easiest procedure is to determine the difference in the pre and post scores. The set of differences is shown in Table 9-D.

Table 9-D

Individual	A	B	C	D	E	F
Difference	-7	2	-5	-3	-6	-2

Because we are taking the difference, Pre score - Post score, the values are negative when the score improves and positive when the score declines. For the set of differences we can obtain a mean, standard deviation and standard error. For our example these values are:

Mean ($\overline{X}$)	-3.5
Standard Deviation (S)	3.27
Standard Error (SE)	1.34

To determine t we use the following formula:

$$t = \frac{\overline{X}_d}{SE_d}$$

where the subscript d stands for "difference" since the mean and standard deviation are calculated using the difference between the paired observations (i.e. they are based on the 6 values in the last row of Table 9-D).

Substituting the appropriate values for the mean and standard error gives us our t value.

$$t = \frac{-3.5}{1.34} = -2.62$$

We look up this t value in a t Table for a two tail test with an Alpha Risk of .05 and N-1 df. Our sample size is 6 so we have 5 df. It is tempting to believe the sample size is 12, after all there are 12 values. However, note that we have only 6 individuals in the data set so the N is 6 not 12.

The critical t value, according to Table 2 of the Appendix, is 2.57. The calculated t value of -2.62 falls in the rejection region (remember in a two-tail test, the critical t values are plus/minus values) so we conclude that there is a statistically significant difference and the program does, in fact, improve performance. If the data had been analysed by a computer program you would have received a printout that looked similar to the one below.

t-Test: Paired Two Sample for Means		
	Pre Score	*Post Score*
Mean	74	77.5
Variance	105.2	78.7
Observations	6	6
Pearson Correlation	0.951749	
Hypothesized Mean Difference	0	
df	5	
t Stat	-2.620908	
P(T<=t) one-tail	0.023523	
t Critical one-tail	2.015049	
P(T<=t) two-tail	0.047046	
t Critical two-tail	2.570578	

These results may give you more information than you want, but the data you need are readily available. If we used the Critical Value Method, we see there is a t of 2.57 ("t Critical two-tail") which must be interpreted as ± 2.57 because we are doing a two tail test. The actual t is –2.62 ("t Stat"). The actual t is more extreme that the critical t so there is statistical significance. The Probability Value Method gives us a p value of .047 ["P(T<=t) two-tail"] which is less that the Alpha Risk of .05 so again there is statistical significance.

One more problem will be worked which illustrates the logic one typically goes through when encountering the results for a typical t Test. Here is the problem:

Before deciding which of two brands of disposable syringes to order, a study was done on the percentage of defects each company had. When the study was analysed it produced the following results:

t-Test: Two-Sample Assuming Unequal Variances

	Co. A	Co. B
Mean	1.533648	1.76546
Variance	0.052598	0.303855
Observations	12	24
Hypothesized Mean Difference	0	
df	33	
t Stat	-1.77563	
P(T<=t) one-tail	0.042509	
t Critical one-tail	1.307737	
P(T<=t) two-tail	0.085017	
t Critical two-tail	1.69236	

If you use an Alpha Risk of .10, in which of the following cases would you have a statistically significant result? In working this problem remember that the comparison is between the percentage of defects and the product with the fewest defects is the superior product.

1] One tail test in which the alternative hypothesis was that Co. B had a better product than Co. A

2] One tail test in which the alternative hypothesis was that Co. A had a better product than Co. B.

3] Two tail test

The answer to part 1 is that there is not a statistically significant difference. For a one tail test you must first examine the difference between the means to see if the result is consistent with the alternative hypothesis. In this case the alternative hypothesis is that Co. B has the "better" result. Since we are measuring the per cent of <u>defective</u> syringes the better result occurs for the company with a <u>lower</u> number of defects. Since, in the trial, Co. B had the <u>higher</u> mean (i.e. "worse" result) the result is incompatible with the alternative hypothesis and there can not be a statistically significant difference.

Part 2 also involves a one tail test so we begin by examining the difference between the means to see if the result is consistent with the alternative hypothesis. In this case the mean for Co. A represents a "better" result (i.e. Co. A has fewer defects) and the result is compatible with the alternative hypothesis. Note that the one tail critical t is 1.31 without any sign. However, since the difference between the means is consistent with the alternative hypothesis, we give the critical t value the same sign as the calculated t value. In this example the calculated t has a negative value (-1.72) so we assign a negative value to the critical t value for a one tail test and that value becomes -1.31. Since the calculated t (-1.78) is more extreme that the critical t (-1.31) there is a statistically significant difference at the .10 level.

We could also have used the Probability Value Method to determine statistical significance in part 2. We note that the p-value for the one tail test (.04) is less than the Alpha Risk (.10) which means that the observed result occurs <u>less than</u> 10 percent of the time. We said in advance that we'd take a .10 chance of concluding there is a difference when one does not exit (our Alpha Risk) and since the actual result is even more rare that that standard, we declare statistical significance at the .10 level.

The answer to part 3 is that there is statistical significance. There are again two ways we know that this is the case. The Critical Value Method requires us to compare the critical t value for a two tail test (shown in the table as "t Critical two-tail") with the calculated t values (shown in the table as "t Stat"). Remember that for a two tail test you interpret the critical t value as a ± value since two tails must be considered. The calculated t (-1.78) is more extreme than the negative critical t value (-1.69) and therefore there is a statistically significant difference at the .10 level. We use the expression "more extreme" since there are two critical t values. We could say "greater than the positive value of the critical t or less than the negative value of the critical t", but it seems easier to simply say "more extreme than the critical t values".

The Probability Value Method to determine statistical significance for the two tail test requires us to take note of the p-value for the two tail test (shown in the table as "P (T<=t)two-tail)". The p-value is given as 0.085 and since it is less than the Alpha Risk (.10), there is statistical significance.

Confidence Limits Using t

At this time, it is worthwhile to point out that there are other applications using t values that we need to be familiar with. We learned how to do a confidence interval when the population standard deviation was known. If we do not know the population standard deviation we can use the sample standard deviation and the appropriate t value in the confidence limit formulae. The following formulae will give you the upper and lower limits.

$$LL = (t_l * SE) + \overline{X}$$

and

$$UL = (t_u * SE) + \overline{X}$$

where SE is $S/\sqrt{N}$, t_l is the t value for the lower limit and t_u is the t value for the upper limit.

For example, what if a sample of records were collected on the annual number of fire alarms received for every 100,000 homes in your state. The Director of Operations wants to know the 98% confidence interval for the number of alarms received. There were 61 fire departments in the sample and the mean number of alarms received was 62.7 with a standard deviation of 14.4. What do you tell the Director?

You first need to transform the "confidence interval of 98%" into a t value. The degrees of freedom are 61-1 = 60 and the 98% confidence limit means we want a t that cuts off a total of 2% in the tails so that 98% of the area is isolated in the center. The 2% corresponds to an Alpha Risk of .02. Remember we must split that proportion so half (.01) is in each tail leaving .98 in the center. Go to the t Table (Table 2 of the Appendix) and find the cell for the row, df = 60, and the column for a one-tail test with a value of .01. The t value you will find is 2.39 and the negative value of that t becomes the lower limit t and the positive value of that t becomes the upper limit t.

We are given $\overline{X}$ (62.7) and can easily calculate SE since we are given S (SE = $S/\sqrt{N}$ = $14.4/\sqrt{61}$ = 1.84). We can now find our confidence interval.

$$LL = (-2.39 * 1.84) + 62.7 = 58.3$$

and

$$UL = (2.39 * 1.84) + 62.7 = 67.1$$

We are 98% confident that the mean number of fire alarms in the state falls between 58.3 and 67.1.

Summary

We already knew that we could use the Z test to compare a sample mean against a population mean if we knew the population standard deviation. However, knowing the population standard deviations is unlikely. Fortunately, we can use the t Test to compare the sample mean by using the sample standard deviation mean in place of the population standard deviation. If N is large we saw that there was little difference between the Z and t Test results.

We also learned that we can use the Two Sample t Test to compare one sample mean against another sample mean. If we were doing one tail tests we learned we needed to examine the difference between the means to see if the result was compatible with the alternative hypothesis. Requiring the actual result to be consistent with the alternative hypothesis makes sense since when you declare statistical significance, you reject the null hypothesis and accept the alternative hypotheses.

We also discovered that there are different types of t Tests - one for paired observations and another one for independent samples. The independent sample t Test can be done with an assumption of equal standard deviations or unequal standard deviations. We also learned that we can construct a confidence interval when the population standard deviation is not known by using the sample standard deviation and t values.

A decision tree to help decide when to use the Z or t Test, as well as the type of test to use, is provided on the next page. To use the tree, start with the diamond called "Standard Deviation Known?" and follow the questions and answers until you reach one of the conclusions (e.g. if the standard deviation is known and the "No. of Sample Means" is 2 your conclusion is that the Two Sample Z Test should be used.

Decision Tree for Selecting the Z or t Test When There are 1 or 2 Means to Compare

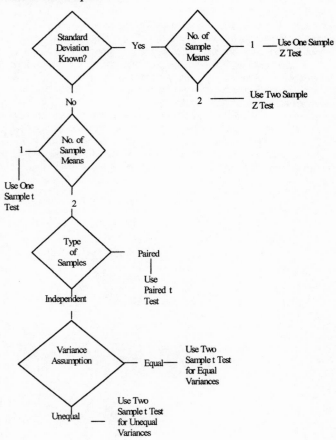

PROBLEMS

1. When doing a One Sample t Test, would you have statistical significance under the following conditions? Assume the alternative hypothesis requires a positive difference.
    a. Alpha Risk=.05, 2 tail test, N=6, calculated t = 2.40
    b. Alpha Risk=.01, 2 tail test, N=11, calculated t = 3.20
    c. Alpha Risk=.01, 1 tail test, N=21, calculated t = 2.30
    d. Alpha Risk=.05, 1 tail test, N=61 calculated t = 2.25

2. When doing a Two Sample t Test with equal variances, would you have statistical significance under the following conditions? Assume the alternative hypothesis requires a positive difference.
    a. Alpha Risk=.05, 2 tail test, $N_1=5, N_2=6$ calculated t = 2.40
    b. Alpha Risk=.01, 2 tail test, $N_1=5, N_2=6$ calculated t = 3.20
    c. Alpha Risk.01, 1 tail test, $N_1=18, N_2=14$ calculated t = 2.30
    d. Alpha Risk=.01, 1 tail test, $N_1=13, N_2=9$ calculated t = 2.50

3. You read an article and find that the average value of cars in your county is $8,750 based on a sample of 21 residents. You wonder if your county residents' investment in cars is better or worse than the state average. You decide to test this question out using the t Test.
    a. Is this a one or a two tail testing situation?
    b. What are the null and alternative hypotheses?
    c. If you do not want to come to the wrong conclusion more than 5 times in a 100, what is your Alpha Risk?
    You make a few calls and find that a state registry of all cars shows that the average car investment is $9,350. You perform the t Test and find t = 2.21.
    d. Is there a statistically significant difference between your county compared to the state as a whole
    e. Why did you come to that conclusion?
    f. How would you state your conclusion?

4. You believe that Europeans drink more than Americans and want to see if your belief is correct. You are uninterested in the possibility that the Americans may drink more than Europeans. You find that a survey was taken of 19 US residents and 13 European citizens on the number of meals in which alcoholic beverages are consumed in a 30 day period. The mean values are 11.2 meals for Americans and 15.3 meals for Europeans. The survey sponsors report that a t Test resulted in a t of +1.78.
    a. In the t formula was $\overline{X}_1$ the Americans or Europeans?
    b. If $\overline{X}_2$ is the American mean and $\overline{X}_1$ the European mean, which of the following Alpha Risks would result in a statistically significant difference?   .10 .05 .01 .001

5. It is generally believed that non-fraternity students do better academically than students who belong to fraternities. The Fraternity Council disputes this belief claiming that fraternity students receive better grades. You are hired to resolve the issue.

   a. Is this a one or two tail testing situation? Why?

   b. If you follow convention, what Alpha Risk should you use?

   At random, 12 fraternity and 10 non-fraternity students are selected and the mean GPAs calculated. The fraternity GPA is 2.61 and for the non-fraternity group the mean is 2.74. Your statistician performs the t Test and finds t = 1.85.

   c. If you set the alpha level at .05, is there a statistically significant difference?

   d. If you used much larger sample sizes, would the chance of a statistically significant difference increase, decrease or stay the same?

   e. Explain the rationale for your answer to part d.

   f. If you were only interested in knowing if non-fraternity students do better than fraternity students, would you change the conclusion you reached in part c?

6. A survey was taken at 16 state police barracks and the number of complaints received was recorded (see below). The Commissioner of Public Safety in the state had set a goal that the average number of complaints should not exceed 15.

No. of Complaints			
14	17	11	15
23	13	22	22
18	27	16	30
13	20	8	19

   a. What is the value of the sample mean and standard deviation?

   b. If a one tail test, using an Alpha Risk of .05, is used - is there a statistically significant difference between the sample mean and the Commissioner's goal?

7. A fitness test, in which a higher score indicated better fitness, was given to a random sample of officers in the police department and the fire department. The results of an analysis of the data are shown below:

   t-Test: Two-Sample Assuming Equal Variances

	Police	Fire
Mean	18.2222222	13.88889
Variance	27.1944444	27.61111
Observations	9	9
Pooled Variance	27.4027778	
Hypothesized Mean Difference	0	
df	16	
t Stat	1.75602648	
P(T<=t) one-tail	0.04910229	
t Critical one-tail	1.33675712	
P(T<=t) two-tail	0.09820459	
t Critical two-tail	1.74588422	

   a. How many subjects participated in the study?

   b. If you use an Alpha Risk of .10, in which of the following cases would you have a statistically significant result?

     1] Two tail test

     2] One tail test in which the alternative hypothesis was that the police department did better than the fire department.

     3] One tail test in which the alternative hypothesis was that the fire department did better than the police department?.

   c. Why did the researcher use this version of the t Test?

8. A psychology professor wanted to know if location made any difference in terms of exam grades. The professor randomly assigned students to the front or back row of his classes and mid way through the course administered an exam. The results of the analysis are given below:

t-Test: Two-Sample Assuming Unequal Variances

	Front	Back
Mean	82.9047619	78.89473684
Variance	21.69047619	70.54385965
Observations	21	19
Hypothesized Mean	0	
df	27	
t Stat	1.84075624	
P(T<=t) one-tail	0.038336213	
t Critical one-tail	1.703288035	
P(T<=t) two-tail	0.076672427	
t Critical two-tail	2.051829142	

   a How many subjects participated in the study?

   b. If you use an Alpha Risk of .05, in which of the following cases would you have a statistically significant result?

     1] Two tail test

     2] One tail test in which the alternative hypothesis was that the front row students did better than the back row students.

     3] One tail test in which the alternative hypothesis was that the back row students did better than the front row students?.

   c. Why did the professor use this version of the t Test?

9. The county decided to test to see if tires from two different manufacturers were different in terms of tire wear. The tires from each manufacturer were placed on the same car and after 5,000 miles the tire wear was recorded. The results of the analysis are shown below:

t-Test: Paired Two Sample for Means

	Mfrg A	Mfrg B
Mean	3.045633	2.602999
Variance	1.784686	1.843735
Observations	20	20
Pearson Correlation	-0.13776	
Hypothesized Mean Difference	0	
df	19	
t Stat	0.974269	
P(T<=t) one-tail	0.171082	
t Critical one-tail	1.327728	
P(T<=t) two-tail	0.342165	
t Critical two-tail	1.729131	

   a. How many cars were used in the study?
   b. If you use an Alpha Risk of .10, in which of the following cases would you have a statistically significant result?
      1] Two tail test
      2] One tail test in which the alternative hypothesis was that the tires of manufacturer A were better than those of manufacturer B.
      3] One tail test in which the alternative hypothesis was that the tires of manufacturer A were worse than those of manufacturer B.
   c. Why was this version of the t Test used?

10. For a sample of a normally distributed variable with mean equal to 60 and a sample size of 25:
    a. Calculate the 95% confidence interval assuming 60 is a population mean, the standard deviation is known and SE = 4.
    b. Calculate an 80% confidence interval assuming 60 is a sample mean, the standard deviation is unknown and SE = 5.

11. What is the difference between a type 1 and a type 2 error?

12. What is the difference between hypothesis testing and estimation?

Excel Instruction

There is no built-in One Sample t test in Excel. However, Excel provides a few functions that make the calculations reasonably easy to do. The functions we can use are TDIST and TINV.

Let's use Excel to work the problem where we wanted to know if graduate textbooks cost more than undergraduate texts. To use the Excel functions, we need the following information: the calculated value of t, the sample size and whether we are doing one or two tail testing. In the example, t is 1.333, N = 36 and df = 35. Since we assumed our interest is in knowing whether graduate books cost more or less than undergraduate, we have a two tail testing situation. Also remember that we set our alpha risk at .05.

On an Excel worksheet find an empty cell, click on the function icon, select "Statistical" and then click on "TDIST". Complete this step by hitting OK. Your screen looks like this:

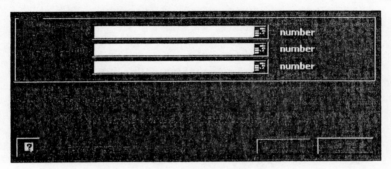

We must supply three items of information. Their names and explanations are given below:

Name	Explanation:
x	the calculated t value
Deg freedom	N-1
Tails	Enter 1 or 2 depending on the testing situation

In our case the values are:

Name	Data to Enter
x	1.333
Deg freedom	35
Tails	2

Complete this step by selecting OK and Excel returns the value .19 (after rounding). There is a .19 probability that we could have obtained a mean of 34 when sampling from a population which has a mean of 32. Since .19 is larger than .05 (the alpha risk) we can not reject the null hypothesis. The evidence isn't strong enough that graduate book costs are different than undergraduate book costs.

It is important to note that Excel requires you to always enter t as a positive value. If you are doing a one tail test and your sample mean turns out to be in the opposite direction of what you expect, then you know there will not be a significant difference and there is no need to do the test. For instance if the graduate book mean was $30, not $34, we would have a negative numerator ($\bar{X}$ -μ) and therefore a negative t. There is no way we can reject the null hypothesis when our sample means is less than the established mean.

We can also find the critical value for t without using the t tables. To do this we use the function TINV. We need to know our alpha risk and our df to use TINV. In the book cost problem, the alpha risk is .05 and N is 36. We select the function icon and then "Statistical". From the list provided we click on "TINV". After hitting OK, our screen looks like this:

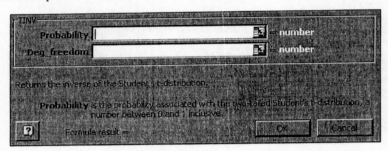

We must supply two items of information. Their names and explanations are given below:

Name	Explanation:
Probability	the Alpha Risk
Deg freedom	N-1

In our case the values are:

Name	Data to Enter
Probability	.05
Deg freedom	35

Excel returns the value 2.03, which is the same as the value listed in the t table. You may ask why didn't TINV request the number of tails as well. Excel always returns a two tail critical t with TINV. However, you can obtain a one tail critical t by doubling the alpha risk and inserting that value as your probability value. You may recall that in a one tail test all the alpha risk is in one tail and we split our alpha risk into 2 halves to do a two tail test. Hence, the one tail test cuts off an area double that of the two tail test. In our example if we want the critical t for a one tail test, we use .10 as the probability value in our input. When you do this, Excel returns the value 1.69, which is the critical value for a one tail test. Again if you check the t table in the Appendix, you will find that the listed value is also 1.69.

To do a Two-Sample t Test we need our raw data on a worksheet. Let's assume we want do determine if there is a difference in male and female pay. Start a new worksheet. Copy the SEX data to cells A4 through A42 with employee 1's sex in cell A4. Copy the Pay data from cell B4 through B42 with employee 1's pay in cell B4. Use Data / Sort to arrange the information so all the Female records precede the Male records. Your last female record should be in cell A25 after the sort.

You select "Tools" from the Menu Bar, "Data Analysis" from the dropdown menu and from the listing select "t-Test: Two Sample Assuming Equal Variances". After hitting "OK" you should be at a screen that looks like this:

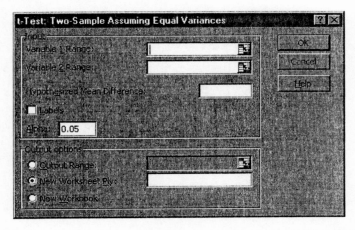

Your "Variable 1 Range" is the location of the Female Pay values, which is B4:B25. Your "Variable 2 Range" is the range of the Male Pay values, which is B26:B42. The "Hypothesized Mean Difference" is 0. Leave the "Labels" entry blank. The "Alpha" value should be set at 0.05, but if it isn't, type in .05 for the "Alpha" value because we want to use a .05 Alpha Risk. For the "Output options" select a "New Worksheet Ply" and name it "Pay Difference". When you've done all the above hit OK and your output on the worksheet, Pay Difference, should look like this:

t-Test: Two-Sample Assuming Equal Variances

	Variable 1	Variable 2
Mean	36.45455	35.76471
Variance	112.9769	147.5674
Observations	22	17
Pooled Variance	127.9350	
Hypothesized Mean Difference	0	
df	37	
t Stat	0.188867	
P(T<=t) one-tail	0.425614	
t Critical one-tail	1.687094	
P(T<=t) two-tail	0.851229	
t Critical two-tail	2.026190	

The interpretation of the output was given in the text and will not be repeated here.

For a two sample t test, under the assumption of unequal variances, all the steps are the same except you chose the "t Test: Two Sample Assuming Unequal Variances" option from the "Analysis Tools" listing provided on the "Tools" / "Data Analysis" menu. In our example of the difference in male and female pay, the results from a two sample t test assuming unequal variances should look like this:

t-Test: Two-Sample Assuming Unequal Variances

	Variable 1	Variable 2
Mean	36.45455	35.76471
Variance	112.9769	147.5674
Observations	22	17
Hypothesized Mean Difference	0	
df	32	
t Stat	0.185593	
P(T<=t) one-tail	0.426968	
t Critical one-tail	1.693888	
P(T<=t) two-tail	0.853936	
t Critical two-tail	2.036932	

Again no explanation of the output is needed since that information was provided in the text. However, note that in this case, the df calculated by the computer program is 32 and does not correspond to the df formula, $(N_1-1)+(N_2-1)$, which is used for the t Test version that assumes equal variances.

To illustrate the Paired t test we will use the example given in the text comparing Pre Scores and Post Scores. Copy the table for that data set to a worksheet where the headings

Individual, Pre Score and Post Score appear in cells A1, B1 and C1 respectively. Enter the actual data for the individuals (e.g. A in cell A2, 65 in cell B2 and 72 in cell C2.). Select "Tools" from the Menu Bar, "Data Analysis" from the dropdown menu and from the "Analysis Tools" listing select "t Test: Paired Two Sample for Means". After hitting "OK" you should be at a screen that looks like this:

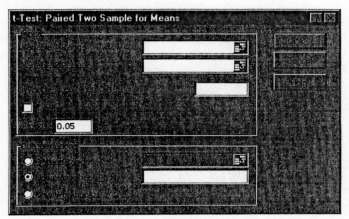

Your "Variable 1 Range" is the location of the Pre Score values and in this case we will include the heading so type B1:B7 in the box. Your "Variable 2 Range" is the range of the Post Scores and since we want headings, type C1:C7 in the second box. The "Hypothesized Mean Difference" is 0. If the "Labels" box is empty click on it so a check mark appears - we included labels in our input statement and this is the way Excel is informed of that fact. Type in .05 for the "Alpha" value if it isn't there because we want to use a .05 Alpha Risk. For the "Output options" select a "New Worksheet Ply" and name it "Score Difference". When you've done all the above, hit OK and your output on the worksheet, Pay Difference, should look like the one in the text. You may have to adjust the width of some columns to have all the information displayed. The explanation of the output information for this example was also included in the text.

A confidence limit for a mean, when the standard deviation is unknown, must be done by hand. The appropriate formulas are given in the text. You can obtain t_l and t_u using TINV, but you will need to write an Excel equation to obtain your actual upper and lower limits.

The Two-Sample Z Test was also discussed and there is an Excel procedure to perform this Z test. Go to the Menu Bar and select "Tools". From the drop down menu select "Data Analysis". From the "Analysis Tools" listing pick the last one – "Z Test. Two Sample for Means". You will now have the following screen before you.

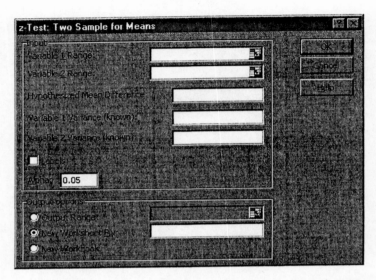

The "Variable 1 Range" refers to the cell location for one of the comparison groups and the "Variable 2 Range" refers to the location of the other comparison group. You activate the "Labels" box if you included labels as part of your range statements above. Enter the variance (i.e. the square of the standard deviation) for each group in the boxes indicated. Alpha should be set at .05 and indicate where you want your result to appear in the "Output options" area. The format for the output is shown below. It is similar to a t Test output and should be self-explanatory.

z-Test: Two Sample for Means

	Variable 1	Variable 2
Mean		
Known Variance		
Observations		
Hypothesized Mean Difference		
z		
P(Z<=z) one-tail		
z Critical one-tail		
P(Z<=z) two-tail		
z Critical two-tail		

Here are some exercises you can do to practice the Excel skills described in this chapter. For each problem assume the data are samples from a larger population.

1. Determine if the mean for special awards is different at a statistically significant level from a mean of 4.0. Conduct you test with an alpha risk of .01.
2. Compare the ages of the male and female employees. Use a .05 alpha risk and test only to see if females are older.
   a. What do you conclude?
   b. What is the 95% confidence interval for the female ages?
   c. Since the variances are quite different, redo the test using the "t Test: Two Sample Assuming Unequal Variances" option - Do you come to a different conclusion?
3. From your database compare the June PC proficiency scores to those of July. Do a two tail test using an alpha risk of .02.
   a. Is there a statistically significant difference?
   b. What is the standard deviation for the July scores?
   c. If you used a .01 alpha risk, would you reject the null hypothesis?
   d. What is the one tail critical t value for part a?

Chapter 10

Difference Between Two or More Means: Analysis of Variance

Introduction

In the last chapter we learned to use the t Test to find if there was a difference between two sample means when the standard deviation was unknown. We will learn that when there are more than two means to compare, we need to use a procedure called the "Analysis of Variance". The result of an Analysis of Variance only tells us that there are differences somewhere among the set of groups whose means were tested. We use a test, such as the Tukey-Kramer test, to identify which of the group means are different from each other. The chapter will also explain the difference between practical and statistical significance.

Experimentwide Error

In a typical experiment we compare two groups to see if their means are different. In this situation there is only one test performed to see if a difference exists. We typically use an a Alpha Risk of .05 when we make this test. Under this arrangement there is a .05 chance of claiming statistical significance when none exist. Statisticians refer to the .05 probability, when it applies to the entire experiment, as the **experimentwide error**. An experimentwide error rate of .05 means there is a .05 chance that the experiment will result in a conclusion that there is a statistically significant difference when one does not exist. Note that when only one comparison is made, the experimentwide error is the same as the Alpha Risk.

Nonetheless, it is sometimes more efficient to test more than two groups in one experiment. We may have two new promising drugs and would like to test both of them against a placebo. It's more efficient to run one study with three groups than conducting two studies with two groups each. In the two study arrangement, the placebo group would be compared against one drug group in the first study and a second placebo group would be compared against the other drug group in study two.

When we analyse the experiment with all three groups included, we want to know if the result for the first drug is different than the placebo group result, but we also want to know if the second drug is different than placebo. In the three-group scenario, we would need to perform two statistical tests. The temptation is to do two t tests in that circumstance, but there is a problem with that approach. If both tests are done with an Alpha Risk of .05, the experimentwide error rate is greater than .05. Why this is so can be seen when we answer the question: What is the probability of at least 1 statistically significant difference when an Alpha Risk of .05 is used for two comparisons. The answer can be found by the binomial probability formula (see chapter 4). At least one statistically significant result means one or both of our comparisons to placebo end up as statistically significant. Our binomial formula, when there is one statistically significant finding, has an n = 2 (the number of trials), r = 1 (the number of statistically significant results) and p = .05 (the probability of getting a statistically significant result).

The binomial formula and calculation is:

$$\frac{n!}{r!*(n-r)!} * p^r * (1-p)^{(n-r)}$$

$$\frac{2!}{1!*1!} * .05^1 * .95^1 = .095$$

The probability of obtaining 1 statistically significant result is .095. However there is also the chance that both comparisons will be statistically significant and that outcome has n=2, r=2 and p=. 05. The binomial calculation in this case is:

$$\frac{2!}{2!*0!} * .05^2 * .95^0 = .0025$$

The probability of obtaining 2 statistically significant results is .0025. The two probabilities must be added together to give us the overall chance of at least one statistically significant finding. When we do the addition (.0950+.0025) we find there is a .0975 of getting at least 1 statistically significant result. The .0975 is the experimentwide error rate for our three treatment experiment in which two statistical tests are conducted. Statisticians consider this error rate to be too high and argue that you should control the experimentwide error rate so that it is at the conventional .05 probability level. For this reason, instead of performing a series of t tests, when more than two groups are being tested, statisticians use a different test - it's called the Analysis of Variance.

Analysis of Variance

The acronym ANOVA is frequently used for the Analysis of Variance. As noted above ANOVA is used when testing hypotheses involving three or more groups. It is also an alternative to the t-test when testing just two groups. That means you will get the identical p-values from a two sample independent group t test as you will with the ANOVA test. The simplest form of an ANOVA is the **One Way ANOVA**. Sometimes the term **Single Factor ANOVA** is used as an alternative name. Typically in a One Way ANOVA, we examine only a single treatment factor and try to determine if there is a difference among the various treatments given.

Mathematical Derivation of a One Way ANOVA

We will spend a little time showing the mathematical derivation of the One Way Analysis of Variance. However, the use and interpretation of the Analysis of Variance doesn't require this knowledge. The Analysis of Variance, just like the t Test or the Z Test, ends up with a calculated test statistic that is compared to a critical value. In this case the test statistic is called the F Ratio and if the calculated F Ratio is greater than the critical F Ratio, then there is statistical significance. However, lets begin with the mathematical derivation.

It will be useful to use a small data set which has three groups and four observations within each group. We will label the three groups in our data set as the Red, White and Blue group.

Red	White	Blue
23	28	29
25	29	30
28	30	34
28	25	33

The mean for each group and the overall mean are displayed below:

Red	White	Blue	Overall
26.0	28.0	31.5	28.5

You may recall that the formula for a sample variance is:

$$S^2 = \frac{\Sigma(X - \overline{X})^2}{N - 1}$$

The numerator, $\Sigma(X - \overline{X})^2$, may be viewed as a "sum of squares". Note that the mean is subtracted from each individual value in the data set, the differences are squared and then summed. If we ignore the group designation and treat the data set as 12 independent observations we can calculate a sum of squares which we will call the "**Total Sum of Squares**". In our example the Total Sum of Squares is 111.

The Total Sum of Squares can be partitioned into two parts, usually called 1] the "Between Groups Sum of Squares" and 2] the "Within Groups Sum of Squares". The Between Groups Sum of Squares is a weighted sum of squares for the three groups. To find the Between Groups Sum of Squares we first determine the difference between each group mean and the overall mean. We then square the differences (see below).

Statistic	Red	White	Blue
Group Mean Difference from Overall	-2.50	-0.50	3.0
Squared Group Mean Difference	6.25	0.25	9.0

The Between Groups Sum of Squares is called a weighted sum because, as you will see, we utilize the sample size for each group in the calculation. To arrive at the weighted sum, each squared group mean difference is multiplied by the group sample size (see the Weighted Value row in the table below). The last step involves adding the three weighted values together. We will expand our table to show these additional calculations. We find that the Between Groups Sum of Squares is 62.

Statistic	Red	White	Blue
Group Mean Difference from Overall	-2.5	-0.50	3.0
Squared Group Mean Difference	6.25	0.25	9.0
Sample Size	4	4	4
Weighted Value (calculation)	25=(6.25*4)	1=(0.25*4)	36=(9.0*4)
Between Groups Sum of Squares		62=(25+1+36)	

The Within Groups Sum of Squares is obtained by finding the sum of squares for each group relative to the group mean. In the case of the Red group, the mean is 26 and the differences between the first observation and that mean is −3.0. The difference values for all three groups is shown below:

Red (Mean=26)	White (Mean=28)	Blue (Mean=31.5)
-3.0	0.0	-2.5
-1.0	1.0	-1.5
2.0	2.0	2.5
2.0	-3.0	1.5

We next square those differences and sum them for each group. The table below shows the squared values and the totals for each group.

	Red	White	Blue
	9.0	0.0	6.25
Squared	1.0	1.0	2.25
Values	4.0	4.0	6.25
	4.0	9.0	2.25
Sum:	18.0	14.0	17.0

The last step to determine the Within Sum of Squares is to add the group-values together. When we do this we obtain a value of 49 for the Within Groups Sum of Squares $(18+14+17) = 49$

The next phase involves calculating the degrees of freedom for each of the Sum of Squares. The pertinent formula and results are shown below:

Source	Formula	Degrees of Freedom (DF)
Between	No. Groups -1	$3-1=2$
Within	$(N_1-1)+(N_2-1)+(N_3-1)$	$(4-1)+(4-1)+(4-1)=9$
Total	$N-1$	$12-1=11$

Unfortunately, we're still not done. We need to divide the respective Sum of Squares by the Degrees of Freedom. The Statistic that result from this calculation is called the "Mean Square".

Source	Sum of Squares	DF	Mean Square (MS)
Between Groups	62	2	31.0
Within Groups	49	9	5.44
Total Group	111	11	10.1

Our final step is to calculate the ratio of the Between Groups Mean Square to the Within Groups Mean Square. That ratio is given a name – the F ratio. In our example the calculation is:

$$F \text{ Ratio} = \frac{31.0}{5.44} = 5.70$$

In the next section we'll see what we do with this statistic.

Interpreting an ANOVA

Technically we do not need the information for the Total Group (the last row in the above table) for our interpretation. However, note that the Sum of Squares and DF values for the Total Group is in fact the sum of the values for Between Groups and Within Groups and it therefore serves as a quality control check on the calculations.

It should be obvious that if the means for the three groups are similar, the Between Groups Sum of Squares and its Mean Square (MS) will be small. The **Within Groups Sum of Squares** represents the variation among individual observations within each of the three groups. The Within Groups Mean Square is often thought to represent random variation that is inherent in an ANOVA. Thus, the Within Groups Mean Square is a kind of normal state or baseline variance that can be used to test for unusual treatment group means. Under the null hypothesis of no treatment differences, the Between Groups and the Within Groups Mean Squares will be the same. However, if the groups are really different then the Between Groups Mean Square becomes large compared to the Within Groups Mean Square. If we divide the Between Groups Mean Square by the Within Groups Mean Square we produce a ratio. The ratio of the Between Groups and Within Groups Mean Squares, called the F ratio, expressed algebraically is:

$$F \text{ Ratio} = \frac{MS_{Between}}{MS_{Within}}$$

The F ratio is used in ANOVA to determine if there is statistical significance between the groups. In fact sometimes people refer to the Analysis of Variance procedure as the "F test". In our example the F Ratio is:

$$F \text{ Ratio} = \frac{31}{5.44} = 5.70$$

Sometimes the term F value is used instead of F ratio. Statisticians have compiled a distribution of the probability of occurrence for F values for different sized studies. In ANOVA testing, size is a function of the number of treatments used and the number of observations made (e.g., the number of individuals tested). It is usual to refer to such a distribution as an F distribution. Thus, one can find from the F distribution, the likelihood of obtaining the calculated F ratio for a study.

The value of 5.70 is our calculated F and we would look up the critical F value in an F Table (See Table 3 in the Appendix). The degrees of freedom for Between Groups is the numerator degrees of freedom and the degrees of freedom for the Within Group is the denominator degrees of freedom. In our example the numerator degrees of freedom is 2 and the denominator degrees of freedom is 9. Table 3 in the Appendix only lists critical values for an Alpha Risk of .05, but many statistical books will provide other Alpha Risk alternatives. In our case the critical F is 4.26 and the calculated F (5.70) is larger so there is statistical significance.

ANOVA by Computer Program

Prior to computers, the mathematical procedure for ANOVA was time consuming and the results presented in a table that showed various statistical values such as Sum of Squares, Mean Square, etc. That format is still used by most computer programs that do an ANOVA today and we can use that output to appreciate the information generated when an ANOVA is performed using a typical computer program.

We'll work with an example to see how the analysis is done and the results presented. We will assume that we have given three separate groups (Groups A, B and C) different appetite suppression treatments. The groups are then compared on the basis of weight loss. We will use an Alpha Risk of .05 and the data that we want analyzed is given below.

WEIGHT LOSS DATA		
A	B	C
19.9	16.0	9.5
23.5	15.0	16.5
28.5	9.0	17.6
18.0	8.1	18.0
18.8	14.0	6.4
19.5	21.0	10.9
16.8	16.5	9.5

Typical computer output for this type of problem is shown below.

SUMMARY

Groups	Count	Sum	Average	Variance
A	7	145	20.71429	16.14476
B	7	99.6	14.22857	19.94905
C	7	88.4	12.62857	21.65238

Statistical Methods for Researchers

ANOVA

Source of Variation	SS	df	MS	F	P-value	F crit
Between Groups	256.6743	2	128.3371	6.667304	0.006811	3.555
Within Groups	346.4771	18	19.24873			
Total	603.1514	20				

The first table titled "SUMMARY" is easy to follow since it contains summary statistics for each group. We are familiar with each of the statistics displayed so no further explanation is needed.

The second table is labeled "ANOVA" and is the is the typical way results are given. The first column of the ANOVA table identifies the three sources of variation – Between Groups, Within Groups, and Total. The next three columns present data that are involved in calculating the F ratio. The SS entry is the Sum of Squares value, df is the degrees of freedom and MS the Mean Square. It's a good policy to determine the df before you run the computer program. This step serves as a useful check on the accuracy of the computer data input. For the source "Total", the df should be the number of observations in the experiment minus 1. We had 21 observations so the df for the Total row of the table is 20. The df for the source "Between" is the total number of groups minus 1. In our case we have three groups so the 2 for the Between groups df is correct The df for the Within Groups source is just the difference between the total df and the between groups df. In our example the Within Group df is 18 (20-2). When using a computer program, if the total and between groups df are not the same as those you calculated in advance, re-check your data input statement.

The next column, MS (which is the abbreviation for Mean Square) gives the result of dividing the SS entry by the df entry. The MS numbers are the Between Groups and Within Groups variances and, as noted above, they are used to calculate the F ratio.

The calculated F ratio is provided in the column headed F. If the populations for the groups being compared are really no different (i.e., the null hypothesis is true) the Between Groups MS will tend to be the same as the Within Groups MS. Statistical theory shows that the F-ratio will be about 1.0 when there is no difference between the group means being compared. In fact, to obtain a ratio of exactly 1.0 requires the Between Groups MS to be the same as the Within Groups MS. On the other hand if the groups means are very different the F-ratio will be quite a bit larger than 1.0.

The data in the ANOVA table allows you to use either the Critical Value Method or the Probability Value Method to determine whether there are statistically significant differences among the groups being compared. To apply the Critical Value Method we need to know the critical F value and the actual F value. The last entry in the table is called "F crit" and as you may guess it is the critical value of F. Since our observed F is 6.67 and the critical F is 3.55 it is clear that we have statistical significance at the .05 alpha level.

In the computer printout table, the p-value needed for the Probability Value Method is in the column headed P-Value. If the p-value is low enough (≤ .05 if one is using an Alpha

Risk of .05) the experimenter concludes that there is a difference somewhere among the groups being compared. In other words, the experimenter is saying that there is sufficient evidence that the group means are not all from the same population. If the probability for the computed F-ratio is larger than .05, the null hypothesis of equal means would not be rejected. In this circumstance there is no statistical significance and one would conclude that the evidence isn't strong enough to claim a difference between the groups. In our example the p-value is .006811. That value is less than .05 and we would declare that there is a statistical difference between the groups.

Differences Between Groups

More analysis is required to determine how the groups differ from each other. There are three comparisons to consider (i.e. is Group A different than Group B, is Group A different than Group C, and is Group B different than Group C). A class of tests known as **multiple range tests** is frequently used to do the additional group analysis. One of the tests available is called the Tukey-Kramer procedure. This test is considered a *post hoc* test since it is introduced <u>after</u> the ANOVA is conducted. Also be aware that because ANOVA compares variances rather than differences between group means, it is possible to find a statistically significant ANOVA result and no individual group differences.

The Tukey-Kramer procedure allows us to test all possible mean comparisons between the three groups to see which ones are associated with statistical significance and which are not. First the differences between the means for the 3 groups are determined. In our case we get the following group differences:

Group A - Group B = 20.7-14.2 = 6.5
Group A - Group C = 20.7-12.6 = 8.1
Group B - Group C = 14.2-12.6 = 1.6

We next need to determine the critical difference for the three sets of differences. This is done by calculating what is called a "Critical Range". The formula for the critical range is:

$$\text{Criitical Range} = Q * \sqrt{\frac{MS_{within}}{2} * \left(\frac{1}{N_1} + \frac{1}{N_2} \right)}$$

Q has to be obtained from a special table for Studentized Ranges (See table 4 in the Appendix). A portion of such a table is shown on the next page. Before we can determine Q we need to determine the following 2 numbers:

1] Number of groups being compared. When we use table 4, this number will be considered the "Numerator df"

2] The total sample size minus the number of groups being compared. When we use table 4, this number will be considered the "Denominator df"

In our case the numbers we want are:

1] Number of groups being compared = 3. Hence, 3 is our Numerator df.

2] Total sample size - number of groups being compared = 21-3=18. Hence, 18 is our Denominator df.

Denominator	Numerator degrees of freedom		
degrees of freedom	2	3	4
1	18.00	27.00	32.80
2	6.09	8.30	9.80
3	4.50	5.91	6.82
.			
.			
.			
17	2.98	3.63	4.00
18	2.97	**3.61**	3.98
19	2.96	3.59	3.96
.			
.			
.			
30	2.89	3.49	3.84
40	2.86	3.44	3.79
60	2.83	3.40	3.74

Note that in this table, the columns are for the Numerator df values and the rows are for the Denominator df values. You identify the cell where Numerator df = 3 and Denominator df = 18. That cell has the value 3.61 (which is in bold so you can locate it) and 3.61 is the Q value we need for the critical range formula.

Another value we need for the Critical Range calculation is MS_{Within}. The MS_{Within} value is found in the MS column of the ANOVA table for the source Within Groups. That value is 19.24873 which can be rounded to 19.25.

We also need to know N_1 and N_2. The N refers to the sample size of each group being compared. Our first comparison is for Group A versus Group B. Both groups have an N of 7 so for our case N_1 and N_2 are both 7. We can now substitute the above numbers into our Critical Range equation:

$$\text{Criitical Range} = 3.61 * \sqrt{\frac{19.25}{2} * \left(\frac{1}{7} + \frac{1}{7} \right)}$$

The result of the calculation gives us a critical Range value of 6.00. If the Group A versus Group B difference is this large or larger, we have a statistically significant difference. A group differences less extreme than 6.00 is not statistically significant. Since all groups have an N of 7, the critical range calculation shown above also applies to the other comparisons.

We previously calculated the group differences which were found to be 6.5, 8.1 and 1.6. In our case, two of the three group differences are more extreme than 6.00. Only the comparison between Group B and Group C results in a difference of less than 6.00. Thus we would conclude that there is a statistically significant difference between Group A and Group B, between Group A and Group C, but there is no statistically significant difference between Group B and Group C.

Note that there is a critical range <u>each</u> time we compare one group to another. In our example all groups had the same N so the critical range we calculated applied to each of the three group comparisons (A vs. B, A vs. C, and B vs. C). However, had the N for group 1 been 8 rather than 7, then the critical range formula would have required an N_1 of 8 not 7. Furthermore, Q would be based on a denominator df of 19 (22-3) rather than 18 (21-3) and in this case Q becomes 3.59. The resulting critical range calculation would be:

$$Criitical Range = 3.59 * \sqrt{\frac{19.25}{2} * \left(\frac{1}{8} + \frac{1}{7}\right)}$$

The calculations result in a value of 5.76 and the two comparisons using group A would use this critical range standard.

Before we leave the topic of the Analysis of Variance, we need to mention that like the Two Sample t Test, there is an assumption that the population variance for each group is the same. Again that assumption is necessary to combine all the groups to get a pooled Within Group Mean Square. If the group sample sizes are about equal, as is true in our case, the results will not be seriously affected by differences in the group variances. Furthermore, in our situation, if we inspect the variances in the Summary table of the ANOVA output, we can see they are relatively the same. Hence, it would be safe to conclude that the equal variance assumption is satisfied. If you run into a case with unequal group sample sizes and fairly large difference in the group variances, you may need to consult a statistician to see what should be done.

Two Way ANOVA

Note that in our One Way ANOVA, we were only interested in testing to see if there were differences among treatment groups. There are cases where we are interested in testing more than one factor in an experiment. For example, we may be interested in knowing if the treatments we are comparing are equally effective in a hospital setting and a home setting.

To test both the treatment factor and the setting factor in the same experiment, we use a procedure called a **Two Way ANOVA** which is also called a **Two Factor ANOVA**. In a Two Way ANOVA we can learn about the effects of a second factor as well as the treatment factor. In our example we can learn about the effect of treatments and the effect of settings. This design is also valuable because you can see if there is a unique relationship between the two factors. For example, we can determine if the relative difference between placebo and antibiotic treatment is the same in a hospital setting as it is in a home setting. Sometimes the Two Way ANOVA is included in a class of studies called factorial or block designs. The treatments (factor 1) are viewed as being encased in blocks represented by the environmental settings (factor 2).

Imagine a study in which we want to compare a new preparation against an old preparation, but we are also interested in knowing whether males and females have similar responses to the two agents. In this case we have a treatment factor (a new preparation or an old preparation) and a gender factor (male or female). Essentially, we have four groups – males on the new preparation, females on the new preparation, males on the old preparation and females on the old preparation.

Let us assume we have collected data on males and females that receive either a new preparation or an old preparation. Both preparations are designed to relieve headaches. The results following treatment are displayed in table 10-A below. The measurement is the number of headaches in a 6-month period.

Table 10-A

Group	Male	Female
New	18	24
Preparation	20	22
	23	20
	15	16
Old	21	18
Preparation	27	20
	29	26
	25	24

Note that the sample size for each of our groups is 4. There are four males and four females who receive the new preparation and there are also four different males and four different females who receive the old preparation. A Two Factor ANOVA will tell us if there is a significant difference between the old and new preparation and whether there is a significant difference between males and females. However, this procedure has one additional piece of information. It can tell us if the relative difference between the new preparation and the old preparation in males is similar to the difference between the old and new preparation in females.

On the next page the computer print-out, after applying a Two Factor ANOVA to the above data set, is displayed. Note that the title identifies the test as "Two-Factor With Replication". The "With Replication" clause means that there are at least two observations for each factor. If there is only one observation, a different ANOVA test (Two-Factor Without Replication) is needed, but that procedure is not presented in this text.

Anova: Two-Factor With Replication

SUMMARY	Male	Female	Total
New Preparation			
Count	4	4	8
Sum	76	82	158
Average	19.0	20.5	19.75
Variance	11.33	11.67	10.5
Old Preparation			
Count	4	4	8
Sum	102	88	190
Average	25.5	22.0	23.75
Variance	11.67	13.33	14.21
Total			
Count	8	8	
Sum	178	170	
Average	22.25	21.25	
Variance	21.93	11.36	

ANOVA

Source of Variation	SS	Df	MS	F	P-value	F crit
Sample	64	1	64	5.33	0.04	4.747
Columns	4	1	4	0.33	0.57	4.747
Interaction	25	1	25	2.08	0.17	4.747
Within	144	12	12			
Total	237	15				

The first set of tables are summary tables and give descriptive statistics for the groups. The first summary table lists the results for those receiving the new preparation and shows that the four males had a mean response if 19.0 whereas the four females had an average response of 20.5. The table also shows the "Total" result (i.e. the results after combining the male and female data) and shows that the mean for this merged group is 19.75 on the new preparation.

The second summary table gives the results for males and females receiving the old preparation. The mean male response was 25.5 and the mean female response was 22.0. Together the males and females had a mean response of 23.75 on the old preparation.

The third summary table shows the pooled results for males and females respectively (i.e. it combines, males and females regardless of the type of preparation given). The eight males on the new preparation had a mean response of 22.25 compared to a mean response of 21.25 for the eight females.

The last table gives the ANOVA findings. The format of the table containing the raw data (see Table 10-A on page 278) determines what factor is considered the "Sample" and what factor is the "Column". The "Column" source is the easiest to identify, it refers to the variable that appears in the last 2 columns of Table 10-A. Those columns are for the male and female data. Therefore, the "Column" reference in the print out on page 279 is for the gender factor. "Sample" refers to the other factor which, in this case, is the treatment factor. We will skip the "Interaction" row for just a moment. The source "Within" refers to the expected variation that would occur if all groups had essentially the same response. It is equivalent to the source "Within Groups" in a One Way ANOVA. In some computer programs the term "Error" is used rather than "Within".

"Interaction" is a new term that measures the consistency of the effects from the two main factors – treatment and gender. In other words, interaction tests to see whether the relative difference between the two preparations in one gender is consistent with the relative treatment effect in the other gender. If the new preparation did much better than the old preparation in males, but worse than the old preparation in females, there would be an inconsistency and that inconsistency would be detected by the interaction information of the print-out. For each Source of Variation, we are given the same statistics presented in a One Way ANOVA (i.e. SS, Df, etc).

Although one is tempted to begin an interpretation of a Two Way ANOVA with the Sample row (i.e. the effect of the treatment factor), you should begin with the interaction term. If there is an inconsistency in your results, there may be little meaning associated with the main factor effects.

You can view the interaction effect as a test of parallelism. The results of our study are displayed below in graphic form (see Figure 10-A on the next page). The means for the four groups are plotted: males on new preparation (19.0), males on old preparation (25.5), females on new preparation (20.5) and females on old preparation (21.5). Note that the graph highlights the relative treatment effects for males on the left-hand side and for females on the right hand side. By drawing a line between the male responses and the female responses on the new preparation and another line between the male and female responses on the old preparation, we can tell visually whether the results are consistent. If the lines are parallel then there is consistency, but if there is incompatibility then the lines diverge.

Our lines have some divergence, but it's not great. In reality, almost all lines will diverge to some degree so we rely on the statistical results displayed in the ANOVA table above to tell us if the divergence is statistically significant or not. If there is statistical significance for interaction then you generally ignore the main factors (Treatment and Gender given in the Sample and Column rows) and discuss your results in terms of the inconsistency depicted in a graph such as Figure 10-A. On the other hand, if interaction is not statistically significant, you then examine the results for the main factors.

Figure 10-A

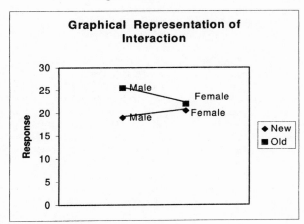

In our example we do not have a statistically significant interaction (the interaction p-value of .17 is greater than .05) so we examine the two main factors. There is a statistically significant difference for treatment (p-value = .04), but not for gender (p-value = .57). We, therefore, conclude that the new preparation is superior because it causes fewer headaches (19.75 versus 23.75) and there is no difference between the males (mean = 22.25) and the females (mean = 21.25) in respect to the frequency of headaches.

Matching Subjects

Before we leave our discussion of ANOVA, it should be pointed out that there are even more sophisticated designs than a Two Way ANOVA. For example, we introduced the subject of paired designs for the t test in chapter 9. The use of the same subject or matched subjects is often deemed desirable because the equivalence of the treatment groups is achieved without relying on randomization. The administration of multiple treatments to each subject is, in theory, a very desirable practice because it clearly enhances the chance of having comparable groups. There is also an ANOVA variation that allows an experimenter to compare treatments when all subjects receive each treatment - it is called an ANOVA **Repeated Measurement design**. Again this approach will be saved for a more advanced course. However, it should be pointed out that repeated use of the same subject for multiple treatments adds to the duration of a study and it may also not be feasible. The feasibility issue is based on the need for a subject to return to his or her pre-treatment state before the second or third treatment is given. The early treatment may cause a relatively permanent change and the subject is, therefore, in a different state when the second treatment is given. This altered state obviously affects the kind of response that can occur with the second treatment. To get around this problem we can use matched subjects rather than the same subject in a paired design. Thus, we could use three matched subjects rather than using the same person three times in an ANOVA Repeated Measures design.

To match subjects we need to place them in categories in which they share common traits (e.g. gender, age etc.). Be aware however, that it is very hard to match subjects on more than a limited number of characteristics. Usually you can match on only a few characteristics. For example we might want to match subjects on the basis of sex, marital status and educational level. For sex there are only two conditions - male or female, but for marital status we would have at least two (married, not married) but more likely four classes (adding widowed and divorced). We now have eight possible categories - Males who were never married, males who are married, males who are divorced, etc. If we plan to give three different treatments in our study we need sets of three people in a category. For example in the category "male and married" we need at least three subjects. If we end up with four or five male subjects who are married, we still can use only three of them in our study. If there are five subjects in the group we must drop two of them. However with six subjects we again can use all subjects assigning two to each of the three treatments. To summarize, all of our 8 categories with less than three subjects (or a multiple of three) result in a loss of subjects.

Things become even more difficult when we include our third variable - educational level. We have many options on what levels to use, but let's assume we will use only three - those who have less than a Bachelor's degree, those with a Bachelor's degree and those with more than a Bachelor's degree (i.e. a graduate degree). We now must expand from our 8 categories to 24 (2 sexes * 4 marital status grouping * 3 educational levels). With 24 groupings we have an even greater risk that we will have more categories with fewer than three (or a multiple of three) subjects. It can be seen that the matching operation becomes more complex with a greater loss of subjects as we add more classes or variables. Thus, experimenters often forgo matching and use designs which 1] randomize all subjects and 2] examine other variables (control variables) that can check for group equivalence in order to reduce the chance of subject selection bias.

Even if all control variables show test group equivalence, there is still no assurance that bias will not be introduced by other variables, whose presence or affect we do not know. We are forced, therefore, to assume that any subject selection difference between groups didn't affect the outcomes. The onus is on the experimenter to look for treatment group incomparability and to defend the assumption of group equivalence. However, no matter what we do there is no guarantee that every group differences can be looked for and identified, so the assumption of group comparability is a reservations or qualifications for studies which use independent groups.

Practical Versus Statistical Significance

Here is a good point to introduce an important concept. We will use the Z formula to show the difference between practical and statistical significance. Remember that we declare statistical significance when the calculated p is less than the Alpha Risk, usually .05. We reject the null hypothesis and say that there is a statistically significant difference. This act of deciding, on the basis of the p-value, that there is a difference between experimental conditions represents **statistical significance**.

In addition to statistical significance there is another way to look at the difference. Is the difference observed of any practical value? It is this view of the difference that we call "practical significance". **Practical significance** means that the difference between experimental conditions is meaningful. If we look at the difference and say: "If this is truly the case then I would certainly prefer one group over the other" then we have practical significance. However, a statistical test of a difference that we believe has practical significance may end up with a p-value for which statistical significance can not be claimed. In that situation there is practical significance, but not statistical significance. The reverse can happen as well. A difference may turn out to be associated with a small p-value and statistical significance declared, but the difference is so small that we see there is no practical distinction between the groups being compared. It should be clear that there can be disagreement between statistical and practical significance.

We might see this situation better by examining the Z test for comparing a sample mean against a standard (μ). The Z formula we use is:

$$Z = \frac{\overline{X} - \mu}{SE} \text{ where } SE = \frac{\sigma}{\sqrt{N}}$$

What if we were searching for a new teaching method that will increase mathematical proficiency. The mean for the current method is well established at 30. The population standard deviation is 10. Assume we would change to a new method if the scores rose by 10% or more. That is, a rise of 3 points or more would cause us to switch to the new method.

A new method must therefore result in scores of at least 33. That 3-point change has practical significance - it prompts us to make a change.

Assume we take a sample with 16 observations and the mean is 33.5. Since this result is over 33 we have practical significance. But do we have statistical significance? We can use a hypothesis testing setting with an Alpha Risk of .05 and a one tail test (we are only interested if the new method is an improvement).

The critical Z value needed is the one that cuts off the top 5% of the standardized normal curve and that value is Z = 1.65. If our calculated Z equals or exceeds that number we have statistical significance. To find the calculated Z we first need to find the standard error:

$$SE = \frac{10}{\sqrt{16}} = 2.5$$

Now we can find our Z value:

$$Z = \frac{33.5 - 30}{2.5} = 1.40$$

Since the calculated Z is less than the critical Z we do not have statistical significance although we have practical significance.

Let's see what happens under a different circumstance. Suppose in our second case everything remains the same except we have 400 rather than 16 subjects. Our standard error is now:

$$SE = \frac{10}{\sqrt{400}} = 0.5$$

We can find the calculated Z value for this situation:

$$Z = \frac{33.5 - 30}{0.5} = 7.00$$

Since the calculated Z is greater than the critical Z we have statistical significance. We also have practical significance. This is an appealing situation – we like it when our statistical significance is also associated with a meaningful difference.

We'll take a look at one more situation. In case 3 we will keep everything the same as case 2 except the sample mean. In case 3 our sample mean is 31.0. This is clearly well short of the 33 we need to declare a practical difference. However, we also must see whether we have statistical significance. Our SE remains 0.5 and when we now calculate Z we get:

$$Z = \frac{31 - 30}{0.5} = 2.00$$

The calculated Z is greater than the critical Z so there is statistical significance, but as noted above the actual difference is trivial. This is an unfortunate situation because we have statistical significance without practical significance. As a matter of fact a study in which there is an incompatibility between practical and statistical significance is often viewed as an unsatisfactory study.

The table below presents the data from the 3 cases described above.

Case	$\overline{X}$	σ	N	Z	Practical Significance	Statistical Significance
1	33.5	10	16	1.40	Yes	No
2	33.5	10	400	7.00	Yes	Yes
3	31.0	10	400	2.00	No	Yes

Statisticians would argue that in the first study too few subjects were used and in the third study too many subjects were used. This discussion points out that one should not blindly accept the presence or absence of statistical significance as being the only important result of a statistical test. If there is no statistical significance it is still prudent to look at the actual difference and decide if there is practical significance. You may lack statistical significance simply because the N is too small to find an important difference. On the other hand, even if there is statistical significance, don't assume it's an important difference till you examine the actual difference reported. If that difference is small, and doesn't represent a truly useful difference, then treat the result warily. The difference, even if it is statistically significant, may not have any useful value.

The ideal case is when you have both statistical and practical significance. Statisticians realize that if you use large enough sample sizes you can show that almost any difference, even the most trivial one, is statistically significant. However, that doesn't mean you have an important difference. What can happen is that just by increasing N we can increase Z. A lackluster mean difference is evident and yet, if N gets large enough, the p-value will go from a non-significant to a significant status.

We should also appreciate the importance of the standard deviation which is in the denominator of the Z calculation. Let us look at what happens to Z when we reduce the standard deviation (σ) and hold the ($\overline{X}$ - μ) difference and N constant. It doesn't matter what numbers we use for these elements since the mathematical effect of increasing or decreasing Z will be the same. For our example we'll use a mean difference of 8.0 and an N of 25.

Case	Mean Difference	N	σ	SE	Z
1	8.0	25	20	4.0	2.00
2	8.0	25	15	3.0	2.67
3	8.0	25	30	6.0	1.33

When we compare case 1 to 2, note that we reduced σ from 20 to 15 and Z rose to 2.67. However, when σ is then increased from 15 to 30 (case 3), the value of Z falls precipitously (from 2.67 to 1.33). It should be clear that changes in the standard deviation have a direct effect on Z and, therefore, on the chance of declaring statistical significance. A smaller standard deviation comes about when we improve the internal validity of an experiment. As we improve our experimental setting there is less variation due to extraneous factors. Reducing the standard deviation is considered desirable because it also means we can get by with a smaller N, thereby reducing cost and time.

Summary

In this chapter we learned that when there are more than two sample means, the Analysis of Variance (ANOVA) is used rather than repeated t tests. The use of ANOVA is based on concerns regarding the experimentwide error rate associated with performing repeated t tests. ANOVA tells us if there is a difference somewhere among the set of means. To find if the differences between any two sample means is statistically significant we learned to use the Tukey-Kramer multiple range test. We also realized that just because you have statistical significance it doesn't mean you have practical significance and you can have a difference between means that is important without having statistical significance.

The sheet that follows shows an updated version of the decision tree given in chapter 9. The revised decision tree incorporates the Analysis of Variance procedure. To use the tree, start with the diamond called "Standard Deviation Known" and follow the questions and answers until you reach one of the conclusions (e.g. if the standard deviation is not known and the No. of samples is 3 you use the ANOVA procedure). If the number of factors is 1 and the One Way ANOVA result indicates that there is statistical significance, you then utilize the Tukey-Kramer Test. Note that the title of the decision tree includes the expression "and the Variables are Interval Measurements". In the next chapter we will see that there are different tests to compare groups which utilize the ordinal and nominal measurements.

Decision Tree for Selecting Statistical Tests When There are More than 2 Means to Compare and the Variables Are Interval Measurements

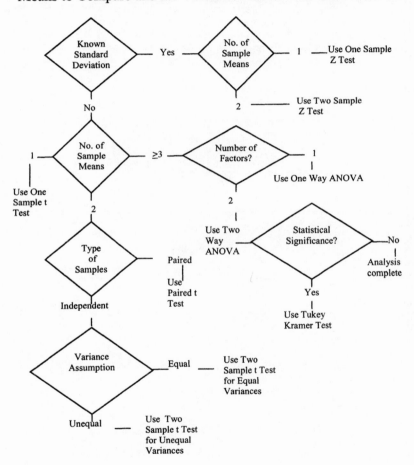

PROBLEMS

1. Due to a large number of accidents, city bus drivers were required to take a safety improvement course. Three different instructional approaches were used and the drivers were assigned at random to one of the methods. After completing the course the driver's records were tracked for the next year and an accident index obtained. The group using method A had the lowest accident score, Group B was in the middle and group C did the worst (i.e., they had the highest score). An Analysis of Variance was conducted on the accident scores and resulted in the following table.

Anova: Single

SUMMARY

Groups	Count	Sum	Average	Variance
A	15	183	12.2	11.88571
B	15	228.8373	15.25582	10.70917
C	15	279.4548	18.63032	11.35232

ANOVA

Source of	SS	df	MS	F	P-value	F crit
Between	310.3715	2	155.1858	13.71416	.000026	3.21993ε
Within	475.2608	42	11.31573			
Total	785.6323	44				

a. If the Alpha Risk was .05, is there a statistically significant difference among the groups?
b. If the Alpha Risk was .01, is there a statistically significant difference among the groups?
c. If the F value in the above table was 3.25, would you accept or reject the null hypothesis?
d. If the F ratio had been 1.00 what would we conclude?
e. If the analyst decided to use the Tukey-Kramer multiple range test, why would he need to calculate only 1 critical range.
f. What Q value would you use to calculate the critical range?
g. What is the critical range?
h. What groups are different from what other groups at the .05 significance level?

2. The police chief wonders if the five precincts that make up his city are equivalent in respect to the number of tickets issued per shift. An analyst conducts a study in which she collects data from 13 shifts randomly selected from each of the five precincts. The results of her analysis are displaced below.

ANOVA: Single Factor

SUMMARY

Precincts	Count	Sum	Average	Variance
1	13	633	48.7	27.1
2	13	738	56.8	24.5
3	13	725	55.8	40.0
4	13	655	50.4	39.1
5	13	682	52.5	26.9

ANOVA

Source of Variation	SS	df	MS	F	P-value	F crit
Between Groups	616.0923	4	154.0231	4.885247	0.001776	2.525212
Within Groups	1891.692	60	31.52821			
Total	2507.785	64				

a. The next decision for the analyst is whether or not to proceed with the Tukey-Kramer multiple range test. What result from the analysis tells her whether to proceed with that test?

b. Since there are 5 precincts, there are 10 two-group comparisons (e.g. one group comparison is precinct 1 v. precinct 2, another is group 1 v. group 3, etc.). You no doubt recognized that we know there are 10 comparison since this is a combination problem where n=5 and r=2. Of the 10 two-way comparisons, which ones are associated with statistical significance at the .05 significance level? You will need to calculate the critical range to answer this question.

c. Why did the analyst need to calculate only one critical range?

d. The police chief knows that in precinct 1 the mean number of tickets issued/shift is 48.7, but he would like to know what the standard deviation is. Tell him what the standard deviation is.

3. What is the difference between practical and statistical significance?

4. From a survey of 120 consumers, preferences were found for a number of items.
    a. Cable news reports were preferred by 32 consumers. What is the 90% confidence interval assuming σ is .35?
    b. American made automobiles were favored over foreign made cars by 102 people. What is the 98% confidence interval assuming σ is unknown? Try solving this problem using t rather than Z.

5. Refer to the study described in the chapter that concerned an old and new preparation for the treatment of headaches. Assume the study was replicated at another time and the results of that second study are displayed below:

Anova: Two-Factor With Replication

SUMMARY	Male	Female	Total
New Preparation			
Count	4	4	8
Sum	73	97	170
Average	18.25	24.25	21.25
Variance	6.25	8.9167	16.786

Old Preparation			
Count	4	4	8
Sum	102	90	192
Average	25.5	22.5	24
Variance	11.667	13.667	13.429

Total		
Count	8	8
Sum	175	187
Average	21.875	23.375
Variance	22.696	10.554

ANOVA

Source of Variation	SS	df	MS	F	P-value	F crit
Sample	30.25	1	30.25	2.9877	0.1095	4.7472
Columns	9	1	9	0.8889	0.3644	4.7472
Interaction	81	1	81	8	0.0152	4.7472
Within	121.5	12	10.125			
Total	241.75	15				

How would you interpret the results form this study?

6. The mean score on a vocabulary test for a sample of 100 college freshmen is 135. The known population standard deviation is 40. The researchers are curious to know if there is a difference between this result compared to past scores they have observed. For the following questions, give the critical Z value and state whether there is a statistically significant difference.
   a. Using an alpha risk of .02 and a past student mean score of 120?
   b. Using an alpha risk of .05 and a past student mean score of 125?
   c Using an alpha risk of .10 and a past student mean score of 140.

7. What option, under the experimenter's control, is available to increase the chance of getting a statistically significant result?

8. How does improving the internal validity of an experiment affect the likelihood of getting a statistically significant result?

9. What statistical test do you use under the following conditions:
   a. The standard deviation is known, and you want to compare the results of two sample means.
   b. The standard deviation is not known and there are four sample means.
   c. There are two independent sample means, the standard deviations are not known, but the sample standard deviations are comparable.

10. What is the difference between independent and paired samples?

11. What is the advantage of the central limit theorem?

12. For the following problems assume that the population mean is 49, the known standard deviation is 12 and the sample size (if needed) is 9.
   a. Use the critical value method to determine if an individual observation of 58 is statistically significant compared to the population mean. Perform the test using a .05 Alpha Risk and a 2 tail test.
   b. Use the critical value method to determine if a mean value of 58 is statistically significant compared to the population mean. Perform the test using a .05 Alpha Risk and a 2 tail test.
   c. Redo part a, but this time assume that the standard deviation value of 12 is a sample standard deviation.
   d. Redo part b, but this time assume that the standard deviation value of 12 is a sample standard deviation.

Excel Instruction

Excel has a built-in function to do the Analysis of Variance (ANOVA). One of the examples discussed in the text dealt with weight loss data. Enter the data for that example on a worksheet. Place the heading and data for the A group in cells A1 to A8. Place the heading and data for the B group in cells B1 to B8 and the heading and data for the C group in cells C1 to C8. From the Main menu select "Tools" followed by "Data Analysis". Select "ANOVA: Single Factor" from the "Analysis Tools" listing. You should have before you a screen that looks like this:

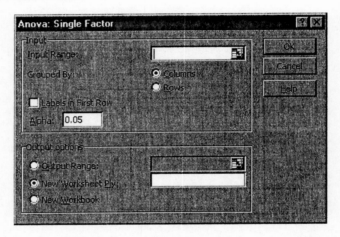

For the input range you must include the complete data set - i.e. the data for all groups so there is only one input range. In addition each group must be in a separate column and the groups must be in adjacent columns. In our example the range we want is "A1:C8". Row 1 is included because we want to include the headings in our analysis output. Our data are grouped by columns so we do not need to make any changes to that portion of the dialog box. If the "Labels in First Row" box does not have a check mark in it, click on the box so it does. Leave "Alpha" at 0.05. For the "Output options" leave the dot in the "New Worksheet Ply" circle and give the new sheet a name - "ANOVA" would seem appropriate. After you hit OK you should have output consistent with that shown in the text. You may have to alter the width of some columns to have all the information show. The ANOVA output was reviewed in the text and won't be repeated here.

We still need to worry about the case where we obtain a statistically significant difference and want to know what groups are different from the other groups. To answer this question we will perform the Tukey-Kramer test. Obtain Q from the Studentized Range table in the Appendix (Table 4). After Q is obtained you just have to calculate the critical range cited in the text. You then compute the individual group differences against that critical range to find which ones exceed the range. Those that exceed the range are declared statistically significant at the .05 level. In our example the Q value we obtained was 3.61 so our full equation, written in one step would be:

$$=3.61*SQRT((19.25/2)*((1/7)+(1/7)))$$

The result of the calculation is 6.00 and any difference between groups means this large or larger would be declared statistically significant at the .05 level.

There is also a procedure in Excel to do a Two Way ANOVA. We will use the raw data for the Two Way ANOVA that compared the New and Old preparation in Males and Females. First copy that data onto a worksheet beginning at cell A1. From the Main menu select "Tools" followed by "Data Analysis" and then select "ANOVA: Two-Factor With Replication". You should have before you a screen that looks like this:

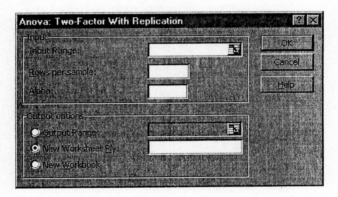

For the input range you must include the complete data set - i.e. the data for all groups including the labels. Note that your raw data is organized in rows and columns. The rows are considered the replications and there must be the same number of rows for each group. There is a Two Way ANOVA procedure that allows unequal replications, but that version is not available in Excel

In our example the range we want to enter is "A1:C9" where column A contains labels, column B the data for males and column C the data for females. Remember Row 1, which also contains only labels, must be included in the input statement. The next box requests the number of "Rows per Sample". In our case we have 4 rows per sample so we enter that number. The other boxes are easy to complete – we identify our Alpha Risk (.05) and where we want our output to appear. When you complete these steps you will obtain a print-out like the one in the text. How to interpret the ANOVA output was reviewed in the text and won't be repeated here.

Before we leave Excel there is one last function that needs to be explained. As noted in chapter 9, the two sample t test has two options - one for equal variances and one for unequal variances. It was also mentioned that you could do a statistical test for the equal variance assumption. Excel has a function to perform that test.

Let us assume we want to test the variances for the male versus female pay problem presented in chapter 9. Create a worksheet that contains the pay information sorted by gender – i.e. all female results together followed by all the male results. The statistical test we want is activated by choosing "Tools" from the Main menu, followed by "Data Analysis". In the "Analysis Tools" listing select "F Test Two Sample for Variances". You will now be looking at the following screen:

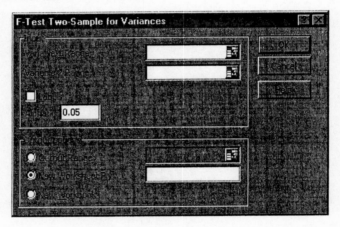

Our input ranges need to be presented in a certain order. Because of the way the test is done, the variable with the larger variance must be variable 1 in the input. Males have a larger variance than females so we enter "B26:B42" for the "Variable 1 Range". We are assuming all the pay values are in column B starting in Row 4. We enter "B4:B25" as the range for variable 2. We did not include labels so be sure that the "Labels" box is blank and again we'll test at the conventional .05 level so make sure that the "Alpha" box has "0.05" in it. We can place the output on our current worksheet so identify an empty area on the sheet (the output takes up 3 columns and 10 rows). In the "Output options" area click on the circle for "Output Range" so it contains a dot. Identify the cell in the upper left-hand corner of the output area you have chosen in the adjacent box. After hitting OK you should have the following results on your screen.

F-Test Two-Sample for Variances

	Variable 1	Variable 2
Mean	35.76471	36.45455
Variance	147.5674	112.9769
Observations	17	22
df	16	21
F	1.306174	
P(F<=f) one-tail	0.279168	
F Critical one-tail	2.156263	

We compare the calculated F, identified as "F" in the table against the critical value for F (identified as "F Critical one tail". The calculated F value is 1.31 (after rounding) and it is less than the critical F of 2.16 (after rounding) so the null hypothesis is accepted and we conclude that there is no statistically significant difference in respect to the variances. The entry "F(F<=f) one tail)" is a p-value identifying the probability that this result would occur if the null hypothesis were true. In other words there is a .28 probability that the observed differences could occur if, in fact, the variances were equal. Since that value is greater than .05 one would not reject the null hypothesis. Remember that the group with the largest variance must be entered as the first variable. If the group with the smaller variance is entered as the first variable the F value will be less than 1. If this result occurs, reverse the order in which the variables are entered.

Here are some exercises you can do to practice the Excel skills described in this chapter. For each problem assume the data are samples from a larger population.

1. Compare the groups with different educational levels in regard to their years of service.
    a. If you use an alpha risk of .05 what would you conclude?
    b. How large would the F ratio have to be for you to change your conclusion?
    c What assumption are you making regarding the variances of the three groups?
2. Three different groups were given exercise programs at three levels of difficulty (low medium and high). The results are shown below.

LOW	33	43	42	61	45	40	38	52	64	55	21
MED	55	61	53	55	78	58	50	48	35	42	61
HIGH	68	54	70	50	83	70	75	62	56	44	65

    a. If you use an alpha risk of .05 what would you conclude?
    b. Calculate the critical range - what differences are there among the groups?

Chapter 11

Non-Parametric Tests

Introduction

In earlier chapters we studied procedures that required the interval scale of measurement. In this chapter we will primarily examine another set of tests which are designed to be used with nominal and ordinal variables. These procedures, frequently called non-parametric tests, can be used to find out how closely two variables are related. Additional procedures such as the Chi Square Test and the Wilcoxon Rank Sum Test will allow us to determine if there are differences between groups.

Parametric Versus Non-Parametric Tests

The best place to begin is by separating statistical procedures into two main classes - the parametric tests and the non-parametric tests. The Z Test, t Test and ANOVA are all parametric procedures. What **parametric tests** have in common is that they require the data to be based on the interval scale and, as we saw, they require certain assumptions such as normality and equal variances to name a few. **Non-parametric tests** require far fewer assumptions (e.g. the underlying population distribution does not have to be normal) and they are used for variables that are in the nominal and ordinal class. Although the non-parametric procedures may sound more appealing based on the above descriptions, they have definite disadvantages compared to the parametric techniques. The parametric tests are more powerful in that they generally can give you more information and you usually can obtain statistical significance with a smaller sample size than that needed to do a comparable non-parametric test.

Correlation Coefficient

We will first look at a non-parametric procedure to calculate a correlation coefficient, but before we do that we need to gain an understanding of what a correlation coefficient is. A **correlation coefficient** measures the degree of relationship between two variables. The symbol r is used to represent a correlation coefficient. An r of 0.0 indicates that there is no relationship between two variables. Positive r values occur when there is a direct relationship between the variables (i.e., increases in one variable are associated with increases in the other variable). Negative r values occur when there is an inverse relationship (i.e., increases in one variable are associated with decreases in the other variable). The maximum r is +1.00 and the minimum r is -1.00.

It is customary to refer to one variable as the X (or independent) variable and the other variable as the Y (or dependent) variable in a simple correlation problem. As noted above, a positive correlation occurs when X increases and there is also an increase in Y. In terms of a graph of a positive correlation, as you move from left to right the data points tend upward (see Figure 11-A).

Figure 11-A

An example of a positive relationship would be hours worked (X) and wages earned (Y). Note that in a correlation graph, each individual is represented by a single symbol. The symbol is located at the intersection of the person's X and Y value. A perfect positive correlation occurs when all the data points are on a straight line that angles upward.

In contrast to a positive correlation, a negative correlation has a relationship in which as X increases Y decreases. A graph of a negative correlation shows the data points falling as one moves from the left to the right along the X axis (see Figure 11-B on the next page). An example of a negative correlation could be the relationship between money spent on a safety program (X) and the number of accidents that occur (Y). As more money is invested in safety the accident rate tends to decrease is an example of a negative correlation. A perfect negative correlation is one in which all the data points form a straight line that angles downward.

Figure 11-B

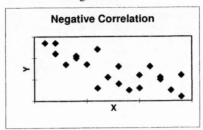

There is a third type of relationship that should be mentioned - the one that occurs when there is no correlation. In this arrangement when X increases the data points form no consistent pattern. A graph of this kind of relationship is given in Figure 11-C. There is no clear pattern and no trend is evident. An example of a relationship without any correlation is the sum of people's social security numbers (X) and their height (Y).

Figure 11-C

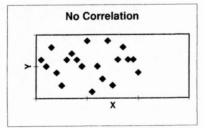

Spearman Rank Correlation Coefficient

One of the best ways to understand what a correlation coefficient represents is to see how one is calculated. For this purpose, we will use a non-parametric procedure because it is quite informative. The correlation procedure we will demonstrate is the Spearman Rank correlation coefficient. The symbol for the correlation coefficient is a lower case r and in the case of the Spearman correlation coefficient there is also a "S" subscript added so the full notation is r_S.

For a correlation coefficient it is necessary that each observation in the data set have a value for two variables. The strength of the relationship between the two variables is then determined. The example we will use is from a data set that consists of students' score on a math test and a self-rating regarding their attitude toward mathematics.

The Spearman procedure is used when both of the variables are ordinal measurements or one is ordinal and one is interval. You cannot use the Spearman procedure with a nominal scaled variable - in a moment you will see why. The math attitude rating is based on a 5 point scale that uses the following ratings:

1. Hate
2. Dislike
3. Tolerate
4. Enjoy
5. Love

The math test score is the percentage of correct answers on a math test. The math test score is an interval scale measurement, but the math attitude rating is an ordinal scale measurement and thus the Spearman method is appropriate.

From a student data base, a sample of 7 students was drawn. For purposes of illustration a sample size of 7 is sufficient, but in practice it's best to use sample sizes quite a bit larger. It's dangerous to generalize, but a sample size of at least 20 would be a reasonable recommendation.

The 7 students in the sample are identified by a single letter and their respective math attitude and test score is shown in Table 11-A.

Table 11-A

Student	Attitude	Score
A	3	68
B	3	74
C	4	79
D	4	89
E	1	74
F	2	95
G	4	84

A numerical value is shown for the attitudes (e.g. student A "tolerated" math which, as noted above, was assigned a rating of 3 and that is what is shown in the table). Unfortunately in the small sample there were no students who "loved" math as can be seen by the absence of any 5 ratings.

The order of the ratings for the ordinal variable can be important because it can affect how you express your findings. Note that the math attitude ratings go from a low number to a high number as the attitude becomes more favorable. The math test scores are also ordered in that the highest score represents the best performance. In this case we will find a positive correlation if math scores increase as attitudes become more positive. If we had coded our attitudes with a 1 for "love" up to a 5 for "hate" we'd end up with a scale that assigned lower scores to more positive attitudes. Consequently, if math scores increased, as the attitude toward math got stronger, the correlation would turn out to be negative. It's more reasonable to think of the most favorable attitudes being positively correlated with higher math scores so the order of the numeric values assigned to the ratings seem appropriate in this example.

We have seven students so we will want to have the students ranked from 1 to 7. A display of the ranks we need to assign is given below:

Rank:	1	2	3	4	5	6	7

To assign ranks to the attitude variable we will first arrange the 7 attitude ratings in order - from lowest to highest. When this is done for the 7 students, the order looks like this (the attitude rating for each student is also shown in the second row):

Student:	E	F	A	B	C	D	G
Attitude	1	2	3	3	4	4	4

What we do next is to assign the ranks to the students. The first two cases are easy - student E gets the rank of 1 and student F the rank of 2. When we get to the next two students, A and B, we see they both have the same attitude value, a rating of 3. We are at the point where we need to assign rank 3 and then rank 4, but who should get what rank? The fairest approach is to assign them the same rank. They essentially are tied for the rank 3 and the rank 4 and it's the average of those two ranks which we need to assign to each student. The average of these ranks is 3.5 ([3+4]/2=3.5) and we will assign each student that rank. At the next level (i.e. assigning rank 5) we find three students with the same attitude value - students C, D and G each have an attitude rating of 4. We again resolve the tie by assigning each student the same rank - in this case the average of ranks 5, 6 and 7. The average rank is 6 ([5+6+7]/3=6) and students C, D and G are each given the same rank of 6.

For the variable math attitude we have created a ranking table that looks like this:

Student:	E	F	A	B	C	D	G
Attitude	1	2	3	3	4	4	4
Rank	1	2	3.5	3.5	6	6	6

We now move to the other variable, math score, and order the students' scores. We end up with the following arrangement.

Student:	A	B	E	C	G	D	F
Score:	68	74	74	79	84	89	95

We begin assigning ranks and student A with the lowest scores receives the rank of 1. The next two students are tied - for the ranks of 2 and 3 so we assign them the mean rank which is 2.5 ([2+3]/2=2.5). We resume our assignment with rank 4 and that rank is given to student C who had a math score of 79. The rest of the assignments are straight forward and we can now add the ranks to the other information and obtain our ranking table for the math scores:

Student:	A	B	E	C	G	D	F
Score:	68	74	74	79	84	89	95
Rank:	1	2.5	2.5	4	5	6	7

We now incorporate the ranks we've obtained for each student to our original data table and that produces Table 11-B (see below).

Table 11-B

	Raw Data		Rank	
Student	Attitude	Score	Attitude	Score
A	3	68	3.5	1
B	3	74	3.5	2.5
C	4	79	6	4
D	4	89	6	6
E	1	74	1	2.5
F	2	95	2	7
G	4	84	6	5

Before we begin the actual calculations, a worthwhile action is to make a scatter graph of the rank data. A reasonable question to ask is why? If we can see the shape of the relationship (see Figures 11-A to 11-C) we'll know what kind of correlation to expect - should it be positive, negative or close to zero? Our graph will give us a good idea of the type of relationship we have. If your calculated result is inconsistent with the graph, you better re-check your work.

The graph we have in our situation is shown on the next page (see figure 11-D). Because of the small sample size it's hard to see any specific pattern, but the general impression is that there may be a slight positive relationship.

In terms of deriving the correlation coefficient the first step was to determine the ranks for both variables which we already did. Logic would tell us that if the variables were highly correlated, the students would rank much the same on each variable. We can measure how close the ranks on the two variables are by obtaining the difference between them. Small differences would indicate a similarity in the ranks of the variables. For person A the difference in ranks is 2.5 (3.5-1). We continue getting the difference in ranks for each student.

Figure 11-D

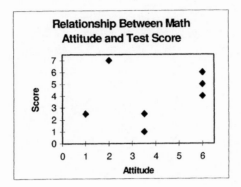

When we are done calculating the differences, we sum the ranks to see the cumulative effect. We find we have a problem - the sum is 0 (see Table 11-C)

Table 11-C

| | Raw Data | | Rank | | |
Student	Attitude	Score	Attitude	Score	Difference
A	3	68	3.5	1	2.5
B	3	74	3.5	2.5	1
C	4	79	6	4	2
D	4	89	6	6	0
E	1	74	1	2.5	-1.5
F	2	95	2	7	-5
G	4	84	6	5	1
				SUM:	0

No matter how similar or different the ranks may be, we will always find that the sum of the ranks is 0. We ran into this dilemma once before - when we were trying to find a measure of variation. In that case we ended up eliminating the problem by squaring each difference and then adding up the squared differences. That is the same thing we do for the Spearman correlation coefficient. Our table can be expanded once again to add the columns for Squared Difference (see Table 11-D on the next page).

Note that the sum of the differences must be 0. If it's not 0 you made a mistake in your rankings so always include the sum of the differences as a way to check the accuracy of your arithmetic.

Table 11-D

Student	Raw Data Attitude	Raw Data Score	Rank Attitude	Rank Score	Difference	Squared Difference
A	3	68	3.5	1	2.5	6.25
B	3	74	3.5	2.5	1	1
C	4	79	6	4	2	4
D	4	89	6	6	0	0
E	1	74	1	2.5	-1.5	2.25
F	2	95	2	7	-5	25
G	4	84	6	5	1	1
				SUM:	0	39.5

The algebraic symbol for the sum of the squared differences (39.5) is ΣD^2 and is used in the calculation of the Spearman correlation coefficient. The only other item we need to calculate r_S is N, the number of students in our sample. That number is 7 so we can now calculate r_S using the formula below. Note that two 1's and a 6 are included in the formula. The two 1's and the 6 are fixed values and are used whenever you apply a Spearman correlation coefficient procedure.

$$r_S = 1 - \frac{6 * \Sigma D^2}{N * (N^2 - 1)}$$

Substituting the values for ΣD^2 and N into the formula gives us:

$$r_S = 1 - \frac{6 * 39.5}{7 * (7^2 - 1)} = 1 - \frac{237}{336} = 1 - .71 = +0.29$$

A correlation of +0.29 is considered to be a weak to moderate r. It indicates that as attitude goes up there is a tendency for scores to go up.

A perfect positive correlation would require the same rank for each student on the two variables (see figure 11-E on the next page). In that event the differences in the ranks for each student will be 0. The squared differences (ΣD^2) will also be 0 and we will obtain a value of +1.0 for our calculated r_S.

$$r_S = 1 - \frac{6 * 0}{7 * (7^2 - 1)} = 1 - \frac{0}{336} = 1 - 0 = +1.0$$

Figure 11-E

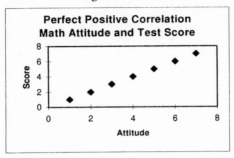

If math attitude and math score were inversely related the ranks would be in exact reverse order (e.g., the student with a rank of 1 for math attitude would have a rank of 7 on the math score and the student with a rank of 2 on math attitude would have a math score rank of 6, etc.). If we made a graph of the data points we'd have a straight line, all the data points would be on the line and the line would be sloping downward. The picture is that of a perfect negative relationship and in fact your calculated r_s would be -1.0 (See figure 11-F).

Figure 11-F

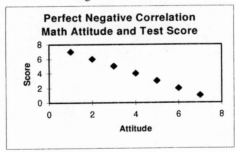

The calculations involved above illustrate the basic logic behind a correlation coefficient, but the Spearman correlation coefficient is limited to situations in which the lowest level scale for a variable is ordinal. If both variables involve interval scales a parametric procedure called the **Pearson correlation coefficient** is used. The symbol for a Pearson correlation coefficient is simply a lower case r without a subscript (r). The formula for the Pearson correlation coefficient is:

$$r = \frac{\sigma_{xy}}{\sigma_x * \sigma_y}$$

We are familiar with the statistics in the denominator – they are the standard deviations for the X and Y variables. The numerator (σ_{xy}) is a new statistic and is called the covariance of X and Y. The covariance is calculated in the following manner:

$$\sigma_{xy} = (X_1 - \overline{X}) * (Y_1 - \overline{Y}) \cdots + \cdots (X_n - \overline{X}) * (Y_n - \overline{Y})$$

If you study this formula you may see that the **covariance** is based on the extent that the X and Y variables move in a similar or dissimilar fashion. If both the X and Y variable are large, relative to the group X and Y mean, we could say they are following a similar pattern. Similarly, if both the X and Y variable are small, relative to the group X and Y means, we could again say the variables follow a similar pattern. Conversely, if one variable decreases as the other variable increases, we would then say that the variables follow dissimilar patterns. Note that the covariance is created by multiplying the deviations from the X mean (X-$\overline{X}$) by the deviations from the Y mean (Y-$\overline{Y}$) for each individual in the data set. The sign of these multiplications is revealing because it determines whether there is a positive or negative correlation. Note that the denominator for r will always be a positive number since the standard deviations for X and Y are always positive values. Hence the sign of the correlation coefficient is determined by the numerator, the covariance of X and Y. This is more apparent if we create a graph with four different quadrants depending on whether the X and Y deviations are positive or negative. This arrangement is depicted in the graph below.

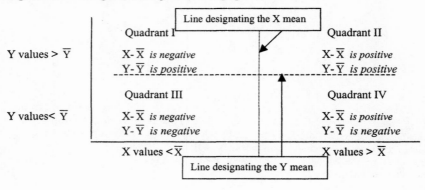

In Quadrant I the X deviations are negative and the Y deviations are positive. The correlation equation requires multiplying the two deviations together for each data element. When the deviations are multiplied together the result is a negative number. In a similar fashion we can see that multiplying the deviations together in Quadrant II produces positive numbers, in Quadrant III we also obtain positive numbers and in Quadrant IV the numbers will be negative. A positive correlation occurs when the majority of points lie in Quadrants II and III – (i.e. the points follow a pattern that flows upward as we move from left to right). On the other hand, a negative correlation occurs when the majority of points lie in Quadrants I and IV (the points follow a downward pattern as we move to the right). Hopefully this explanation illustrates that the Pearson correlation formula is a reasonable way to express the relationship between two variables. Although the formula for the Pearson correlation is provided, the arithmetic is better done by a computer program - usually a few simple commands are all that is needed.

You interpret a Pearson r the same way as a Spearman r_S. The advantage of a Pearson correlation over a Spearman is that the Spearman must convert the original data to ranks. In that process important information is lost since the rank values only indicate that a given student had a favorable or unfavorable attitude toward math. Had we been able to use an interval scaled variable we would be able to take advantage of how much more favorable an attitude was since interval scale measurements have equal intervals between their scale units. However the two techniques tend to give very similar results. Had the math attitude scale been an interval scale, and we used the Pearson procedure, our r would be +.25 which is very similar to the +.29 obtained with the Spearman method.

Although there is no standard way to interpret r, one way this can be done is given in Table 11-E below.

Table 11-E Interpretation of Correlation Coefficients

Correlation Coefficient	Interpretation
+1.00	Perfect Positive Relationship
+.99 to + .75	Very Strong Positive Relationship
+.74 to + .50	Strong Positive Relationship
+.49 to + .30	Moderate Positive Relationship
+.29 to + .10	Weak Positive Relationship
+.09 to + .01	Trivial Positive Relationship
0.00	No Relationship
-.01 to -.09	Trivial Negative Relationship
-.10 to -.29	Weak Negative Relationship
-.30 to -.49	Moderate Negative Relationship
-.50 to -.74	Strong Negative Relationship
-.75 to -.99	Very Strong Negative Relationship
-1.00	Perfect negative Relationship

The choice of the Pearson or Spearman technique depends on the type of variables you have. However, as mentioned in chapter 6, the category a scale is placed in is not always rigidly fixed. There are instances where an ordinal scale has properties that allow it to be used as if it were an interval scale. However, that judgment is often best made by a professional statistician. If you run into a situation where you have an ordinal variable and believe the results of a parametric procedure would have value, it would be reasonable to discuss the situation with a statistician to see if your ordinal variable may be treated as an interval variable.

Regardless of the size of a correlation, it is critical to appreciate the fact that a correlation, even a very strong one, does not necessarily mean the two factors are related in a cause and effect fashion. Correlation is a measure of **association** and association is not the same as causation. For example, a researcher may find that there is a positive correlation between years of college and conservative political beliefs. Such a correlation would not mean a college education causes conservative attitudes − in fact, it is more likely that college instruction emphasizes liberal rather than conservative viewpoints. The correlation between education and conservatism may be due to a third variable that is related to both variables and is behind the positive correlation. For example, more students may attend college because of their financial capability and conservatism may also appeal to wealthy people. It is the fact that both of the correlated factors are related to financial wealth that creates their positive correlation.

One last observation - to determine the correlation between two variables, you need data for <u>both variables</u> for each individual in your data base. If either variable is missing the individual must be rejected from the analysis. There are methods to estimate missing data, but a review of those methods is better reserved for a more advanced course.

Chi Square Test

In an earlier chapter we used the Z Test to compare two proportions. We can obtain an equivalent result by using the Chi Square Test. Instead of examining proportions, the **Chi Square Test** compares the frequencies that make up the proportion. The main difference between the two techniques is that the Chi Square Test does not require the assumption of normality. If you recall, the Z procedures for individual values required that the variable be normally distributed, but when the test involved group means the central limit theorem lessened the need for the normality assumption. Nevertheless, because the Chi Square Test does not require the normality assumption it is placed in the non-parametric class of tests.

Two By Two Table − Comparing Two Groups

The symbol for the Chi Square Test is X^2. To illustrate the use of this test method, assume we want to determine the probability that two groups have the same proportion of members who behave in a certain fashion. Our example will look at the difference in voting patterns for Democrats and Republicans. More specifically we will compare the proportion of Democrats and Republicans who voted in the last election in a hypothetical community. The data base we will use is a random sample of registered voters in the county. From the records we select 30 Democrats and 70 Republicans for our sample. For the selected individuals we review the voting record to see if they voted in the last election. Assume the count is done, but we are only told that of the 100 people, 40 did not vote and 60 voted in the last election. At this point we are not given the party breakdown.

We can construct a table of what we know. That table looks like this - we have marginal totals (i.e. 40, 60 and 30, 70) for the rows and columns, but the interior cells are blank.

Voted	Dem	Rep	Total
No			40
Yes			60
Total	30	70	100

Now if the null hypothesis is true, what frequencies would there be in the interior cells? To find this out we will first determine the proportion of people for each cell of the table and then the expected number for those cells.

We start with the assumption that there is no difference in the voting intensity of Democrats and Republicans and ask, under that assumption, what would be the <u>expected proportion</u> of Democrats who voted? We can then ask comparable questions for each of the 3 other cells in our table. We will then record our results in a table we'll call the Expected Proportionality Table.

If it doesn't matter what political party you belong to, your likelihood of not voting is based on the overall proportion who did not vote. We can see from the marginal row total that 40 of the 100 individuals in our sample did not vote. Hence the proportion of Democrats who would be non-voters is .40 (40/100). We can use the identical reasoning to determine the proportion of Republicans who would not be expected to vote and their proportion of non-voters is also .40. The identical proportion makes sense since we are working under the assumption there is no difference between the two groups.

We next determine the proportion of people in each party who would be expected to vote, under the assumption that there is no difference between Democrats and Republicans. There are a total of 60 people who voted out of 100 so we would predict that .60 proportion of the people in each party would have voted, if there were no difference between the parities. We can place these values in an Expected Proportionality Table (See below).

Table 11-F
Expected Proportionality Table

Voted	Dem.	Rep.
No	0.40	0.40
Yes	0.60	0.60

The proportionality table indicates that if there is no difference between groups, then the proportion of Democrats and the proportion of Republicans who do not vote is .40 and the proportion who do vote is .60.

Given these proportions we next need to find the <u>expected number</u> of Democrats and Republicans who would vote or not vote. From the table that gave the marginal totals, note that we have 30 Democrats. To determine the number of Democrats who did not vote, on the assumption their likelihood of voting is .40, we simply multiply the 30 by our .40 proportion. The result is shown below.

$$\text{Democrats not voting} = 30 * .40 = 12$$

We would expect, under the null hypothesis, that 12 Democrats did not vote. Note that we represented our answer as a whole number because it turned out to be exactly 12 with no decimal places. In calculating expected frequencies it is best to express your answer to at least one and preferably two decimal places. If you round to the nearest whole number you can get imprecise figures for your final result.

We obtained the frequencies for our table in two steps - determining the expected proportion and then the expected number. There is also a single mathematical formula that is frequently used to directly obtain the frequencies - it is:

$$\text{Expected Cell Count} = \frac{\text{Row Marginal Total} * \text{Column Marginal Total}}{\text{Grand Total}}$$

Essentially this is the same formula as the formula we used above. For the cell "Democrats not voting" we obtain:

$$\text{Expected Cell Count} = \frac{30 * 40}{100} = 12$$

We can use this formula to determine the expected number of Republicans who did not vote. Try to do this in the space below.

For your answer you should have gotten the number 28. Your calculation should have been:

$$\text{Republicans not voting} = \frac{70*40}{100} = \frac{2800}{100} = 28$$

Using the expected cell count formula, we would find that the expected number of Democrats who voted is 18 and the expected number of Republicans who voted is 42. We can now fill in the interior cells of the table and we'll give this table a name - we'll call it the Expected Frequency Table. In a table such as ours (sometimes it is called a 2 by 2 table because there are 2 columns and 2 rows) we really only need to calculate one cell frequency and we can get the other cell counts by subtraction. If we only had the cell count for the expected number of Democrats who did not vote, we can subtract that value from the total number who did not vote (that number is the marginal row total of 40). If we do the subtraction we obtain the expected number of Republicans who did not vote (40-12 = 28). Also the marginal total for the total number of Democrats is 30 so we can subtract 12 from this number to find out the expected number of Democrats who voted. It is 18 (30-12 = 18). Since there are a total of 70 Republicans and the expected number who did not vote is 28, the expected number who voted is 42 (70-28 = 42). However, even though you need to calculate the expected value of only one interior cell, it is wise to calculate the value for two cells using the expected cell count formula. In that way you can reduce the chance you made a calculation error in the first use of the formula. Our Expected Frequency Table looks like this:

Table 11-G
Expected Frequency Table

Voted	Dem	Rep	Total
No	12	28	40
Yes	18	42	60
Total	30	70	100

The Expected Frequency Table tell us what would happen if the null hypothesis were true. We contrast this expectation with what actually happened. We now will share the observed voting patterns in the next table which we will term the Observed Frequency Table.

Table 11-H
Observed Frequency Table

Voted	Dem	Rep	Total
No	16	24	40
Yes	14	46	60
Total	30	70	100

We now compare the expected and observed results using the Chi Square formula. The formula is:

$$X^2 = \frac{\Sigma(f_o - f_e)^2}{f_e}$$

where:

f_o = observed frequency for a cell (table 11-H)

f_e = expected frequency for that cell (table 11-G)

The Σ sign indicates we should sum the results of the calculation for each cell - in our case for 4 cells.

We will calculate the value for the upper left hand cell (the cell for Democrats not voting). The result is shown below:

$$X^2_{cell\ 1} = \frac{(16-12)^2}{12} = \frac{4^2}{12} = \frac{16}{12} = 1.33$$

Observe that we recorded our answer with 2 decimal places - as noted above you should always add a little more detail to your calculation to avoid a lack of precision in your final answer. For each of the cells of our table we repeat the above calculation. What do you get when you use the Chi Square formula for the Republicans who did not vote cell?

You should have 0.57 as your answer. The calculation is:

$$X^2_{cell\ 2} = \frac{(24-28)^2}{28} = \frac{-4^2}{28} = \frac{16}{28} = 0.57$$

We continue using the formula to obtain the values for the cells in row 2. Those values are 0.89 and 0.38 for Democrats and Republicans respectively.

As indicated by the overall formula we now sum (Σ) the 4 values to get our X^2 value. X^2 is 3.17 (1.33+0.57+0.89+0.38). The X^2 value is a test statistics just as Z, t and F were. We find whether the probability associated with a X^2 value of 3.17 is statistically significant from another table constructed by statisticians. It is called a Chi Square Table and the table is given in the Appendix - Table 5.

Review the table and note that there are no negative values for X^2 so our rejection region is always in the upper tail. If you go back to the X^2 formula you will realize that, if the null hypothesis is true, X^2 will technically be 0 since the expected and observed frequencies should be the same. However, we know that due to sampling variation we can have some deviation in our observed frequencies even if there is no difference between Democrats and Republicans.

To use the Chi Square Table we need to know the degrees of freedom. The degrees of freedom are found by using the following formula:

$$Degrees \ of \ Freedom = (No. \ Rows - 1) * (No. \ Columns - 1)$$

In our example we have 2 rows and 2 columns so our df is:

$$Degrees \ of \ Freedom = (2 - 1) * (2 - 1) = 1 * 1 = 1$$

There is 1 degree of freedom for our X^2. We again need to set a level of significance (i.e. the Alpha Risk) and in this case we will follow convention and use the .05 level. We can now use the Critical Value Method and from Table 5 of the Appendix, we can find the critical X^2 value. If the computed X^2 value equals or exceeds the critical X^2 value, then we have statistical significance. If the computed value is less than the critical X^2 value, then we must conclude that the evidence isn't strong enough to claim a difference in the voting intensity of Republicans and Democrats. To find the critical X^2 value we enter the Chi Square table at the row for df = 1. Along the top of the table the alpha levels are listed for two tail tests. Since we are using a .05 level we find the .05 column and for the df = 1 row we read off the X^2 value. In our case it is 3.841 - we reject the null hypothesis and declare statistical significance if our calculated X^2 is greater than or equal to 3.841. The calculated X^2 is only 3.17 so we do not have a statistically significant difference between Democrats and Republican for voting intensity.

You can also do the Chi Square Test for frequency tables larger than 2 by 2. The same steps are involved - you determine the expected frequency for each interior cell and calculate the X^2 with the same formula used above. In looking up the critical X^2 value remember you need to use the df formula shown above to calculate the correct df number.

The Chi Square Test looks for any difference between the proportions for the two groups being compared and hence the Chi Square Table cites critical values for a two tail test. However, if you wanted to do a one tail test you still can use the Chi Square Table. If, for example, you wanted to do a one tail test at the .05 level of significance, you could use the column for the .10 level of significance to find your critical value. Remember, in a one tail test we are placing all our risk in one tail rather than spreading it equally among two tails. Therefore the .05 level for a one tail test has the same critical values as a two tail test for the .10 level of significance. If you wanted to do a one tail test with alpha at .05 and there was 1 degree of freedom you would look in the Chi Square Table using the column for the .10 level and the row for 1 df. Your critical value would then be 2.706. Just as in the t test, you need to be careful though to make sure the difference in the two groups being compared is consistent with the alternative hypothesis.

One Way Table - Comparing One Group to a Theoretical Distribution

The Chi Square Test can also be used to see if the observed frequency distribution is consistent with a theoretical one. In this case the theoretical values become our expected frequencies.

In horse racing, bettors use different systems and a fan may wish to see if there is merit in the post position system. The assumption is that horses closest to the rail or post have an advantage (i.e. they have less distance to run). Before betting on such a system one can determine if there is merit in it. A person could look over recent results and see if the horses closest to rail are more likely to win. A large number of races involving 8 horses could be reviewed where the horse closest to the rail is in post position 1 and the one furthest away is in post position 8. Results from 136 races could look like this:

Table 11-I

Post Positions	1	2	3	4	5	6	7	8
No. of Wins	21	19	18	25	17	10	15	11

Under the null hypothesis each post position would have an equal chance of winning the race - i.e. each post position would have an equal number of winners. We need to convert that proposition into expected frequencies. Since the expected frequency for each post position has an equal chance of winning, each position has an expected frequency of 17 (136/8=17). We now have our expected frequency and can present our data in a one-way table (see the first two rows of Table 11-J). We can also calculate the X^2 value (see last three rows of Table 11-J) which gives us a calculated X^2 of 10.22..

Table 11-J

Row No.	Post Position	1	2	3	4	5	6	7	8	Total
1	Observed Frequency	21	19	18	25	17	10	15	11	136
2	Expected frequency	17	17	17	17	17	17	17	17	136
3	Observed-Expected	4	2	1	8	0	-7	-2	-6	0
4	$(\text{Observed} - \text{Expected})^2$	16	4	1	64	0	49	4	36	174
5	$\dfrac{(\text{Observed} - \text{Expected})^2}{\text{Expected}}$	0.94	0.24	0.06	3.76	0	2.88	0.24	2.12	10.22

The df formula for a one-way table is:

Degrees of freedom = Number of cells - 1

We have 8 cells in our table so df is 7 (8-1). We can find our critical X^2 value from the X^2 table (Table 5 of the Appendix). If we do our test using an alpha risk of .05, the critical X^2 is 14.1. Since our calculated X^2 is 10.22, and it is less than the critical X^2 we can not reject the null hypothesis. Therefore we conclude that there is not enough evidence to claim that there is a difference in the proportion of times a horse wins when stating at different post positions.

Using Chi Square as a Test for Independence

We could have also set up this problem with a slightly different hypothesis. We could have started with a null hypothesis that says there is no relationship between post position and winning the race. In other words, post position and winning the race are independent events. The term **"Test for Independence"** is used when we structure our hypothesis like this. The alternative hypothesis is that there is a relationship between post position and winning a race. Note that in a X^2 test for independence we are stating the hypotheses in a different manner than when we do a test of proportions. The mathematics are the same - only the hypothesis is different. Had we done a X^2 Test for independence rather than proportions we would again reject the null hypothesis and, in this case, conclude that "Post position and winning a race are independent events." In our first X^2 example about voting intensity among Democrats and Republicans we could also have done a test of independence. Because that result also was not statistically significant we would have concluded that party affiliation and the likelihood of voting are independent events.

Chi Square and Sample Size

Whether you are using the Chi Square Test for a two-way table, a one-way table or as a test of independence, a note of caution concerning sample size is in order. When you are dealing with small expected frequencies you need to make what is called a "continuity correction". In the X^2 formula, the correction is made by subtracting .5 from each (fo-fe) difference before you divide by fe. Furthermore if you have any table with expected frequencies less than .5 you should not do the X^2 Test. An alternative test, the **Fisher Exact Test,** would be used in that case. We will not present that test, but most statistical computer programs will include the Fisher Exact Test.

Wilcoxon Rank Sum Test

A comparable non-parametric test to the Two Sample t Test is the **Wilcoxon Rank Sum Test**. The name is informative because it tells us something about what the test will be based on - it will use the sum of ranks to create the test statistics, just as the Spearman Rank Correlation procedure relied on ranks to determine a correlation coefficient. Also be aware that there is a test called the Wilcoxon Paired Ranks Test and it should not be confused with the Wilcoxon Rank Sum Test.

The Wilcoxon Rank Sum Test is a very suitable test when one is using variables based on ordinal measurements. To perform the Wilcoxon Rank Sum Test, you replace the observations in each sample with the rank for that observation. Therefore, we combine the observations from the two samples and the rank of 1 is given to the smallest value. A rank of 2 is assigned to the next smallest value regardless of what sample the observation belongs to. You continue assigning ranks until all observations in the combined data set have a rank.

A T score is then calculated for each sample. For our discussion, we will identify the samples as Sample 1 and Sample 2. A T_1 score is the total sum of the ranks assigned to the observations in Sample 1 and a T_2 score is the total sum of the ranks for the observations in Sample 2. Do not confuse this T (upper case) with the t (lower case) Test discussed in chapter 9. There is no connection between the two expressions.

Two additional calculations are called for. We need to calculate the mean of the ranks for each sample. We will use the designation μ_1 for the mean in Sample 1 and μ_2 for the mean in Sample 2. The formula for μ_1 is:

$$\mu_1 = \frac{N_1 * (N_1 + N_2 + 1)}{2}$$

and for μ_2, the formula is:

$$\mu_2 = \frac{N_2 * (N_1 + N_2 + 1)}{2}$$

We also need the standard deviation (σ) for the complete set of ranks. The formula is:

$$\sigma = \sqrt{\frac{N_1 * N_2 * (N_1 + N_2 + 1)}{12}}$$

The last step is to calculate a Z value using the formula:

$$Z = \frac{T_1 - \mu_1}{\sigma} \text{ or } Z = \frac{T_2 - \mu_2}{\sigma}$$

For large sample sizes, the statistic T is normally distributed so that the creation of a Z value is considered appropriate. However, if the sample sizes are 10 or less you should not use the Z statistic. There are separate tables for situations in which the samples sizes are 10 or less so the Wilcoxon Rank sum Test can still be used, but not using the Z value. In our work we will only work with relatively large sample sizes so the Z statistic will be satisfactory. We already know how to interpret a Z statistic so we have all the background we need to perform a Wilcoxon Rank Sum Test.

To demonstrate the use of the test we'll use the following data set:

Table 11-K

East Coast		West Coast	
2.21	1.88	2.91	2.50
2.22	2.20	2.61	1.95
2.60	1.58	2.97	3.28
2.48	2.40	2.74	1.59
2.19	1.96	3.82	1.76
2.81	1.78	2.49	3.67

The numbers in the table are yield rates for savings accounts at a random sample of banks on the east and west coasts. Normally one would use a parametric test on the data because the variable is based on the interval scale. However, a review of the data showed a non-normal distribution - so to avoid violating the normality assumption a non-parametric analysis was conducted.

There was an interest in seeing if the rates on the two coasts were similar or different so a two tail test situation was chosen. The .05 level of significance was also felt to be appropriate. We will refer to the east coast banks as sample 1 and the west coast banks as sample 2.

The data first needed to be pooled and ranks assigned to the combined data base. Table 11-L on the next page shows the ranks assigned to the set of 24 banks and it also identifies the T_1 and T_2 values.

T_1 is the total of the ranks assigned to the east coast banks and T_2 is the total of the ranks assigned to the west coast banks. If there are ties then the average rank is assigned to each observation. Summing the east coast ranks gives us a total of 115 and the sum of the west coast ranks is 185. You can check to make sure you have done the ranking process properly by adding T_1 and T_2 and comparing that sum to:

$$\frac{(N_1 + N_{2)}) * (N_1 + N_2 + 1)}{2}$$

For our example $N_1 = 12$ and $N_2 = 12$ so the value we need to check our total sum of ranks against is:

$$\frac{(12 + 12) * (12 + 12 + 1)}{2} = \frac{24 * 25}{2} = 300$$

Table 11-L

Location	Yield	Rank	East Ranks	West Ranks
East	1.58	1	1	
West	1.59	2		2
West	1.76	3		3
East	1.78	4	4	
East	1.88	5	5	
West	1.95	6		6
East	1.96	7	7	
East	2.19	8	8	
East	2.20	9	9	
East	2.21	10	10	
East	2.22	11	11	
East	2.40	12	12	
East	2.48	13	13	
West	2.49	14		14
West	2.50	15		15
East	2.60	16	16	
West	2.61	17		17
West	2.74	18		18
East	2.81	19	19	
West	2.91	20		20
West	2.97	21		21
West	3.28	22		22
West	3.67	23		23
West	3.82	24		24
Totals $(T_1 \& T_2)$			$115(T_1)$	$185(T_2)$

Our sum of ranks for the two samples is also 300 (115+185=300) so we apparently made no error in assigning ranks. Of course, there could have been off setting errors but that is pretty unlikely.

We now calculate μ and σ. We only need to calculate either μ_1 or μ_2 and it doesn't matter which one is computed. In our example we will use the μ for sample 2. The value for μ_2 is:

$$\mu_2 = \frac{12 * (12 + 12 + 1)}{2} = 150$$

and for σ the calculation is:

$$\sigma = \sqrt{\frac{12 * 12 * (12 + 12 + 1)}{12}} = \sqrt{12 * 25} = \sqrt{300} = 17.3$$

The last step is to calculate Z and we compute the following value for Z:

$$Z = \frac{185 - 150}{17.3} = +2.02$$

We now compare the calculated Z value against the critical Z values, which for a two tail test at the .05 level of significance, is ±1.96. Our calculated Z (+2.02) is more extreme than the critical Z value (+1.96) so we conclude that there is a statistically significant difference for yield rates between east and west coast banks.

The formulae for the Wilcoxon Rank Sum Test looks unusual but, because we have equal sample sizes and no ties in our example, it is possible to show that the Wilcoxon formulae are equivalent to the Z Test formula we used in chapter 9. The Z formula we are referring to is the One Sample Z Test:

$$Z = \frac{\overline{X} - \mu}{SE}$$

To apply this formula we use the ranks as our data set, and simply calculate the appropriate statistics (μ, $\overline{X}$ and SE) from that set. The mean of the full set is μ (300/24 = 12.5). The mean of ($\overline{X}$) for sample 2 is 15.417 (185/12 = 15.417). The SE is calculated in the usual fashion where the sample standard deviation (7.07) is divided by the square root of N (24). The SE turns out to be 1.443. You then apply the Z formula.

$$Z = \frac{15.417 - 12.5}{1.443} = \frac{2.917}{1.443} = 2.02$$

The Z value we obtain is 2.02, the same as the one we obtained with the Wilcoxson formula. The point of this last digression was to show that although the Wilcoxon Rank Sum formulae looked unusual, they produce perfectly reasonable results which are consistent with the results we would expect had we used the more conventional Z formulae. However, when the sample sizes are unequal, or there are ties among the ranks, it is necessary to use the Wilcoxon Rank Sum formulae because the conventional Z formula will not produce an accurate result. The adjustments needed to use the conventional Z Test are mathematically tedious and it is much easier to simply apply the Wilcoxon Rank Sum formulae.

We will not review the Wilcoxon Paired Rank Test, but want to emphasize that it is the non-parametric alternative to the paired t test. Furthermore, the Kruskal-Wallis Test is the non-parametric alternative to the ANOVA. These tests procedure can be found in many statistical texts that focus on non-parametric procedures.

Choice of a Statistical Test/Procedure

Generally you must use the test/procedure that matches the type of measurement variable you are using. The table below summarizes the tests we have reviewed by the type of measurement variable involved.

Type of Variable	Test/Procedure to Use
Interval	Z Test, t Test, ANOVA, Pearson r
Ordinal	Wilcoxon Rank Sum Test, Spearman r
Nominal	Chi Square Test

As noted in chapter 6, the type of measurement can be viewed in a hierarical fashion with interval scaled variables at the top and variables that are nominal at the bottom of the ranking. It is possible, but inadvisable to treat a variable that is classified at a higher level on the measurement hierarchy as a lower level variable. For example you could use the Wilcoxon Rank Sum Test on interval level data, but as previously noted the non-parametric procedures are designed for ordinal and nominal level variables and they are less efficient than the methods developed for interval level data.

Summary

In this chapter we learned that statistical procedures can be divided into parametric and non-parametric categories. The non-parametric methods we studied included the Chi Square Test for nominal variables and the Wilcoxon Rank Sum Test for ordinal variables. We also learned that the strength of the relationship of two variables can be demonstrated by the correlation coefficient. For ordinal variables, the Spearman Rank correlation method is used and for interval variables the Pearson correlation method is used. A decision tree is provided on the next page to help chose the appropriate test described in this chapter. Note that the tree includes reference to the decision tree for interval measurements described in Chapter 10. To use the tree you start with the "Type of Problem" diamond. As an illustration, if you wanted to find a difference between groups in which the date were based on a nominal variable. the tree indicates that you use the Chi Square Test.

Decision Tree for Selecting Statistical Procedures for Different Types of Variables

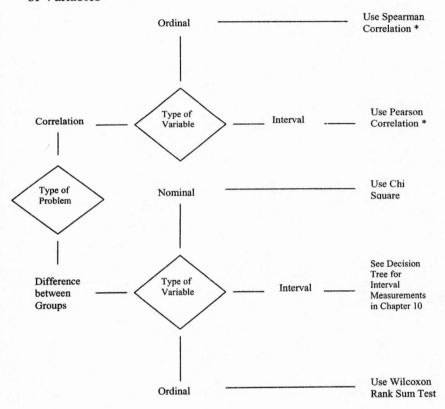

* The Spearman procedure is used when one variable is ordinal and the other is interval or both variables are ordinal. The Pearson procedure is used when both variables are interval.

Statistical Methods for Researchers

PROBLEMS
Unless otherwise specified assume you are doing two tail tests with an Alpha Risk of .05.

1. Do a Chi Square Test comparing Democrats to Republicans on their past voting records using the following results:
>   Democrats not voted: 43
>   Democrats voted: 43
>   Republicans not voted: 33
>   Republicans voted: 61

   What is your conclusion?

2. At your place of work a survey is taken among management and non-management personnel in respect to whether or not they experience job satisfaction in their jobs. The results are:
>   Management with satisfaction: 72
>   Management without satisfaction: 32
>   Non-Management with satisfaction: 158
>   Non-Management without satisfaction: 108

   a. If you were doing a test of proportions, what is your conclusion?
   b. If you were doing a test of independence what is your conclusion?

3. The religious beliefs of 140 residents in urban areas and 140 residents in suburban settings were obtained. The findings are shown below:
>   Urban with Conservative religious beliefs: 63
>   Urban with Moderate religious beliefs: 41
>   Urban with Liberal religious beliefs: 36
>   Suburban with Conservative religious beliefs: 59
>   Suburban with Moderate religious beliefs: 59
>   Suburban with Liberal religious beliefs: 22

   Is there a statistically significant difference in the results?

4. A genetic trait is caused by a recessive gene which has 1 chance in 4 of occurring. At the only hospital in your community a sample of 86 newborn babies showed that 28 of the newborns had the recessive gene. The Health Commissioner wants to know if the sample evidence is sufficient to declare that the community has an unusual proportion of people with the recessive gene. He asks that you use a .10 alpha risk and let him know if there is a statistically significant difference. What can you tell him?

5. Two employment consultants review the resumes of 11 MPA graduates and 14 MBA graduates regarding their appeal to health care organizations. The two ratings were pooled and the sum of the ratings are presented below:

MPA		MBA	
10	30	22	13
6	16	17	21
25	14	9	21
27	26	23	6
15		7	27
12		24	21
18		19	3

Using the Wilcoxon Rank Sum Test determine if there a statistically significant difference between the groups?

6. Contestants in a singing competition were rated as High, Medium or Low by two independent evaluators. Their ratings are given below.

Contestant	Evaluator A	Evaluator B
A	High	Medium
B	Medium	Low
C	Low	Low
D	Low	Low
E	High	High
F	Medium	Medium

   a. What is the correlation between the two judges?
   b. What kind of reliability in being measured in this situation?

7. Thirteen instructors teach both graduate and undergraduate courses at the local university. The Dean of the undergraduate school and the Dean of the graduate school rated each teacher. The <u>rank</u> assigned to the instructors by the deans is given below. How strong is the relationship between the deans' rankings?

Instructor	Undergraduate Dean Ranking	Graduate Dean Ranking
Alpha	7	2
Beta	8	1
Chi	3	9
Delta	11	7
Epsilon	5	4
Gamma	6	11
Iota	1	10
Mu	13	13
Omega	2	8
Sigma	9	3
Tau, B	10	6
Tau, F	4	12
Zeta	12	5

8. What is the difference between parametric and non-parametric tests?

9. A poll indicates that 312 citizens among the 485 surveyed support their town's acquisition of land for a new park. Because of the large sample size, the sample standard deviation may be used as the population standard deviation in this problem.
   a. What is the 80% confidence interval for the percent of town residents supporting the town park?
   b. Do you think the measure will pass when it's presented to all the voters? Why?

10. A study is done comparing a new drug against the standard treatment. A total of 31 subjects were given the new drug. The new drug caused a better response ($\overline{X} = 127.0$) compared to the historical average for the standard treatment which was 121.6. The standard deviation for those receiving the new drug is 18.
    a. What is the calculated t value?
    b. If a two tail test is done using an alpha risk of .05, what would you conclude?
    c. If the researchers are only interested in knowing whether the new drug is better than the standard drug what is the conclusion using a .05 alpha risk?

11. An ANOVA with an Alpha Risk of .05 produced the following analysis:

Anova: Single Factor

SUMMARY

Groups	Count	Sum	Average	Variance
A	14	617.0368	44.07406	71.52217
B	11	498.2203	45.29276	80.34314
C	15	683.4849	45.56566	64.63323
D	12	466.3196	38.85996	49.49484

ANOVA

Source of Variation	SS	df	MS	F	P-value	F crit
Between Groups	362.1506	3	120.7169	1.820694	0.15598	2.7981
Within Groups	3182.528	48	66.30267			
Total	3544.679	51				

   a. What group has the highest standard deviation?
   b. Is there a statistically significant difference between the groups?
   c. What is the probability that these results are due to random variation?
   d. If you used an Alpha Risk greater than .05, would you be more likely or less likely to declare statistical significance?

12. A survey is to be taken and the pollsters want to be 95 per cent certain that the margin of error is not greater than ±3.5 per cent. There is no background information on what kind or response to expect to the questions they will ask, but they expect only 1 of 7 people who are sent the questionnaire to fill it out and return it. To how many people should they plan to send questionnaires?

13. Based on a sample of 41 middle class investors, it is reported that their average performance in the stock market is a gain of $158 per $1,000 investment. The sample standard deviation is $205.
    a. What is the 90% confidence interval for the average gain for all middle class investors?
    b. Is there a statistically significant difference (2 tail with alpha = .05) between the sample mean and an average gain of $125?

Excel Instruction

There is no function in Excel to determine the Spearman Rank Correlation Coefficient, but the arithmetic is straightforward and it can be done manually with little trouble in Excel.

Excel has a function for the Pearson correlation coefficient. From the Main Menu click on the Function icon and then select "Statistical" from the drop down menu. From the "Function Name" list, select either "PEARSON" or "CORREL" - the functions are essentially identical. Click on "OK" and you will be presented with the following dialog box.

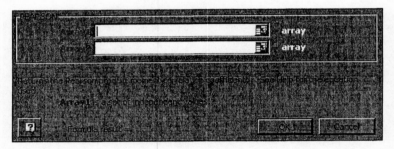

Array1 refers to the location of one of the variables and Array2 refers to the location of the other variable. If variable 1 is in cells D3 through D14 and variable 2 is in cells E3 through E14 you enter D3:D14 for array1 and E3:E14 for array2. Excel returns the correlation coefficient.

In doing a Chi Square problem, Excel will do most of the work, but you also have to make a contribution. You need to enter the observed frequency counts in a worksheet and you also must calculate the expected frequencies. The table of observed frequencies for the problem we worked in the text is reproduced below with the addition of labels for the rows and columns of a worksheet.

Row	A	B	C	D
			Column	
15	Observed Frequency Table			
16	Voted	Dem	Rep	Total
17	No	16	24	40
18	Yes	14	46	60
19	Total	30	70	100

To calculate the expected frequencies we would simply need to write the expected cell count formula in an appropriate portion of the worksheet. Let's place the expected cell frequencies for the category "Democrats not voting" in cell B20. We therefore would enter the following expression in cell B20: "=B19*D17/D19" and the result (12) would appear in that cell. We could also calculate the expected frequencies for the other 3 cells and place these results in cells B21, C20 and C21 and by doing that our worksheet would now look like this:

		Column		
Row	A	B	C	D
15		Observed Frequency Table		
16	Voted	Dem	Rep	Total
17	No	16	24	40
18	Yes	14	46	60
19	Total	30	70	100
20		12	28	
21		18	42	

Our contribution is now done and we can let Excel do the rest of the calculations. Go to a cell where you want the probability value for our Chi Square Test to appear - Cell D21 would do nicely. Select the Function icon and "Statistical". From the "Function Name" list choose "CHITEST". After clicking 'OK", the following dialog box appears.

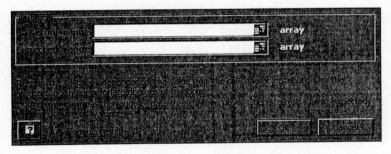

The "Actual range" refers to our location of the observed frequencies so enter "B17:C18" in that box. The "Expected range" refers to the location of our expected frequencies so we enter "B20:C21" in the second box. When we hit "OK" the two tail probability value for our example, .0748 appears in Cell D21. If we were testing at the .05 level of significance we would reject the null hypothesis since .07 is greater than .05. Had we wanted to know the Chi Square value for this problem we can use a function called "CHIINV", which is another Excel function icon. The dialog box for CHIINV is shown below.

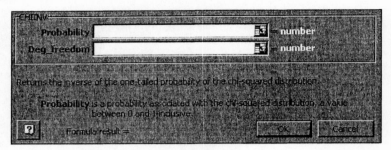

The "Probability" entry is our calculated probability value so we enter "D21" or ".0748" in the first box. The degrees of freedom for our 2 x 2 table is 1 so that number is entered into the second box, "Deg freedom". After hitting "OK" Excel returns our calculated X^2 value of 3.1746. We also do not need to use the X^2 table to find a critical X^2 value because Excel can furnish that number for us. Move to a new cell and again select the "CHIINV" function, but this time the probability entry will be the Alpha Risk. Following convention we will use ".05" so enter that value in the Probability box. The "Deg freedom" is still 1 so we enter that number in box 2. After clicking on "OK", Excel return the critical X^2 value of 3.8415. Since our calculated X^2 value (3.17) is less than the critical X^2 value (3.84) we conclude that the differences are not statistically significant at the .05 level.

There is one additional Chi Square function available in Excel. If you have calculated a X^2 and want to know the probability associated with the result you can use the function "CHIDIST". The dialog box for "CHIDIST" is:

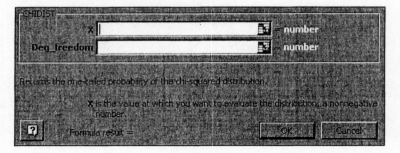

The "X" is your X^2 value. In our example X^2 is 3.17 so we would enter that value in box 1. The "Deg freedom" entry is again 1 so that number would be entered in the second box.. Excel returns the probably value of .0748 - the likelihood that our observed frequencies could have occurred by chance.

The last problem we will work on is the Wilcoxon Rank Sum Test. Unfortunately most of the work needs to be done by us, but there is an Excel feature that can reduce some of the work and it also has applications to other situation so it's worth knowing. In a Wilcoxon Test a major step is assigning and summing the ranks. As you may recall we calculate a T1 and T2 value, which are the sum of the ranks for sample 1 and sample 2 respectively. We will show how this can be done efficiently in Excel.

In our example, we compared East and West coast banks. The example we will work with assumes we used labels of "Location" and "Yield" in Cells A1 and B1 and then entered the raw data in those columns. We begin by sorting the data set. You select both the Location variable (i.e. East or West coast bank) and the Yield variable for the sort. Yield is then used as the primary sort variable. Specify an ascending sort order. When you have finished, the beginning of the worksheet should look like that shown in columns A and B below.

Row	A	B	C	D	E
	Location	Yield	Rank	East Ranks	West Ranks
1					
2	East	1.58	1	1	
3	West	1.59	2		2
4	West	1.76	3		3
5	East	1.78	4	4	
6	East	1.88	5	5	
7	West	1.95	6		6

(Column header spans A–E)

We haven't shown all the entries, but enough so you can see what the result of the sort will look like. We next add a column C to our table which we will use to enter the ranks from 1 to the last number which in our case would be 25. Remember, instead of manually entering 1, 2, 3 etc., you can use the Edit/Fill/Series commands to assign those ranks quickly. However, be aware that if there are ties, you will need to calculate a mean and substitute that value for the automatically assigned rank.

After the ranks are assigned we need to obtain the rank sum for the two locations (East and West). We begin this operation by assigning column D for the East ranks and column E for the West Ranks. To have only the ranks for the east banks entered in column D and the ranks for the west banks entered in column E, we use a new Excel function called "IF". The format for the "If" function is:

= If(State a condition, Action if true, Action if false)

Note that there are 3 parts to this function, separated by commas. Part 1, "condition", refers to the content of a referenced cell. By "action" we mean what value or words do we want to appear in the designated cell (i.e. the cell containing the If function). If there is a match you enter the value/words you want in part 2. If there is no match you enter the value/words you want in part 3.

We highlight cell D2 (our designated cell) and note that the "condition" we want to test is whether the rank of 1 belongs to an east or west coast bank. We determine this by noting whether the entry in Cell A2 is "East". If it is "East" then the rank of 1 should be assigned to the East column and specifically to cell D2. Hence our "condition" is whether Cell A2 contains the entry "East". We can now begin to write the first part of the If function which will be:

$$=If(A2="East",$$

Note that in the If function, Excel requires names to be in quotation marks (i.e. "East") and you complete the condition statement with a comma.

We next must instruct Excel what to do if the condition is true. We want all the ranks for the east locations in column D so if the condition is true we want the rank, which is in cell C2, to be copied into cell D2. The second part of the If function is, therefore, simply "C2" What we are saying is that if the condition is true the action we want taken is to place the value of Cell C2 into D2. To complete the function we must also indicate the action to be take if the condition is false. In our case we want do not want any value to appear so we enter " " for the last part of the function (i.e. in between the two quotation marks you enter a space) . Our complete expression which is to go into cell D2 is therefore the following:

$$=IF(A2="East",C2," ")$$

In our example, the entry in cell A2 is East so the condition is true and the entry in cell C2, (which is a 1) should appear in cell D2. We can now copy our If function to the other cells in column D and when we do this the numbers which are shown above in column D will appear.

We also want to perform the same operation for the west coast banks so we need another If function. In this case the If function that is written in Cell E2 should read:

$$=IF(A2="West",C2," ")$$

In other words we want the rank recorded in Cell C2, copied to Cell E2 providing the location (Cell A2) is West. In our example the condition is false (the A2 entry is not "West") so we create a space by using the " " expression. The space is treated as a blank or no response in subsequent calculations. We again copy the If function in Cell E2 to all the other cells in Column E. We can use the Σ function to obtain a sum of the ranks for the east coast banks as well as the west coast banks. The rest of the calculations require using mathematical operations we've already learned. When you eventually calculate the Z value, remember that you can use the NORMSINV function, which we studied in chapter 7, to find the critical Z value. If we did a two tail test with an Alpha Risk of .05, you would write: "=NORMSINV(.025)" for the critical Z in the lower rejection region and "=NORMSINV(.975) for the critical Z in the upper rejection region.

Here are a few problems you can use to test your skills using Excel and your data base. Assume the data base is a sample taken from a larger population.

1. Do a Chi Square Test to determine if there is a difference between males and females in respect to advancement potential using an Alpha Risk of .05.

2. Assume your boss believes that intelligence, as reflected in the test scores, are related to an employees' salary - i.e., the higher the test score the higher the salary. Prepare an XY graph for the two variables. Use the Pearson correlation procedure to find the correlation between these two variables.

3. Use the Wilcoxon Rank Sum Test to examine the hypothesis that females are more sensitive than males based on the sensitivity ratings. Use a .05 Alpha Risk.

4. What is the Spearman correlation coefficient between sensitivity levels and the creativity codes?

Chapter 12

Controls, Research Designs and the Statistical Index

Introduction

Controls are methods and activities that are used to reduce the chances that a research study will produce an erroneous result. There are different ways to use controls in experiments. There can be an experimental control group, a control variable or a controlled experimental setting. We will also look at ways we can go about designing our research projects. We will see that there are designs for estimation as well as hypothesis testing. A number of specific designs will be reviewed in terms of their strengths and weaknesses. In addition we will look at how a statistical index is created. This will lead us to learn how to standardize measurements - i.e. to convert variables that are represented in different units of measurements so they can be combined.

Control Group

If a person has an infection we can treat that infection with antibiotics. If the person is cured after treatment we'd like to say it was because of the antibiotics. However, it is also possible that the person would have fought off the infection without the antibiotic. Perhaps the infection was not due to an organism sensitive to the antibiotic. To rule out those cases in which we experience a treatment success, but it is not due to the agent we used, experimenters like to include **control groups** in their studies. In medical experiments a placebo group is a frequently used control group. The placebo will look and taste just like the experimental medication, but it will be composed of inert substances. To decide if the experimental medication is any good the experimenter compares the success rate of the treatment group to that of the placebo control group.

333

Control Variable

We have seen that randomization can not guarantee that there will be no selection bias in an experiment. Variables that can be used to examine comparability between treatment groups are called **control variables**. In addition, to be considered a control variable, the variable has to have the potential to affect the response of the dependent variable. In our antibiotic experiment there are factors that can influence the outcome variable such as duration of disease or the presence of another disease. These concomitant conditions represent our control variables and it is, therefore, important to collect data about these conditions to see if they are represented in an equivalent manner among the treatment groups. If the baseline means or distributions for a control variable are unequal in the comparison groups, the final results could be biased. Statisticians frequently use a mathematical technique (i.e., Analysis of Covariance) to adjust for inequality among the treatment groups for baseline variables. The **Analysis of Covariance** provides an F-Test for determining statistical significance after adjusting for a variable so that the comparison groups are essentially equivalent at the baseline observation. Unfortunately, the details on how to conduct the Analysis of Covariance will have to be found in a more advanced text.

Controlled Experimental Setting

In the antibiotic experiment described above, we would want our test group to be in settings that reduce the chance that extraneous factors could enhance or reduce the chance for a cure. For example, we would want to be sure the treatments (antibiotic or placebo) were given on time and in the proper dose and we would want the subjects to be in a setting that would not expose them to a new infection. We would also want the subjects to have healthy food, ample rest, etc. Providing an environment where these conditions exist, and where the handling of individuals can be controlled so both experimental groups are treated similarly, is what we mean by a **controlled experimental setting**. We want to regulate the setting so factors, other than those intentionally introduced by the experimenter, do not influence the results. Here our concern is with the internal validity of the experiment.

There is a downside to having too tight of a control - we reduce the external validity. If we conducted our antibiotic trial in a strict hospital setting we technically can not say, with certainty, that we'd get the same results had we used a home environment.

Double Blind Study

Another source of bias that can enter a study is based on the behavior of the subjects and evaluators. A **double blind study** requires that the two most important participants in an experiment (the evaluator and the subject) be unaware of the experimental condition being used. If the experimental conditions are administration of an active drug or placebo then the two critical actors are the person making the observations and the subject receiving the treatment. The concern is that if these individuals know what is being given they will react differently than they would if they are unaware of the treatments being used. The purpose of the double blind techniques is to control bias on the part of the subject and the observer.

Bias is not used here to mean intentional favoritism. Knowledge of the treatment being given may cause an observer to inflate the observation for people receiving an active drug and under report any positive response by subjects receiving placebo. Knowledge of the treatment being administered may also prompt subjects to respond in the way that they think they should respond. For example, people who know they are receiving a placebo may claim less benefit than they would if they had the same response on an active drug.

Frequently reference is made to the **Hawthorne Effect** to explain what can go wrong if the subjects in a study know they are involved in an experiment. The Hawthorne study was performed at the Hawthorne Works, a manufacturing facility for the Western Electric company. The study was done by Elton Mayo and his Harvard University colleagues. What happened? In the most important study for our purposes, two experimental groups were formed and the effect of lighting on work related tasks was tested. The experimental group had light increased or decreased to see how the group's work output was affected. For the control group the lighting was held constant. No matter what the experimenters did however, productivity was increased in both the experimental and control groups. Even when the lighting was lowered, although the results were erratic, productivity still went up. The results demonstrated the fact that awareness that one is involved in a study changes behavior. Although one cannot conceal the fact subjects are participating in a trial, the bias can be equalized by not letting the subjects know what treatment group they are in.

Furthermore, the individual making the observations can also be biased. If the evaluator knows that a person is on placebo he or she may, without being fully aware of it, give lower scores or evaluations then if the subject was receiving an active treatment. The evaluator has expectations about what the active treatment and what the placebo should produce, and his or her reporting is influenced by those expectations.

Most people can understand the power of expected outcomes. In a classic example, teachers were told that some of their classes had students who were the best and the brightest in the school. For other classes, the teachers were told that the students were academically poor and among the worst students in the school. In actuality the two groups of students were a mix of bright, average and poor students. However, the teachers gave higher grades to the classes they believed contained the bright students and lower grades to classes that they believed consisted of poor students.

Unfortunately, it is not always possible to perform double blind studies. Think of a study to test a new surgical technique versus the current surgical method. You can not conceal the surgical techniques from the surgeon. Furthermore, it could be unethical to not tell the patient what surgical procedure would be applied. As a second example, if one wants to test a revised training program against the current program, it may be all but impossible to keep the fact that two programs are being used from the participants and those doing the evaluation.

Research Designs

The way we are going to administer treatments to our subjects, and the time pattern we use to make our measurements of the dependent variable, are what research design is all about. We have a surprisingly large number of ways to design a study. There is no commonly agreed to way to classify study designs. The scheme we will use in this text is to first divide designs into two main categories: 1] estimation designs and 2] hypothesis testing designs.

Estimation Designs

Estimation designs are usually associated with survey research. The two most popular approaches are the cross sectional and the longitudinal study.

CROSS SECTIONAL STUDY. In a cross sectional plan all the data are collected at one time. "At one time" does not necessarily mean exactly at the same time, but it means within a set time frame - perhaps a matter of days or weeks. For example, if we wanted to know the prevalence of a certain asymptomatic virus we could do health exams on a random sample of people over a two or three day period in one county. In that time frame, all our selected subjects would be tested for the presence or absence of the virus and we would use the sample results to estimate the population percentages.

LONGITUDINAL STUDY. In a longitudinal study we collect data periodically over a period of time on a group of people and measure the same traits. As an example, this year we could examine a group of people with a chronic disease in terms of their signs and symptoms and then annually thereafter.

If the same people are examined at each observation period the study is called a **longitudinal panel**. For example a famous health survey, the Framingham Study, makes periodic observations on the same group of people that began the program in the 1960s. If we use different people at each observation point the term **longitudinal time series** is used. A time series example is the selection of a sample of people with each 10 year census to acquire more specific census information on a set of critical variables. The sample chosen is a different group of people in each census.

Hypothesis Testing Designs

There is also another set of designs aimed at testing a hypothesis rather than making an estimation. In a hypothesis testing design one tries to determine if there is an association between an independent variable and a dependent variable. This group consists of two major types of designs: 1] special designs and 2] standard designs.

SPECIAL DESIGNS – THE CASE STUDY. There are two types of special designs. One of the designs is called a case study. The **case study** usually concentrates on evaluating the effect of a change in an important procedure, method or phenomenon within a single organization. The new procedure is examined in depth at a given site. For example we could study one county welfare agency to see how well it instituted a new welfare program. At the agency, evaluators could observe and interview staff and clients on whether the program is doing well. In addition, they could review SOPs, memos, and statistics that measure success or failure. The case study process usually generates large amounts of data from different sources in the organization. Furthermore, the information is specific to the program being evaluated.

SPECIAL DESIGNS – THE CASE CONTROL METHOD. There is also an approach called the **case control method**. It is a plan used frequently in epidemiology work, but has application to other settings as well. In a case control design, the researcher starts with an outcome and looks retrospectively for a cause. Note that in our classical designs we introduce a cause and then look prospectively for the outcome. In a prospective study, the causative agent could be a new antibiotic treatment and the outcome we look for would be a cure. In the case control approach we begin with the outcome (e.g. lung cancer) and look for the cause (e.g., smoking). The way the case study method works is to find a group of people with the disease (the cases) and then identify another group that's very similar to the diseased group except they do not have the disease (the controls). The second group, called the control group, should be equivalent to the group that has the disease in that both have subjects about the same age, with similar concomitant illnesses, etc. Data on variables, other than those used to match the groups, are then examined to see if the groups are comparable. If it is found that the case group has a higher percentage of people with a characteristic than the control group, it suggests that that characteristic may be responsible for the disease. Study replication is exceedingly important in the case control method since type 1 and type 2 errors are hard to control.

The other category of hypothesis testing designs can be referred to as standard as opposed to special designs. The standard design category can be further subdivided into experimental and quasi-experimental designs.

STANDARD DESIGNS - EXPERIMENTS. In an experiment the researcher has control over:

1. What treatments to use.
2. How to assign the treatments to the experimental units.
3. The ways to reduce confounding.

In some situation (e.g. evaluations of social programs), the treatments are pre-defined and the experimenter must accept those choices. However, in a true experiment the experimenter selects the treatments to be administered as well as the treatment regimens (i.e. treatment frequency, dosage, etc.). The experimenter is also free to select one of the randomization methods discussed in chapter 7 for allocating the treatments to the subjects.

Confounding occurs when we have a subject selection bias between the comparison groups. This can occur when there are many people with a characteristic (e.g. a variable that can affect the dependent variable) in one group and few or none with that characteristic in the other group. We will call a variable that is disproportionally represented in the comparison groups an **extraneous variable**. For one treatment group, the independent variable (i.e. the treatment given to the group) and the extraneous variable are considered linked because that group ends up with most of the people with the extraneous variable. Thus, we can't be sure if the treatment effect is due to the independent variable, the extraneous factor or a combination of the two and we have confounding.

STANDARD DESIGNS - QUASI-EXPERIMENTS. In a quasi-experimental design one or more of the three elements listed above are beyond the experimenter's control. The term "quasi" implies as much - a testing situation in which good design principles are only partially satisfied. In a quasi-experimental design the researcher is often limited by practicalities. For example, it may not be appropriate to use a control group because of costs, time or it may simply be against policy to include such a group - e.g., "all" people should be given the "best" treatment available.

Frequently quasi-experimental designs are used in "natural settings" as opposed to what we might call "laboratory settings" (i.e. the environment for an experiment). A natural setting may be an organization that is operating on a day by day basis and needs to function in an efficient and unfettered manner. The number of experimental design features that can be used in a functioning organization is limited. For example, the needs of an organization may require certain groups or locations to be given only one treatment and the other treatment must, therefore, be given at a different location. If we were comparing two ways to care for the homeless, we might have to have one program tried in one city and the alternative program tried in another similar city. The wanderings of the homeless would make it impractical to offer two programs in the same city. Note that there is confounding of the settings and treatments in the above example.

Furthermore, in a natural environment it may not be feasible for the researcher to allocate treatments to subjects on a random basis. Although both treatments may be appropriate for many subjects, one treatment may be contraindicated in certain cases, and the other treatment may be improper for other circumstances. In either event, random assignment of treatments to subjects is impossible. Since treatments groups may be "naturally" selected rather than randomly selected, the possibility of confounding is often unavoidable in the quasi-experimental setting.

On the surface the quasi-experimental design may compare badly to the experimental approach. However, the quasi-experimental design is often used when the experimenter has little or no other choice. To judge those designs harshly could lead to a certain degree of nihilism. It is frequently better to undertake a research project, even if it is not perfect, than to do nothing at all. All designs have some flaws - hidden or obvious flaws - and yet we need to continue to pursue research projects. As noted before, we accept the risk that any given study will give a wrong answer and we count on replications of a study to weed out the erroneous results and reinforce the valid findings.

Design Descriptions

A description of a variety of quasi-experimental and experimental designs are given below. In these descriptions the following notation will be used:

Notation	Meaning
Tr	Tr will stand for treatment and is the point in the study when treatment may be given. An "X" will designate that a treatment was administered and a dash (-) indicate that no treatment was given.
Pre-Obs	Pre-Obs will be the point in a study when a pre-treatment observation can be made. An "O" will designate that the observation is made and a dash (-) indicates that there is no observation.
Post-Obs	Post-Obs will be the point in a study when the post-treatment observation can be made. An "O" will designate that the observation is made and a dash (-) that there is no observation.

DESIGN 1 (POST TEST)

Pre Obs	Tr	Post Obs
-	X	O

Design 1 is sometimes called a Post Test Design because we only take an observation after a treatment is administered. However, it is an example of a design that may not even satisfied a quasi-experimental standard. A treatment is given (X) and then an observation is made (O). In a sense this is what is done when we are not using any kind of comparison. We just do something and then see what happens. There are no controls nor any data to compare our outcome to except possibly our expectation. Design 1 lacks a baseline observation - what a person was like before treatment commenced is not known. An example of Design 1 could be to give a patient an antidepressant and then see the degree of depression the patient has based on the results of a post-treatment assessment. In the Post Test Design the pre-treatment status is unknown so how much help the antidepressant provided can't be quantified.

It could be argued that Design 1 has no place as a legitimate research technique, but there is a set of circumstances that can be used to justify this kind of design. Conceive of a life threatening illness in which there is no available treatment and everyone who has this condition dies. This scenario can occur when a new virulent virus suddenly appears and we have no treatment to counteract it. Hence, there is no outcome except death when one is stricken with this virus. Therefore, if we give a treatment and a subject doesn't die, we would certainly be in a strong position to argue that the treatment made a difference.

Still the main flaw with Design 1 is that it lacks a baseline observation so let's add an observation prior to treatment. We will call this arrangement a Pre and Post Test design and it is illustrated below as Design 2.

DESIGN 2 (PRE AND POST TEST)

Pre Obs	Tr	Post Obs
O	X	O

Here we have the ability to see if there is a change after treatment is given by comparing the post-treatment observation to the pre-treatment observation. However, even if we observe a change we may not know why the change occurred. An example of Design 2 follows: we observe disease symptoms in a person, a treatment is given for the disease, and we then repeat our observation of disease symptoms. Let's assume the post-treatment observation indicates that the disease is no longer present. We are tempted to attribute the elimination of the symptoms to the agent we administered, yet we may have had the same result if we had done nothing. There is a saying: If you take something for your cold it will go away in one week, otherwise it takes seven days.

Assume that in our antidepressant example the follow-up observation showed a lower level of depression than that exhibited at the pre-treatment observation. This may be a sign of statistical regression. If the depression is episodic, the higher score may just be a high point in the cycle and not mean much. The testing effect we studied previously applies to this situation as well. If the dependent variable we observe is prone to improvement through repetition, then it can be associated with a change that may not be attributed to the treatment.

To avoid this latest problem let us add a control group to our design. This will be shown as a new dimension to our schematic - the addition of a Group category. The numerals I and II will be used to represent two separate groups of subjects. We will also eliminate the baseline observation in our example so we can appreciate what occurs when we have a control group, but no baseline reading. We will call Design 3 a "Post Test with Inactive Control Group" design. Note that Group II does not actively receive anything so we add the qualifying term "Inactive" to the design name.

DESIGN 3: (POST TEST WITH INACTIVE CONTROL GROUP)

Group	Pre Obs	Tr	Post Obs
I	-	X	O
II	-	-	O

Without the baseline observation we do not know if the two groups have an equivalent degree of illness before treatment commences. Consequently we may have major problems interpreting the results of the post-treatment observation. Let's assume the groups give equivalent responses at the post-treatment observation point and see what conclusions could be reached. We could conclude that the treatment didn't help since the groups end up at the same level. However, if Group I started off much worse than Group II, it could mean that the treatment was helpful. On the other hand, if Group I started off much better than Group II, and the two groups ended up at the same level, it may mean that the treatment was harmful. Furthermore, the lack of treatment for Group II means the study is not double blind and we may have bias influencing the results on the part of the subjects, evaluators or both parties.

There are steps we can take to overcome the disadvantages identified in Design 2. We can make multiple observation both before and after we introduce our treatment. A design with multiple observations preceding and following a treatment is called an **Interrupted Time Series** and is shown as Design 4. Note that Design 4 does not include a control group.

DESIGN 4 (INTERRUPTED TIME SERIES)

Pre-Obs	Tr	Post-Obs
O,O,O,O,O	X	O,O,O,O

Design 4 is an option experimenters consider when they have access to only a few subjects, but they are in a position to make repeated observation on the subjects before and after treatment. In some respects it's like Design 2, but with repeated observations in the pre and post-treatment phases. The repeated observations provide the opportunity to see if the condition is following a consistent pattern in a given subject. If the response has stabilized in the pretreatment phase, and then changes in a favorable direction after the treatment is given, there is pretty good evidence that the treatment is helpful. Furthermore, if there was improvement right after treatment was given and then that improvement slowly dissipated, and the subject's condition returned to the pre-treatment levels, we may be even more convinced that the treatment had a positive effect, even though it was not a permanent effect. Thus an Interrupted Time Series is an appealing design because it can pick up an unusual change if a stable pattern exists before treatment is given. However, it also has a very serious drawback - it requires multiple observations which may be costly and time consuming.

Another attempt to overcome problems in research studies is to use a paired arrangement. To get around the problem of a lack of group equivalence it is frequently felt desirable to give all subjects both treatments. Frequently this means the treatments are given at two different times. When this is done we have what is called a Counter Balanced design, which is depicted as Design 5.

DESIGN 5. (COUNTERED BALANCED)

Pre Obs	TR	Post Ops	Tr	Post Obs
-	Xa	Oa	Xb	Ob

The notation Xa means treatment A is given to all subjects and the notation Xb means treatment B is given to all subjects. Oa refers to the observation made after treatment A was given and Ob to the observation after treatment B was given. Note that our evaluation is based on the difference between Oa and Ob. If the two observations are similar we could conclude that treatments A and B were equivalent. If the Oa observation point is associated with a better response than the Ob observation point we could conclude that treatment A is better than treatment B. If the reverse result occurred we could conclude that treatment B is better than treatment A. However, in this design, we still have the problem of a testing or maturation effect. If observation Oa shows improvement, is it due to either of these effects or to treatment A? If observation Ob is better than observation Oa, there are a number of possible explanations: 1] treatment B may be better than treatment A; 2] more time has allowed the subject's natural defenses to produce a positive response at Ob (the maturation effect); or 3] the testing effect is operating and has allowed a better response at Ob.

Another problem with Design 5 is that if the first treatment (A) causes a cure then there is no way to know what the second treatment (B) can do because the subject is not in need of treatment and giving B is meaningless. This design requires an assumption that all subjects are in an equivalent state at the time <u>each</u> treatment (Xa and Xb) is administered. It should be clear that because of the many design options that researchers have at their disposal, they must be familiar with the proposed treatments and the outcome assessments to avoid picking a poor design.

We can also introduce randomization into our designs by adding that event to our schematic. We will include a step called "Rand" (for randomization of treatments to subjects) prior to the Tr stage. Let's go back to Design 3 and insert a randomization step. We will also take care of our subject/evaluator bias by adding a second treatment group to this design and assume the trial is done under double blind conditions. When all this is done we have a Randomized Post Test with Active Control Group Design - see Design 6.

DESIGN 6 (RANDOMIZED POST TEST WITH ACTIVE CONTROL GROUP)

Group	Pre Obs	Rand	Tr	Post Obs
I	-	R	Xa	O
II	-	R	Xb	O

In Design 6, by adding a randomization step we have controlled, as well as we usually can, for group comparability, especially if the N is large. Nevertheless, the problem with Design 6 is the lack of a pre-treatment observation (see the criticism for Design 3). Without the pre-observation step we can not have an accurate assessment of the amount of change that takes place at the post-treatment observation point. If the two groups have a similar result at the post-treatment observation, it may mean they were both effective or both ineffective. To remedy the problem we can add a baseline observation before administering treatments and in this case we have a design which we will call a Classical Randomized Design

DESIGN 7 (CLASSICAL RANDOMIZED)

Group	Pre Obs	Rand	Tr	Post Obs
I	O	R	Xa	O
II	O	R	Xb	O

Design 7 takes care of many of the problems we have identified in earlier designs. We would be in a position to compare the change from the pre-treatment observation to the post-treatment observation for each group. If one group showed greater improvement than the other we could conclude that that treatment was best. However, we still have a potential problem with a testing or maturation effect and even though the two groups could be affected in an equivalent manner, we wouldn't want to use design 7 if the testing or maturation effect were possible. Alternatively, we could add a number of pre-treatment observations till the results stabilized indicating that the effects had been dissipated.

Some researchers are interested in seeing if there is a testing effect and they can do this with a Solomon Four Group Design. The design can determine if there is a testing effect as well as a treatment effect and is shown below.

DESIGN 8 (SOLOMON FOUR GROUP)

Group	Pre Obs	Rand	Tr	Post Obs
I	O	R	X	O
II	-	R	X	O
III	O	R	-	O
IV	-	R	-	O

To tell if there is a testing effect, the groups with a pre-treatment observation are compared against the groups without a pre-treatment observation. In other words the results for Groups I and III are combined and compared to the merged results for Group II and IV. If the post-treatment results for Groups I and III are better than those in Groups II and IV, it indicates that there is a testing effect.

To test to see if the treatment was beneficial, treatment Groups I and II are combined and compared to the two groups not receiving the treatment (i.e. treatment Groups III and IV). If the post-treatment results for Groups I and II are better than those in Groups III and IV it indicates that the treatment is efficacious. What we must remember about this design is that it is quite large. Four treatment groups are required and the researcher must decide if it is worth the extra time, energy and effort to obtain the answers.

We have gone on long enough describing the advantages and disadvantages of various research design. But before concluding this subject, let's look at one more design. How can we be sure we have a good experimental model? For instance, if we test a new treatment against a placebo and find no difference, how do we know that the study could have found a difference if one existed. We may have found no difference because the study is insensitive. To determine if the study has sensitivity we will introduce a treatment group to receive the established treatment for the condition being studied. It is assumed that such a treatment exist, otherwise we wouldn't pursue this design.

We now would have three treatment groups - one for the new treatment which we will designate as Xn, one for the placebo group which we will designate as Xp and one for the group which will be given the established treatment which we shall call Xe. We can call this design the Classical Randomized Design with Experimental Model Control.

DESIGN 9 (CLASSICAL RANDOMIZED WITH EXPERIMENTAL MODEL CONTROL)

Group	Pre Obs	Rand	Tr	Post Obs
I	O	R	Xn	O
II	O	-	Xp	-
II	O	R	Xe	O

To evaluate a study using this design, we would first look at the ANOVA result to see if the group on placebo was different than the group on the established treatment (Group II versus Group III]. If a difference is present in favor of the established treatment, Xe, we would have assurance that the model we were using was sensitive - i.e., it can tell the difference between an established treatment and placebo. If a difference didn't emerge, it would indicate that the research model lacks sensitivity. This conclusion would make us wonder if there would be any purpose in continuing the analysis. If the established treatment produces a more favorable result than the placebo treatment, we would then proceed to see if the new treatment was better, equal to or worse than the established treatment. The design seeks to answer a very important question - Can the experiment detect a difference, if one exists?

As you can see by manipulating the steps in our schematic we can come up with many design variations. We only showed some of the most common design possibilities. We also did not examine timing - the duration of time between giving a treatment and measuring the outcome can be critical. If you pick the wrong time to make your observations you may miss the point at which a treatment has produced a positive response or caused a deleterious effect. When observations are made can be critical to the success of a research project.

It should be clear that a researcher must be astute and pick the design that is best suited for his or her purposes - inappropriate designs lead to incorrect conclusions. Furthermore, in addition to identifying a design's strengths and weaknesses, we also must realize that there may be other circumstances that influence the selection of a design. There may be cost limitations or time constraints. There can be cases where the use of placebo may not be justified. For example when the first drug was found to be effective against HIV (Human

Immumodeficiency Virus), it was done in studies comparing it against placebo - you could not add an established treatment drug since none existed. However, subsequent studies with newer drugs could now be designed with an established drug (the drug shown to be effective for HIV), but you could no longer justify using a placebo. On ethical grounds you could not give a subject an inert substance when you had a drug available that showed it was effective for HIV. Excluding placebo from our design options meant that we could not employ Design 9, which is arguably the safest design presented.

Statistical Index

It is often useful to combine several variables to form a single measure of a phenomenon. One variable may be incomplete in trying to represent a complex concept. By combining several variables a more comprehensive representation is possible. There are many commonly used indexes, such as the Dow Jones Industrial Stock Average or the Consumer Price Index (CPI). A **statistical index** is defined as a set of variables which are combined and used to measure a complex concept. Some indexes such as the Dow Jones and CPI take a representative number of items - the items may be stocks (the Dow Jones index) or the cost of goods (the CPI). It is intuitively clear that the average of many items is a better representation of the change in consumer prices than the change of any one item such as a loaf of bread or a gallon of gasoline.

Frequently it is valuable to use an index to represent change over time. For example the CPI is usually determined for a base year and then changes in subsequent years are shown relative to the base year. It is usual to fix the value of the base year index at 100. The value 100 is arrived at in a straight forward calculation:

$$Index = 100 * \frac{Current\ Year}{Base\ Year}$$

Suppose the index has only two items. In the base year, the price of the items are $1.50 and $4.50. Thus, the sum is $6.00. In the first year of an index, the current year and the base year are the same year and the calculation of the index is:

$$Index = 100 * \frac{6.00}{6.00} = 100 * 1.0 = 100$$

If in the next year (i.e. year 2) the prices of the two items are $1.60 and $4.70 the sum of the items is $6.30 and the index is:

$$Index_{Year\ 2} = 100 * \frac{6.30}{6.00} = 100 * 1.05 = 105$$

Unit Free Index

In our illustration, the index used the same unit of measurement for the variables - the price of a commodity. However, there is often an interest in developing an index in which the units of measurement are different. For example, come up with an index for the health of a community - the variables one wants for the concept "health of a community" may be quite diverse. The variable we want to use could be the number of MDs per 1,000 residents, the average time it takes to reach a hospital, the infant mortality rate, the amount of alcohol consumed per person per year, the average number of work days lost due to health related accidents, etc. These variables do not use the same unit of measurement. We simply can not add them up and make any sense out of the sum. There is a need to develop the index on the bases of unit free measurements for our variables. What we want to do is to transform the measurements so the unit of measurement is removed.

This is easier to do then we might at first think - mainly because we have already done transformations that gave us unit free measurements. Recall the Z transformation in which the following formula was used:

$$Z = \frac{X - \mu}{\sigma}$$

Let us assume the variable we are working with is dollars. Each of the elements (X, μ and σ) in the Z formula is expressed in dollars. However, note that what we do is subtract X dollars from μ dollars in the numerator and arrive at a difference value that is still expressed as dollars. However, when we divide that difference by σ, which is also expressed as dollars, we eliminate the unit of measurement. In a sense "dollars" are divided by "dollars" and the result is 1 - i.e. the dollars as a unit of measurement becomes 1 and "dollars" are no longer applicable.

COMBINED Z TRANSFORMATION. To illustrate how an index could be created using what we'll call the **Combined Z Transformation** we will work with a baseball example. Suppose we decide we will select the All Star team, not on the basis of a popularity poll, but on the basis of "merit". The concept for which we want an index is baseball merit. We often rate baseball players in terms of the player's batting average, the number of home runs the player hit or the individual's fielding average. These variables therefore can become the elements we want in our index. There could be more variables in our index, but for illustrative purposes the following ones will suffice:

  1. Batting average (BA)
  2. Fielding average (FA)
  3. Number of home runs hit (HR)

Most would agree that if we just added the numbers together the result would be pretty silly. Before we combine the three variables, we need to transform the numbers so the units of measurement (e.g. times at bat or fielding chances) are eliminated.

In addition, since we are picking an All Star team, we need to calculate an index for each player at each position in the league. We first need to identify all the eligible players at a given position and then calculate the mean and standard deviation for this set of players for each variable. Let us start by selecting the All Star for first base. We are given the following statistics for the mean and standard deviation for all eligible first basemen.

Variable	Mean	Standard Deviation
BA	240	40
FA	.970	.050
HR	11.5	3.0

We will calculate our index for three of the players although there would be many more who would be eligible. We will assume these are the top three players who will be identified as Able, Baker and Charlie.

The statistics for the three players are:

Variable	Able	Baker	Charlie
BA	258	311	345
FA	.965	.995	.985
HR	37	9	12

The Z values for player Able are shown in Table 11-M in which the Z calculation are based on the standard Z formula ($Z=(X-\mu)/\sigma$)

Table 11-M
Player Able Index-Combined Z Method

Variable	X	μ	σ	Z calcualtion	Result
BA	258	240	40	(258-240)/40	0.45
FA	.965	.970	.050	(.965-.970)/.050	-0.10
HR	37	11.5	8	(37-11.5)/8	3.19
				Sum:	3.54

The sum of the Z values (3.54) is the index value for Able. Calculate the index value for Baker and Charlie below:

Your answer should be 1.96 for Baker and 2.99 for Charlie. Since Abel's 3.54 is the highest total, he would be our All Star first baseman.

PERCENTIZING METHOD. An alternative way to transform variables and make them unit free is by means of the percentizing technique. In this case you need to know the best and worst values for the elements in your data set. Let us assume in our case that the relevant statistics for the best and worst players were:

Variable	Best	Worst
BA	345	205
FA	.995	.960
HR	37	1

The formula for the transformation by means of percentizing is:

$$X_a = 100 * \frac{Y_a - Y_w}{Y_b - Y_w}$$

where X_a = Percentized value for player a
        Y_a = Actual value for player a
        Y_w = Actual value for the worst player in the data set
        Y_b = Actual value for the best player in the data set

A percentized value is obtained for each variable and the values summed to get the index value. For Able we would perform the calculations shown in Table 11-N and obtain an index score of 152.1

Table 11-N
Player Able Index - Percentizing

Variable	X	Best	Worst	Pct. Calculation	Result
BA	258	345	205	100*(258-205)/(345-205)	37.9
FA	.965	.995	.960	100*(.965-.960)/(.995-.960)	14.3
HR	37	37	1	100*(37-1)/(37-1)	100.0
				Sum:	152.1

Note that because Able had the most home runs his percentized score for home runs is 100. In the space below determine the index value for Baker and Charlie using the percentizing method.:

Your answers should be 197.9 for Baker and 202 for Charlie. Using the percentizing method, Charlie will be the All Star first basemen.

Weighting

One last comment is in order. The illustrations above assume that all the variables are equally important. In fact, that may not be the case. It may be felt that batting average is more important than the other variables. It may also be believed that fielding average is the least important variable. In that event, you need to weight the variables before you form the final index. If what you were doing was of a critical nature you might want to enlist the help of experts rather than rely on your own opinion to form the weights. However, in either case you need to assign a weight to the variables and incorporate the weighting into the computation of the index.

One way to assign weights is to give a weight of 10 to the most important variable. You then take the next most important variable and give it a weight relative to the variable you judged as most important. If the second most valuable variable is almost as important as the first, it would be given a weight of 9. On the other hand, if the second variable was considered no where near as important as the first, it would be assigned a much lower weight - for example, a weight of 6. You continue giving weights to the variables based on their relative importance. If you can not distinguish between the importance of two variables they may be given the same weight.

The index in which you use weights is frequently referred to as a **weighted index**. The weighted index is calculated using the following formula:

$$Weighted\, Index = \Sigma(W_1 * V_1)$$

where W_1 refers to the weight assigned to variable 1

V_1 is the index value for variable 1

In our example if we decided hitting home runs was the most important variable, we would give it a weight of 10. Assuming batting average was the next most important variable we might assign it a weight of 8. Finally fielding average, which we may not consider of much value, would be given a weight of 5. Our weighted index on the percentized data for Able would be based on the following calculation:

$$Weighted\, Index = (8*37.9)+(5*14.3)+(10*100) = 1375$$

Able's weighted index is 1375. If you calculate the weighted index for Baker and Charlie your answers would be 1328 and 1463 respectively. Based on the weighted index, Charlie would remain the All Star first baseman.

Summary

In this chapter we see that by using controls we can reduce the possibility of obtaining misleading results. Including control groups and control variables can strengthen a study. By following tight standards during the conduct of a study we can also keep extraneous factors from distorting the trial results. Based on this chapter we also should appreciate that there are multiple research designs that can be applied to a research problem. Researchers must be familiar with the strengths and weaknesses of a variety of designs so they can select the plan that will be best for their research project.

In the last part of the chapter we learned how to calculate a statistical index. When the elements in the index use the same units of measurement, the calculation of an index is straight forward. However, we can also use the Z transformation or the percentizing technique to form an index, even if the units of measurement for the components in the index are different.

PROBLEMS

1. a What kind of analysis is appropriate when using treatment A at site 1 and treatment B at site 2?
    b Why is this design necessary?

2. What assumption is required when the same subject receives all treatments (design 5)?

3. Why is matching done and what problem is frequently encountered?

4. Why is group comparability always an assumption in an independent group experiment?

5. Why is the Hawthorn Effect relevant to research methods?

6. What is the difference between cross sectional and longitudinal designs?

7. What is confounding? Give an example.

8. How does the case control approach differ from the usual research design that takes a prospective approach?

9. a What is the advantage of the Randomized Classical Design with Experimental Model Control?
    b When might this design be unfeasible?

10. What is the advantage and what is the necessary subject requirement of the Interrupted Time Series design?

11. For the variables listed below, use the Percentizing technique to find the standardized values for the cities listed.

City	Avg. Per Capita Income	Per Cent On Welfare	Avg. Years Of Education
Appelton	15,000	95.0	18.1
Bakers Ford	16,500	96.8	17.5
Calumet City	12,400	88.2	11.5
Drangeville	20,000	97.4	16.4
Best City:	20,000	98.0	18.5
Worst City:	10,000	78.0	11.5

12. It was decided to create a Vacation Cost Index for a 4 year period based on the factors shown below. Calculate the index starting with an index value of 100 for the first year.

FACTOR	YEAR			
	1	2	3	4
Motel Room Cost	27.85	29.34	30.10	33.24
Price of Gasoline	1.24	1.30	1.28	1.38
Avg. Meal Cost/Day	25.20	26.29	28.78	29.40
Daily Fees Paid	5.80	6.25	6.45	6.45

13. For the variables listed below use the Combined Z Technique to find the standardized values for each of the movie studios listed.

Movie Studio	Academy Award Nominations	Avg. Profit	Public Rating	Pct. Change in Stock Price
Atomic	25	22	18.1	+17
Mammoth	12	15	21.5	+ 2
New Age	31	12	11.5	+20
Cyber	28	7	8.5	+11
All Studio Mean	20	12	18.5	+ 9
" " Std Dev	5	6	10.0	6

14. The administrators at one of the state colleges wants to monitor student assessment of the college using the four characteristics listed in the table below. Data for latest available 5 year period were collected. The characteristics and 5 year data are shown in the table. The minimum and maximum value for each characteristic are also shown in the table.

Characteristic	1993	1994	1995	1996	1997	Min	Max
Pct. students returning	88%	87%	81%	86%	85%	81%	89%
No. of events without a rule violation	27	35	23	27	27	20	35
Student revenue from college sponsored events (in millions)	2.5	2.9	3.2	2.9	3.3	2.4	3.3
Average faculty evaluation score	2.04	2.08	2.02	2.07	2.07	2.01	2.08

a   Calculate an index for the 5 year data starting with an index value of 100 for 1993.

b When asked to weight the variables the following weights were assigned. Recalculate the index based on these weights.

Characteristic	Weight
Pct. students returning	7
No. events without a rule violation	4
Student revenue from college sponsored events	10
Average faculty evaluation score	2

15. Part of a Consumer Safety Program involved a survey of 76 financial institutions in the state. The goal was to see how the newer institutions compared to older institutions in respect to the adequacy of their loan prospectus given to customers. A prospectus was judged as adequate or inadequate by independent experts. The results of the survey are given below:

Type	Adequate	Inadequate
Older Institutions	27	11
Newer Institutions	32	6

Is there a statistically significant difference between the two groups of institutions using an .05 Alpha Risk?

16. A political scientist was interested in the effect of location on the retention of senior governmental officials. Based on public records he compiled the following table. Test the hypothesis that location makes no difference in turnover rate at the .05 level of significance.

Location	Low	Turnover Medium	High	Total
Urban	25	28	28	81
Suburban	30	21	12	63
Rural	20	11	5	36
Total	75	60	45	180

17. An evaluation team was asked to compare how well warranties were understood by consumers when a product was sold by private companies versus products sold by non-profit companies. A random sample of 36 products (18 for each type of organization) was taken. The researchers also obtained data on consumer understanding of the warranty which were assigned ratings of 1 to 8. A 1 was assigned if consumers had no understanding of the warranty and an 8 if there was complete understanding. The following results were obtained.

Non-Profit Organizations		Private Organizations	
8	5	8	1
3	6	6	6
4	7	1	5
7	7	3	5
7	5	3	7
7	3	3	5
7	7	3	2
3	6	7	4
5	7	8	8

a   Is there a statistically significant difference between the organizations at the .05 level of statistical significance?

b.   If you assume the ratings are interval level data would you change your conclusion?

18. Prior to a recent swimming meet the coach of your team rated her divers on a scale of 1 to 10. At the meet four judges rated the divers actual performance on a scale of 1 to 5. The pooled results of the judges ratings are given below along with the coach's evaluations. What is the correlation coefficient?

Diver	Judges' Ratings	Coach's Rating
Agnes	11	8
Betty	15	6
Carol	19	10
Dedra	17	10
Eve	9	4
Fran	13	9
Gloria	14	7
Holly	10	2
Irma	16	8
Jane	20	7

19. The Department of Health tested samples for bacteria which were collected from 16 local restaurants that had been chosen at random. Some bacteria is present in all food tested and the following results were obtained:

Bacterial Count			
160	80	107	83
100	91	71	79
196	122	99	84
111	140	124	129

Because this is a health related matter significance tests are done with a .10 alpha risk. A bacterial count above 95 is considered unacceptable.

a   Is there sufficient evidence for the health commissioner to take some action based on theses results?

b   What was the critical and the calculated values of the test statistic you used?

20. If a sample of 40 union members shows that 45% will vote for a strike, what is the 90% confidence interval?

There are no instructions in Excel for this chapter.

Chapter 13

Regression

Introduction

Regression is a statistical method that is used to show the relationship between a dependent and one or more independent variables. A correlation coefficient also shows the degree of relationship between variables, but regression gives much more information about the nature of that relationship. With correlation we get one value - the correlation coefficient. With regression we get a much larger battery of information including the ability to make predictions. We will examine two models of regression - simple linear regression and multiple linear regression. The assumptions underlying both models will be briefly reviewed. Although regression is designed for interval variables we will learn a technique that allows use of ordinal and nominal variables in multiple regression problems. We will also see how we can measure the contribution for each variable in a multiple regression analysis. The distinction between hypothesis testing and hypothesis generation will also be covered in this chapter.

Simple Linear Regression

In simple linear regression we study the relationship between only two variables. The term "multiple linear regression" is used when there are more than two variables. It is important to distinguish between the two variables because one is considered the independent or X variable and the other, the dependent variable or Y variable. Remember the independent variable is suppose to act on or influence a response in the dependent variable. When we examine the relationship between the independent and dependent variable remember to always treat the independent variable as the X variable and the dependent variables as the Y variable.

Let's begin with a small example of simple linear regression. Our two variables will be IQ and test scores for a group of students. What is the independent variable?

You should have chosen IQ since intelligence should influence how well a person does on a test. Intelligence may not be the only influence, but it seems reasonable to believe that "smarter" people do better on a test. In other words IQ should affect test scores and the cause/effect relationship would not be the other way around.

We'll work with the following data set which provides an IQ and a test score for 6 students. Note that for <u>each</u> student there are two observations.

Student.	IQ	Score
1	98	80
2	98	78
3	110	82
4	110	88
5	122	92
6	122	96

In practice, if you intend to do a regression, you should have more than 6 individuals. Data sets of 20 or more persons would be more reasonable. However, it is quicker to see what is occurring in linear regression when the sample size is small.

First we should plot the values and look at the graph (see Figure 13-A below).

Figure 13-A

We can see, as we move from the left to the right on the IQ axis, the test scores tend to increase. This is the sign of a positive correlation - as the independent variable increases the dependent variable also increases.

Regression Line

We might even draw a free hand straight line through the points such that it splits the points so about half are above and half are below the line. We'll call that line the preliminary regression line. The line is based on our judgment and we will see later that we can draw a much more accurate line based on mathematical calculations. Once we have a regression line in place we can use it to make predictions. Figure 13-B is an enlarged version of Figure 13-A. It also contains 1] grid lines to give clarity and 2] the preliminary regression line.

Figure 13-B

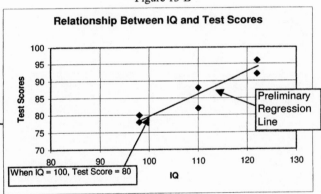

When IQ (the X variable) is 100 we can estimate what the test score (the Y variable) would be. Based on our graph, when IQ is 100 test score is about 80. Hence we would predict that someone with an IQ of 100 should score around 80 on the test. We can also make predictions for other IQ scores. For example what test scores seems reasonable for an IQ score of 105?

A test score somewhere around 84 seems to match up with an IQ score of 105. Although it's not advisable, we could even make predictions beyond the points we have plotted. For example an IQ of 125 appears to line up with a test score of approximately 96.

Any time we draw a straight line, there is a mathematical equation that represents that line. The equation for a line is:

$$Y = a + bX$$

Regression Equation

When the equation for a line is used for a simple regression problem, it may be called the regression equation. The terms in the equation are defined as follows:

Y = dependent variable [test score]
X = independent variable [IQ]
a = **intercept**, the value of Y when X = 0.
b = **slope**, the amount Y changes for a 1 unit change in X

The slope, b, is sometimes called **beta** or the **regression coefficient**, and it is the amount that Y changes for each one unit change in X. In our example how much do test scores (our Y) change for a 1 unit change in IQ (our X)? To determine that value we could find any point on the X axis, record the Y value corresponding to that X, locate a second X that is 1 unit away from the first X, record the Y for that second X and the difference in the two Y readings is the slope. Use of that method may be a little hard in our example because the X axis is in 10 unit increments rather than 1 unit increments. Nevertheless, we can find out how much Y changes for a 10 unit change in X and then divide that change by 10 to give us the change for 1 unit of X. The test score for an IQ of 100 is about 80 and for an IQ of 110 it is about 86. Thus for a 10 unit change in X (100 to 110), Y increases 6 units (80 to 86). This means that for a one unit change in X, Y changes 0.6 point (6/10 = 0.6). The 0.6 is our b, the slope.

We still need to find a, the intercept on the Y axis. We could find the intercept by extending our graph until we expose an X value of 0. We would then record the Y value for an X = 0 and the Y value would be the intercept. In Figure 13-C (displayed on next page) we extended our graph to include an X of 0 and we see that when X is 0, Y is somewhat under 20. A test score of 15 seems like a realistic value for Y.

We can now write an estimated equation for our relationship between test scores and IQ. We estimated a as 15 and b as 0.6 so substituting those values in our equation for a line we get:

$$Y = 15 + 0.6*X$$

Figure 13-C

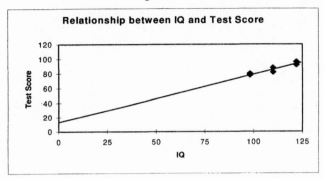

There is a mathematical technique to determine the value of a and b , but the equations are labor intensive and they are not particularly informative so they will not be presented. The equations are available in many statistical texts that include regression methodology. You are advised to use a computer program if you want to do regression - the gain in time and accuracy is an overwhelming advantage. An example of the computer output we would obtain for our regression problem is shown below.

Regression Statistics

Multiple R	0.9375
R Square	0.87890625
Adjusted R Square	0.84863281
Standard Error	2.78388218
Observations	6

ANOVA

	df	SS	MS	F	Significance F
Regression	1	225	225	29.0323	0.005737305
Residual	4	31	7.75		
Total	5	256			

	Coefficients	Standard Error	t Stat	P-value	Lower 95%	Upper 95%
Intercept	17.25	12.80997604	1.346607	0.24935	-18.31626894	52.81627
X Variable 1	0.625	0.115995091	5.388159	0.00574	0.302945331	0.947055

The information we want, the values for a and b, is provided in the last portion of our output. The table has column headings for "*Coefficients*", "*Standard Error*", etc. The first column, which is unlabeled, has a heading for "Intercept" and another for "X Variable 1". Our initial interest is with the first data column - the column headed "*Coefficients*". For "Intercept" the value 17.25 appears - that is the value of a. It is the value of the Y variable when the X variable is 0. Hence, 17.25 is the test score when IQ is 0. It is close to our estimated value of 15, but 17.25 is the mathematically accurate value of a. The coefficient for the "X Variable 1" is given in the table as 0.625. The generic term "X Variable" is used in the print-out, but it's important to realize that the variable in question is the independent or X variable which in our example is IQ score.

The value 0.625 is our slope, the b value in our formula. We estimated the slope to be 0.6, but the 0.625 is the precise value based on mathematical calculations. We can now write the correct equation for our relationship between test scores and IQ. It is:

$$Y = 17.25 + 0.625*X$$

For our example, this is the mathematically correct **regression equation**. We are able to obtain a Y value for any X using this formula. For example if X is 100, we substitute that value in the equation and solve for Y.

$$Y = 17.25 + 0.625*100$$
$$Y = 17.25 + 62.5 = 79.75$$

When X = 100, Y is 79.75.

We also would like to draw a precise regression line. To draw that line we need to know two points that are on the line and then we connect the points. We just found out that one point on our regression line is (100,79.75). The values 100 and 79.95 are called coordinates. The X coordinate (100) always precedes the Y coordinate (79.75) and a comma separates the values. To create our regression line we need one other point, but it's a good practice to use three points rather than the minimal number of two. We'll use X = 120 for our second point and find the value for the corresponding Y coordinate. See if you can determine what Y is when X=120 below.

You should have obtained a Y of 92.25 based on the following calculation:

$$Y = 17.25 + (0.625*120) = 92.25$$

We now have our second set of coordinates (120,92.25). For our third point we will use the mean value of X and the mean value of Y. Those means are 110 for X and 86 for Y. We can use that point (110, 86) because all regression lines must pass through the mean X and mean Y point. Including this point gives us a good check on our arithmetic. The line that passes through the three points we identified, is the regression line. We will extend the line so it goes from our minimal X value (98) to our maximal X value (122). (See Figure 13-D below).

Figure 13-D

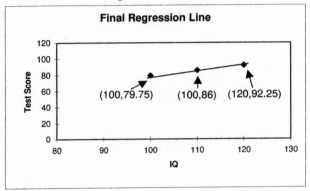

If we use our regression formula we can obtain predicted test scores for the 6 students included in our data base. The original observed score and the predicted score are shown in Table 13-A.

Table 13-A Original and Predicted Score

Student	Original Score	Predicted Score
1	80	78.5
2	78	78.5
3	82	86
4	88	86
5	92	93.5
6	96	93.5

It was mentioned before that it is unwise to make predictions for scores that go far beyond the set of X values in your data set. We do not have data for IQs less than 98 nor beyond 122. The IQ/test score relationship may change if we go too far out from our original set of values so avoid predictions for IQs that are much larger or smaller than those in the original data set. Also we should realize that our test scores could have a maximum value - for instance 100. However, if we predicted a test score for X = 140, we'd get a test score of 104.75. Our common sense tells us that this answer is impossible for a test with a maximum score of 100. Unfortunately, a regression formula has no limits built into it based on common sense. It is, therefore, up to the users to apply their practical knowledge when using regression equations.

Note that the slope value, 0.625, is a positive value. This means that our regression line moves in an upward or positive direction. A negative slope would mean that the regression has a negative or downward direction. If the slope were 0 it would mean the regression line is flat.

Explanation of the Regression Output

In addition to the above information there are other useful items included in the regression output. At this time we won't need to know what all the values mean, but some are useful to our understanding of the relationship between the variables used in a regression. Also note that in the text, we will round values rather than using the values given in the computer print out. Based on the computer print out, a portion of which is reproduced below, we see that the X Variable 1 row (which gave us the coefficient for the slope) also contains a value under the heading "*Standard Error*". That value, 0.116, is the standard error for the slope. In the next column we see the result of a t Test on whether the slope of 0.625 is different from a slope of 0.

	Coefficients	Standard Error	t Stat	P-value	Lower 95%	Upper 95%
Intercept	17.25	12.80997604	1.346607	0.24935	-18.31626894	52.81627
X Variable 1	0.625	0.115995091	5.388159	0.00574	0.302945331	0.947055

The t value is 5.39 and Excel translates that value into a probability in the next column which is headed "P-Value". The p-value for the test is .006 and indicates that the difference between the observed slope of 0.625 and a slope of 0 is statistically significant (testing at an alpha risk of .05). It means it is very unlikely (6 chances in a thousand) that the true slope could be 0 (i.e. a flat line). This is considered a good outcome because we know a slope of 0 means there is no relationship between the variables, but our statistically significant slope in this case means we have a relationship between IQ and test scores. If the t Test fails to produce a statistically significant result, you have to conclude that the relationship between the variables may be non-existent. In the last two columns of the table the 95% confidence interval for the slope is given. We are 95% confident that the population slope is between .30 and .95.

RESIDUALS. Before we discuss other parts of the regression computer output, we will look at other data that can be obtained in a regression. Statistical programs can produce a table, which is often referred to as "Residual Output" (see below).

RESIDUAL OUTPUT

Observation	Predicted Y	Residuals
1	78.5	1.5
2	78.5	-0.5
3	86	-4
4	86	2
5	93.5	-1.5
6	93.5	2.5

A residual means something left over and you will see that that definition of the term is quite reasonable in regression analysis. Note that our residual table lists each observation in the first column - in our case students 1 through 6. For each person a predicted test score is given in column 2. For individual 1 (who incidentally had an IQ of 98) the predicted test score is 78.5. Note that all the predicted points listed in the above table are, in fact, on the regression line - this is always the case. The last column headed "*Residuals*" gives the distance between the observed Y for an observation and the predicted Y. Hence, a **residual** can be defined as the difference between the actual value and the value predicted from the regression model. Mathematically it is:

Residual = Actual Value − Predicted Value

For observation 1 the actual Y was 80 and the predicted Y was 78.5 so the **residual** is 1.5. From this table we can see how much each predicted point differs from the observed point. As a result, this table tells us a lot about how good our predicted regression line fits the actual data. In a perfect fit, the observed and predicted points would be identical. In that sense, the residual is the amount left over from a perfect fit. When we discuss the standard error of the estimate we will discuss a way to use this information.

COMPUTER OUTPUT EXPLANATION CONTINUED. Returning to the explanation of the computer print out, we will continue our explanation with the table at the top of the output. That table is reproduced below:

Regression Statistics

Multiple R	0.9375
R Square	0.87890625
Adjusted R Square	0.84863281
Standard Error	2.78388218
Observations	6

The **Multiple R** value is the Pearson correlation coefficient (r) which we discussed in an earlier chapter. The next entry, **R Square**, is the result you get when R is squared (i.e., $R^2 = r * r$). It is sometimes called the **Coefficient of Determination**. R Square represents the amount of variation in Y that can be explained by X. If $R^2 = .50$ then 50% of the variation in Y is explained by X. If R^2 is small, it means X has not accounted for much of the behavior of Y and indicates that there are many other factors influencing Y. In our example, R^2 is .88 and we can state that "88% of the variation in test scores is explained by IQ". Just as a large r is considered good because it shows a strong association between variables, a large R^2 value is also considered good for much the same reason.

To help appreciate R^2, suppose we were trying to predict test scores on the basis of IQ. If we put the names of all of the people in a hat and predicted that the first name drawn would have the highest test score and the second one drawn would have the next highest, we realize that this would be a silly way to make predictions. The correlation (r) between test scores and IQ scores that would result from this method is, in theory, 0. For any given experiment the answer would not be exactly 0, but on average, after many experiments, the average would approach and could actually reach 0. Consequently R^2 would also be 0 or close to 0 and we would say that 0% of the variation in test scores was explained by IQ. But in our example 88% of the variation in test scores was accounted for when IQ was used. Using IQs to predict test scores is obviously much superior compared to simply making predictions by pulling items from a hat.

Researchers tend to pay more attention to R^2 than to r because they feel R^2 gives them more useful information about the ability of the independent variable to predict the dependent variable. Frequently, researchers calculate the complement of R^2 (i.e. $1- R^2$). For example, in our case $1- R^2$ is $1-.88$ or .12. If IQ accounted for 88% of the variation in test scores then we can reason that other "factors" must account for the other 12%. There are therefore other variables available that could help us make even better test score predictions.

The adjusted R Square value is of little use in a simple regression so we will bypass it. The next item, Standard Error, is also a useful statistic obtained from regression. It often is called the **Standard Error of the Estimate**. Unless there is perfect correlation (i.e. all the observed data points fall on the regression line) the regression equation is not a completely accurate predictor. It is valuable, therefore, to have a statistic that measures how well the observed and predicted points compare. The Standard Error of the Estimate is that measure. Our Standard Error of the Estimate is 2.78 and, it is essentially a standard deviation for the residuals discussed earlier.

The Standard Error of the Estimate has the same numerator as any sample standard deviation (i.e. $\Sigma(X- \overline{X})$, but the denominator is n-2, not n-1. If you recall, back in chapter 1, we discussed a standard deviation. A standard deviation measures how spread out a set of data is. We learned that if we add and subtract 1 standard deviation from the mean of a normally distributed variable, the interval we created will contain about 2/3s of all the variables in the data set. Therefore, we can conclude that about 2/3s of the predicted test scores will be within ±2.78 points of the actual test score. Is that good or bad? That decision must be made by the researcher who is familiar with the data variables being used in the regression. In general, it seems reasonable to have a preference for data sets that are more homogeneous, as opposed to being very diverse. Hence, researchers would like a small Standard Error of the Estimate because the smaller the Standard Error of the Estimate the smaller the deviation between predicted and observed values.

It may help to appreciate the Standard Error of the Estimate if we realize that we could have the same regression line, but it could be obtained from different data sets. In Figure 13-E the graph of the original data set is reproduced, but the regression line has been added and the value of the Standard Error of the Estimate is also provided. The predicted points are on the regression line so our residuals are the vertical distance between an observed point and the regression line. As noted above the Standard Error of the Estimate is the standard deviation of the data set made up of the residuals.

In Figure 13-F we show the graph of another data set with the same regression line. The a (intercept) and b (slope) values are identical to those in the original data set). However, note that the observed values are much closer to their predicted scores in this data set, which we will call data set 2 to differentiate it form the original data set. In this latter case the standard error of the estimate is 1.19 and it is clearly smaller than that of the original data set.

Figure 13-E   Figure 13-F

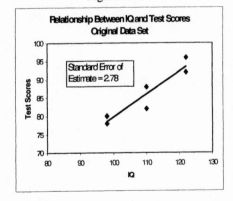
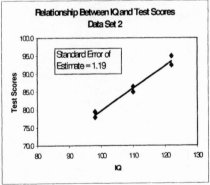

We could also have the same regression line, but with observed points that are farther away from their predicted scores. In this latter case (see figure 13-G on the next page) the standard error of the estimate would be much larger than 2.78 and it, in fact, is 5.07.

Statistical Methods for Researchers

Figure 13-G

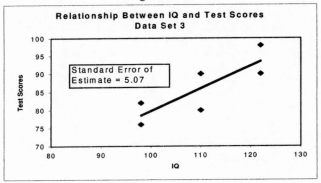

The last entry in the upper table of the regression output is "Observations" and this is simply a count of the number of individuals used in the analysis. In our case we used 6 people so the Observation number is 6.

The middle table is an ANOVA table, but gives us no new information that we haven't obtained from other sources so it can be ignored in simple linear regression.

Assumptions

There are certain assumptions necessary to justify a linear regression analysis. Perhaps the most important one is that the relationship between the variables is, in fact, linear (i.e. a straight line). Statisticians have formal methods to test this assumption, but the methodology goes beyond what can be done in this text. However, even non-professionals can do a rather simple check on the linearity assumption - draw a graph and examine it. Do the observed data points form a pattern <u>other</u> than a straight line? The term "other than a straight line" is used because you are free to assume linearity and only reject that assumption when it is unsupported. In a correlation problem in which r is very small, you would see no pattern and yet it would be correct to apply a linear regression to the data set. On the other hand, if the data points appear to be more consistent with the shape of a curve rather than a line, then the linearity assumption is unfounded. There are ways to do curvilinear regression, but they again go beyond the scope of this text.

Other assumptions involve normality and equal variances. The distribution of Y must be a normally distributed for <u>each</u> X. Furthermore, the variation around the line of regression should be equal for <u>each</u> X value. Typically there are a limited number of Y observations for each X so these assumptions are often impossible to test. One therefore usually proceeds with the regression procedure without testing the assumptions, but it must be realized that the answer could be inaccurate depending on the degree to which the assumptions are not satisfied. A final assumption involves independence which was discussed in a previous chapter.

Regression Conclusion

After conducting a simple linear regression, the conclusions can be presented in a rather succinct paragraph. One usually includes the following comments:

1. Whether the relationship is statistically significant (in simple regression this is based on whether the slope for b is statistically significant). If there is statistical significance, points 2-4 are also included.
2. A quantitative comment regarding how Y changes in respect to X - i.e., what is the direction and how much does Y change for each 1 unit change in X.
3. How well do the predicted points for the dependent variable compare to the observed points. By using the standard error of the estimate, it is possible to comment on how far the predicted points are from the observed points.
4. The amount of total variation in Y that is explained by X using the R^2 value.

An example of how the findings from the IQ/test score relationship could be summarized is given below:

> A statistically significant relationship (p=.006) exists between IQ and test scores. For each one point increase in IQ, the test score increases by 0.625 points. About 2/3s of the predicted test scores are within ±2.8 points of the observed test scores. Approximately 88% of the total variation in test scores is accounted for by IQ.

Hypothesis Testing Versus Hypothesis Generation

Before we move on, it is important to point out that there is a difference between hypothesis generating research and hypothesis testing research. Sometimes we use a statistical procedure, such as regression, to look for independent variables that might be related to a dependent variable. There may be no a priori reason to suspect that the independent variable has a connection with the dependent variable. We do our tests to see what variables may be related. We are exploring, looking for possible relationships. In statistical circles this activity is called **hypothesis generating research**. It differs from hypothesis testing, in that, in hypothesis testing we have established a rationale for believing that a relationship exists and we are testing that belief.

When regression is used in exploratory research we find variables that are candidates for further research. In the exploratory stage, the variables that show a statistically significant relationship to the dependent variable become our candidates. However, in this exploratory step we are only generating hypothetical relationships not testing them. If we test 20 possibly related variables, and use the .05 level of significance, we should expect to find one variable with a statistically significant relationship with the dependent variable by chance. In fact, our 5% risk of a wrong decision means that one of the 20 variables should be associated with statistical significance even when no relationship exists. It stands to reason that if people look hard enough and long enough they will find some relationships - relationships that are due only to chance. The term **fishing exercise** and **data mining** are sometimes used to refer to this searching/exploratory practice. There is nothing wrong with the practice as long as researchers recognize that they are generating a hypothesis and not testing one. The variables found from hypothesis generating research need to be tested with fresh data to see if, in fact, their relationship to the dependent variable is at a statistically significant level.

Multiple Regression

If there is only one independent variable we have simple regression. If there is more than one independent variable we have multiple regression. In regression, we always deal with a single dependent variable.

To see how multiple regression works we'll look at an example. What if our goal is to predict mortality rates in community hospitals. In this case, our dependent variable is the mortality rate. For independent variables we need to select those factors that can influence mortality. An obvious one is age. As age increases the chance of dying also increases. We could collect the average age of people in different communities as our X variable and the mortality rate for each community as our Y variable and then do a simple linear regression. But age is not the only important characteristic. Other factors such as number of physicians in the area and average distance a resident is from a hospital could influence mortality. What we want to do is add variables that are related to mortality so we can develop a better, more complete, model that can predict mortality. We assume that additional variables, if they influence dying, will add precision to our prediction model. Dealing with more than one independent variable means we need to use a multiple regression procedure. As you will see, multiple linear regression is just a straight forward extension of simple linear regression.

As we have learned, the regression equation in simple linear regression is:

$$Y = a + bX$$

For multiple regression we still have one dependent variable (Y), but we can have two or more independent variables (X_1, X_2, etc.). As a result our equation, when we have two independent variables, becomes:

$$Y = a + b_1X_1 + b_2X_2$$

If we have more than two independent variables we just keep adding bX terms. However, you should add more independent variables as long as you believe they contribute to explaining the dependent variable. Do not just "throw in" variables in the hopes that some may work. If you do this you may find spurious relationships. If you are conducting exploratory research then the addition of variables, for which there is no *a priori* reason to believe that they influence the dependent variable, may be tried. However, this approach must be viewed as hypothesis generating and not hypothesis testing. Remember that if a positive connection between a "possible" variable and the dependent variable does emerge, you still need to test that relationship with new data to rule out the possibility that the relationship occurred by chance.

Multiple Regression Explanation

Our interpretation of the elements in the multiple regression equation are consistent, but not identical, with those we learned when doing simple regression. The main difference is the need to expand or qualify an explanation. We will explain the terms based on the multiple regression equation shown above. The terms in the above equation are defined as follows:

Y = dependent variable

X_1 = independent variable 1

X_2 = independent variable 2

a = intercept, the value of Y when both X variables = 0.

b_1 = slope for variable 1, the amount Y changes for a 1 unit change in X_1 controlling for X_2. The term "controlling for" means the effect of X_2 has been held constant when determining the slope for X_1.

b_2 = slope for variable 2, the amount Y changes for a 1 unit change in X_2 controlling for X_1. The term "controlling for" means the effect of X_1 has been held constant when determining the slope for X_2.

If there are more than two independent variables the explanations are modified to take into account the new independent variables such as X_3, X_4, etc.

We can again work with our IQ/Test Score data set by adding a new independent variable. Study time (in hours) is another factor that could have an effect on test scores so we will add it as our second independent variable. We will also add a few new students (a data set of 6 is very small for regression and as we add more variables the effect of small sample size becomes more serious). Our new data set is shown below:

Student	IQ	Study Hours	Test Score
1	98	8	80
2	98	5.5	78
3	110	4.5	82
4	110	7	88
5	122	5	92
6	122	7	96
7	100	6.5	83
8	105	8.5	87
9	115	8	89

We'll let X_1 represent IQ and X_2 represent study time in hours. To write our multiple regression formula we need to know a, b_1 and b_2. To find these values we will rely on the calculations from a computer program which is shown below.

SUMMARY OUTPUT

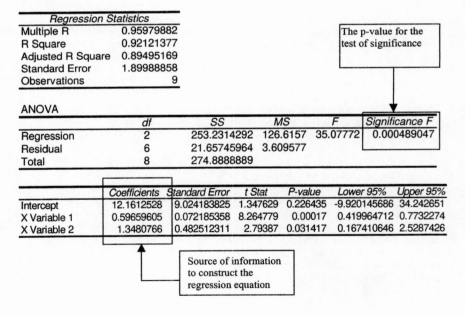

Regression Statistics	
Multiple R	0.95979882
R Square	0.92121377
Adjusted R Square	0.89495169
Standard Error	1.89988858
Observations	9

The p-value for the test of significance

ANOVA

	df	SS	MS	F	Significance F
Regression	2	253.2314292	126.6157	35.07772	0.000489047
Residual	6	21.65745964	3.609577		
Total	8	274.8888889			

	Coefficients	Standard Error	t Stat	P-value	Lower 95%	Upper 95%
Intercept	12.1612528	9.024183825	1.347629	0.226435	-9.920145686	34.242651
X Variable 1	0.59659605	0.072185358	8.264779	0.00017	0.419964712	0.7732274
X Variable 2	1.3480766	0.482512311	2.79387	0.031417	0.167410646	2.5287426

Source of information to construct the regression equation

The last or third table is where we find the elements needed to form the regression equation. In the coefficients column of this table, our interest is in the two independent variables. Because of the way we entered our data, "X Variable 1" refers to IQ and "X Variable 2" refers to study time. We can see that the slope in this model for IQ is .597 and that it is significantly different than a slope of 0 (p value = .00017). In multiple regression the b values are often called **partial regression coefficients** or **partial slopes** because, as noted above, they are calculated after controlling for the influence of the other variables. In this model test scores increase by .597 points for every increase of 1 IQ point. Furthermore, in our equation, b_1 will be .597. The results for the study time are given in the "X Variable 2" row and show that the slope for this variable is 1.35. This slope is also statistically significant because the p value of .03 is less than .05. We can also state that for every hour of study time the test score improves by 1.35 points. In our equation the b2 value will be 1.35. The last thing we need from this table is the intercept, a. From the table we see that the intercept coefficient is 12.16 so we can now write our multiple regression equation:

$$Y = 12.16 + (.597*X_1) + (1.35*X_2)$$

Before leaving this explanation it is important to note that we could have ended up with a non-significant independent variable, but that doesn't mean the variable by itself is insignificant. What can happen in multiple regression is that two or more variables contain the same information. Thus, when multiple regression is done the effect of the variable of interest is calculated in the "presence" of the other variable(s). In the definition of beta that's what the expression "controlling for" means. The net result is that the variable of interest has less effect due to the presence of variable 2 than it would have on its own.

The null hypothesis for multiple regression is tested with an F test which is found in the ANOVA table (see the second table in the print-out titled ANOVA). We can concentrate on the "Significance F" column in that table because the value in that column is the p-value for the test of significance. If the p value is less than .05 it means you have a statistically significant regression. A significant regression means that at least one regression coefficient is not equal to 0. If the regression were not significant there would be little reason to explore the results further. Our p-value is .0005 so we have a highly significant regression.

Contribution of Independent Variables

Many researchers feel that your final multiple regression model should contain only those variables that are useful in predicting the dependent variable. Useful means that the variable makes a statistically significant contribution. Variables that are not statistically significant are usually discarded. A popular method used to determine the contribution of an independent variable is by means of the **partial F-test criterion**. The method entails determining the contribution of each variable based on its regression sum of square.

Be forewarned that the following explanation may sound complex because the data come from so many different sources, but the process is pretty straight forward. Nevertheless, you need to be careful and follow the steps thoroughly. To apply the partial F-test criterion we need to have data from three different regressions – the multiple regression which we have been working with and a regression run for each independent variable which are run as simple linear regression models. From each regression we rely on the data presented in the ANOVA table. The three ANOVA tables are presented as Table 13-B (the multiple regression), Table 13-C (the simple regression run for the IQ variable), and Table 13-D (the simple regression run for the Study Time variable). The data needed to determine the Partial F-test Criterion is identified by the boxed entries in these tables.

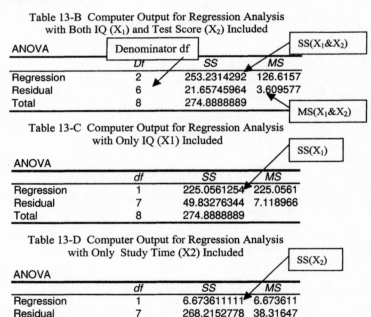

Table 13-B Computer Output for Regression Analysis
with Both IQ (X_1) and Test Score (X_2) Included

ANOVA	Denominator df		SS(X_1&X_2)
	Df	SS	MS
Regression	2	253.2314292	126.6157
Residual	6	21.65745964	3.609577
Total	8	274.8888889	MS(X_1&X_2)

Table 13-C Computer Output for Regression Analysis
with Only IQ (X1) Included

ANOVA			SS(X_1)
	df	SS	MS
Regression	1	225.0561254	225.0561
Residual	7	49.83276344	7.118966
Total	8	274.8888889	

Table 13-D Computer Output for Regression Analysis
with Only Study Time (X2) Included

ANOVA			SS(X_2)
	df	SS	MS
Regression	1	6.673611111	6.673611
Residual	7	268.2152778	38.31647
Total	8	274.8888889	

To do the calculations we also need to introduce and define new data terms which are given in Table 13-E.

Table 13-E Data Needed to Conduct the Partial F-Test Criterion

Data	Definition	Source	Value
$SS(X_1\&X_2)$	The sum of squares for regression when X_1 and X_2 are both included in the regression model	Table 13-B	253.2
$SS(X_2)$	The sum of squares for regression when X_2 is used alone in the regression model	Table 13-D	6.7
$MS(X_1\&X_2)$	The mean square (MS) for residual when X_1 and X_2 are both included in the regression model.	Table 13-B	3.6
$SS(X_1)$	The sum of squares for regression when X_1 is used alone in the regression model	Table 13-C	225.1

To determine if IQ (the X_1 variable) makes a statistically significant contribution we need to do a preliminary calculation. We need to find a value we'll identify as $SS(X_1|X_2)$ which can be defined as the Sum of Squares for X_1 given X_2's presence. The formula is:

$$SS(X_1|X_2) = SS(X_1\&X_2) - SS(X_2)$$

As shown in Table 13-E, the values we need are in Tables 13-B and D and we get:

$$SS(X_1|X_2) = 253.2 - 6.7 = 246.5$$

Our last calculation requires use of the following F formula:

$$F = \frac{SS(X1 \mid X2)}{MS(X1 \& X2)}$$

$SS(X_1|X_2)$ is obtained from the preliminary calculation shown above and $MS(X_1\&X_2)$, as noted in Table 13-E, is obtained from Table 13-B.

$$F = \frac{246.5}{3.6} = 68.5$$

This calculated value for F is then compared to a critical F value found in an F table (Table 3 of the Appendix). If the calculated F exceeds the critical F the contribution of the variable is considered statistically significant. In looking up the critical F value, the degrees of freedom for the numerator is always 1 and the degrees of freedom for the denominator is the same as that given in the Residual row in the ANOVA table for the multiple regression (Table 13-B). In our example the denominator degrees of freedom is therefore 6. The critical F in this case (using the .05 level of significance) is 5.99. Since the computed F far exceeds the critical F we conclude that IQ makes a statistically significant contribution to the multiple regression model.

To determine if Study Time (the X_2 variable) makes a statistically significant contribution we need to repeat the steps above with just a few alterations. The preliminary calculation involves the following formula:

$$SS(X_2|X_1) = SS(X_1 \& X_2) - SS(X_1)$$

where $SS(X_1 \& X_2)$ and $SS(X_1)$ are obtained from Tables 13-B and C. Inserting the appropriate values in our $SS(X_2|X_1)$ formula produces the following result:

$$SS(X_2|X_1) = 253.2 - 225.1 = 28.1$$

Our last calculation requires use of the following F formula:

$$F = \frac{SS(X2 \mid X1)}{MS(X1 \& X2)}$$

$SS(X_2|X_1)$ is obtained from the preliminary calculation and $MS(X_1 \& X_2)$ is obtained from Table 13-B. Inserting the appropriate values produces the following result:

$$F = \frac{28.1}{3.6} = 7.8$$

The critical value for F in this case is also 5.99 since the degrees of freedom are the same as that given for the IQ test. Our calculated F is 7.8 which is greater than the critical F so we conclude that study time also makes a statistically significant contribution to the multiple regression model.

Multiple Regression Explanation (Continued)

Continuing with our explanation of the output from multiple regression we move to the first table which is reproduced on the next page.

We'll begin with the second entry in the table, the R square value. The value .92 indicates that 92% of the variation in test scores can be attributed to our independent variables. The rest of the variation (8%) is due to random variation coming from other sources.

SUMMARY OUTPUT

Regression Statistics	
Multiple R	0.95979882
R Square	0.92121377
Adjusted R Square	0.89495169
Standard Error	1.89988858
Observations	9

Note that the first table entry, **Multiple R**, is the square root of R Square $\left(\sqrt{.9212} = .96\right)$. However, in this case it's not possible to attach a positive or negative sign to the value of Multiple R. Some of the independent variables that go into the regression may have positive slopes and others may have negative slopes so no sign can be attached to Multiple R.

The **Adjusted R Square** attempts to modify R square on the basis of the number of independent variables that are used. As new independent variables are added to a regression, R Square will never decrease even if the new variables are useless. It, therefore, is not possible to tell from R square how much the addition of new independent variables improved the model. The Adjusted R Square modifies R square for the number of independent variables used. In this way it provides another way for you to know whether the addition of independent variables are worthwhile. In our example the Adjusted R Square is .89. If we added a third independent variable to our regression and the Adjusted R Square stayed the same or became less than .89, it would tell us that adding the third variable was of little or no value in predicting test scores.

The Standard Error, or as we prefer to call it the Standard Error of the Estimate, is interpreted the same as it was in a simple regression. The value we have, 1.90, indicates that about 2/3s of the predicted test scores will be within ±1.90 points of the observed test score. Note that in the simple regression example, the standard error of the estimate was 2.78 so in our new regression the predicted points and the observed points are, on average, closer together which is a sign of a better regression fit.

Just as in simple linear regression, we can obtain a display of the residuals with multiple regression. A table showing the residuals values is given on the next page. The table shows the predicted test score (see the *Predicted Y* column) for each observation and the distance that that point is from the observed test score (see the *Residuals* column). We see that observation 1 has an expected test score of 81.4. When this score is compared to the observed score of 80 (which is not given in the table) the distance between the two scores is -1.4 units. As you may recall a residual subtracts off the predicted score from the actual score so in this case our difference is a negative number. As you look through this table you can see that poorer predictions occur with observation 9 and 7 and the best predictions occur with observations 2 and 5.

RESIDUAL OUTPUT

Observation	Predicted Y	Residuals
1	81.4122785	-1.412278481
2	78.042087	-0.042086981
3	83.853163	-1.853162977
4	87.2233545	0.776645523
5	91.6863539	0.313646127
6	94.3825071	1.617492926
7	80.5833557	2.41664432
8	86.2624891	0.737510871
9	91.5544113	-2.554411326

We could also have found the predicted score for observation 1 by using our regression equation. Substituting the observation 1 values for IQ [98] and study time [8] would give us:

$$Y = 12.16 + (.597*98) + (1.348*8)$$

$$Y = 12.16 + 58.51 + 10.78 = 81.4$$

This is, of course, the same as the predicted value from the Residual Output table. The advantage of knowing the regression equation is that we can make predictions about new cases. For example, if a student has an IQ of 112 and studies 6 hours for the test, what is that student's predicted score? See if you can find the answer.

Your answer should be 87.1. We could also ask: "If the student studies one more hour what will be the expected score?" To find the answer, we can use the formula or, since we know the slope for study time, all we have to do is add the value of the slope, 1.3, to the previous answer and we get 88.4.

We will close this section on the explanation of the output from multiple regression with a suggestion on what to include when reporting multiple regression results. For instance, we could report the results as follows:

A statistically significant relationship exists between test scores and the variables IQ and study ($p<.001$). For a given study time, the estimated test score will increase by .597 points for each one point increase in IQ. For a given IQ, the estimated test score will increase by 1.35 points for each hour of study time. Both variables make a statistically significant contribution when predicting test scores. About 2/3s of the predicted test scores are within 1.9 points of the observed test score value. Approximately 92% of the total variation in test scores is accounted for by the variables IQ and study time. The adjusted R square for this relationship is .89.

Note that the summary statement for a multiple regression includes the same elements we used in the simple linear regression statement. However, the multiple regression statement is a bit more complex because there is more than one independent variable.

Dummy Variables

The theory behind regression is based on the use of interval scale measurements, but as it turns out ordinal and even nominal variables can frequently be used in multiple regression. Based on what we have discussed about the unequal distance between the categories of an ordinal scale you need to be careful about your interpretation. Furthermore, nominal variables are especially tricky and must be entered and interpreted with care. In fact, the term **"Dummy Variable"** is used to refer to nominal variables introduced in multiple regression and the term is descriptive because it suggests that a false or specious variable is being used.

Let's work with an example of a multiple regression. Our dependent variable will be salary based on a person's years of experience and sex (the independent variables). Suppose salary is reported in thousands of dollars (i.e., $10,000 is represented as 10.0) and the regression equation for this relationship is:

$$Y = 10.5 + (2.10*X_1) + (3.50*X_2)$$

where: X_1 is Years of experience, an interval scale measurement.

X_2 is Sex, a nominal scale measurement.

In the original data set, the sex variable should be coded as a numeric entry of 0 or 1. It doesn't matter which sex is coded 1 or which is coded 0. For our example, assume we coded female as 0 and male as 1. What is our prediction of the salary for a male who is 10 years old?

Your answer should be 35.0 ($Y = 10.5+[2.10*10]+[3.50*1]$). What is the predicted salary for a female with the same years of experience?

In this case the predicted value is 31.5 (Y = 10.5+[21.0*10]+[3.5*0]). Note that the difference is 3.5 which is the b value for X_2 (sex).

Had we reversed the coding for sex, we would have created a new data set. All the sex codes would be different in the first data set compared to the second set. Consequently we would have produced a new regression equation. In our example the new regression equation would be:

$$Y = 14.0 +(2.10*X_1) -(3.50*X_2)$$

Note that the slope for sex is still 3.5, but now it is a negative slope rather than a positive slope and as a result the last part of the equation requires subtraction rather than addition. If we substitute the data on our hypothetical male with 10 years experience and our female with the same amount of experience what predicted salaries would we get (Remember a male is now coded as a 0 and a female as a 1)?

The male salary would be 35.0 (Y = 14.0+[21.0*10]-[3.50*0]). The female's salary is 31.5 (Y = 14.0+[21.0*10]-[3.5*1]). The salaries are identical to those obtained with the original coding system.

We can also use nominal variables as the Y variable. It's easiest to see how this is done if the variable is a binomial so that will be the case in our first example. In this example, we will start out by asking what is important in the selection of students to an honorary leadership society. The dependent variable in the example is whether or not a person is selected to the society. That variable is a nominal variable with the possible answers of Yes or No. It is important for interpretation purposes that the No response be coded as 0 and the Yes response be coded as 1.

For independent variables, we could use GPA as the X_1 variable, and the number of organizations a student belongs to as the X_2 variable. Assume we use these independent variables, perform a regression and obtain the following regression equation:

$$Y = .05 + (.14*X_1) + (.05*X_2)$$

If GPA = 3.30 and a student belongs to 6 organizations what is Y?

Your answer should be .81. Substituting the given information into the regression equation produces:

$$Y = .05 + [.14*3.30] + [.05*6]$$

$$Y = .05 + 0.46 + .30 = .81$$

We interpret Y as a probability, but with reservations. Remember that Y is whether or not a person is elected to the society. Note that our coding scheme for Y was 0 = No and 1 = Yes. Thus our range of Y values, 0-1, is the same as a probability range. With a GPA of 3.30 and membership in 6 organizations our prediction is that there is a .81 probability that this person will be selected for the society. Now for some cautions - the mathematics of regression doesn't truly treat the data as a probability and therefore predictions greater than 1 can occur. Here's an example: A person with a 3.50 GPA and membership in 10 organizations. What does the regression equation give as the probability prediction for this individual?

Your answer should be 1.04 (Y = .05+[.14*3.50]+[.05*10]). Of course that probability is impossible so we would realize something is wrong. As a rule of thumb you need to be skeptical of results close to 0 or 1 when using a nominal dependent variable.

Here's another case that is illogical - a person with a GPA of 1.00 and membership in no societies. We know from experience that that person would not even be permitted to remain in school. However, the regression equation would produce an answer. What is it?

You should have found that Y = .19 (Y = .05+[.14*1.00]+.[05*0]). There is a .19 chance that the person would be selected for the honorary society. Remember regression is a mathematical process that does not incorporate human common sense.

Summary

In this chapter we found that regression analysis is a powerful tool to determine the nature of a relationship between a dependent variable and an independent variable(s). The regression model we studied requires certain assumptions and at a minimum we must believe that there is a linear or straight line relationship between the dependent and independent variables used in the regression analysis. If there is a single independent variable the analysis is considered simple linear regression, but if there are 2 or more independent variables the multiple linear regression procedure is used.

The results of a regression analysis tells us 1] how much the dependent variable changes for a 1 unit change in the independent variable, 2] the strength of the relationship as measured by r or multiple R, 3] whether the relationship is at a statistically significant level and 4] the amount of variation in the dependent variable that can be accounted for by the independent variable(s). We are also able to write a mathematical equation describing the relationship between the variables, measure the spread of observed responses around a fitted regression line and to make predictions regarding the dependent variable. We also saw that in multiple regression we can test to see if the independent variables are making a statistically significant contribution to the regression model. Although designed for interval scale measurements, we found that we can use dummy variables for measurements based on the nominal scale providing we are circumspect in how we interpret the results. We also learned that if we are doing exploratory research, we must consider the results of regression analysis as hypothesis generating information rather than as a definitive hypothesis testing result.

PROBLEMS
(Unless otherwise specified assume all test are done at the .05 level of significance.)

1. If the regression equation is:
   Y = 212 + 5.5X
   a. At what point does the regression line cross the X axis?
   b. How many units does Y change when there is a 10 unit fall in the X variable?
   c. Is there a positive or negative correlation between the two variables? What data did you use to come to your conclusion?
   d. If a person has an X value of -40, what is the person's Y value?

2. A simple regression problem was carried out to determine the effect of mileage on car maintenance. The results of a regression analysis are given below:

Regression Statistics	
Multiple R	0.623
R Square	0.388
Adjusted R Square	0.369
Standard Error	460.955
Observations	34

ANOVA

	df	SS	MS	F	Significance F
Regression	1	4311609.4	4311609	20.3	8.33E-05
Residual	32	6799347.0	212480		
Total	33	11110956.4			

	Coefficients	Standard Error	t Stat	P-value
Intercept	474.7	179.5	2.645	0.013
miles	0.011	0.002	4.505	8.33E-05

   a. Write the regression equation.
   b. Is there a statistically significant relationship between the two variables? What information did you use to arrive at your answer.
   c. What is the predicted maintenance cost for cars that are driven 80,000 miles and 125,000 miles respectively?
   d. If you decide to drive a randomly selected car 5,000 more miles, how much should you expect to pay in maintenance costs?
   e. What percent of the variation in a car's maintenance cost is accounted for by the miles driven variable?
   f. Within what range do the middle 67% of the residuals probably fall?

3. How does the interpretation of the following variables differ between simple and multiple regression?
   a   Beta
   b   The intercept
   c   R Square
   d   A statistically significant F value

4. The following report is derived from an analysis of college sophomores in which the effect of SAT scores and sex on GPA was examined. Sex was coded as 0 = Male and 1 = Female. What conclusions can you reach based on the report?

Regression Statistics	
Multiple R	0.61851183
R Square	0.38255688
Adjusted R Square	0.31756287
Standard Error	0.28501855
Observations	22

ANOVA

	df	SS	MS	F	Significance F
Regression	2	0.956310468	0.478155	5.886033	0.010248754
Residual	19	1.543475896	0.081236		
Total	21	2.499786364			

	Coefficients	Standard Error	t Stat	P-value	Lower 95%	Upper 95%
Intercept	0.027	0.831	0.032	0.975	-1.713	1.766
Sex	0.200	0.122	1.643	0.117	-0.055	0.456
SAT	0.002	0.001	3.021	0.007	0.001	0.004

5. If you added the variable "study habits" to the regression done in problem 4 and the Regression Statistics changed to:

Regression Statistics	
Multiple R	0.622319
R Square	0.387281
Adjusted R Square	0.285161
Standard Error	0.291706
Observations	22

a.   Is Study Habits a useful variable that improves the model?
b.   What information did you use for your answer to part a?

6. State historical sites can be evaluated in terms of the average number of visitors per week at a site. It is believed that the attendance is a function of the following variables:

X_1 The number of people who live within a short drive (150 miles or less).
X_2 The number of motel rooms in the area.
X_3 The number of days in a year when the weather is rated as "poor".

You are told that R square is .60 and the p-value from the ANOVA analysis is .042. You also receive the following data:

	Coefficients	P-Value
Intercept	14575	.037
X Variable 1	.003	.231
X Variable 2	12.5	.218
X Variable 3	-181	.002

a Write a memo describing your conclusions.
b If there are 50,000 people within a 150 mile radius, 350 motel rooms in the area and 43 poor weather days a year what is the expected weekly attendance?

7. A Residual Output table is shown below.

Observation	Predicted Y	Residuals
1	8.7	4.2
2	10.7	4.7
3	8.3	-2.6
4	7.8	-3.8
5	5.9	1.0
6	9.1	1.1
7	6.8	2.7
8	11.1	2.6
9	7.0	-1.0
10	9.8	-5.3
11	10.8	-3.6

a. For observation 7 what is the predicted value and is it larger or smaller than the actual value?
b. What was the actual score for a person with a predicted score of 9.1?
c. What observation would be considered to have the poorest prediction?
d. What is the standard error of the estimate?

8. Refer to problem 4 and the data given below. Determine if each independent variable (i.e. sex and SAT) makes a significant contribution to the regression model.

Computer Output for Regression Analysis with Only Sex (X_1) Included

ANOVA

	df	SS	MS
Regression	1	0.21492	0.21492
Residual	20	2.28486	0.11424
Total	21	2.49978	

Computer Output for Regression Analysis with Only SAT (X2) Included

ANOVA

	df	SS	MS
Regression	1	0.73712	0.73712
Residual	20	1.76266	0.08813
Total	21	2.49978	

9. What problems might be encountered in a Counter Balanced Design?

10. What is a double blind study? Why is it done?

11. In a longitudinal study, what is the difference between a panel and a time series?

12. What is the experimentwide error if we compare 3 different experimental conditions and do separate t tests for each of the 3 possible paired group differences?

13. Ten candidates for the school's best all around athlete were ranked from high to low (best to worst) by two independent raters. What is the correlation coefficient between the raters?

Candidate	Rater 1	Rater 2	Candidate	Rater 1	Rater 2
Art	5	4	Fred	6	7
Bob	2	3	Gail	3	1
Carol	7	9	Harry	8	8
Don	9	10	Irene	10	6
Ed	1	2	Joan	4	5

14. Allergy tests were done on the left and right arms of a group of 17 volunteers. The area of redness, in square centimeters, was measured on each arm. Half the subjects were given a new agent on the left arm and the established agent on the right arm. For the other half the new agent was given on the right arm and the old agent on the left arm. The results of the tests are shown below. Is there a statistically significant difference?

Subject ID No.	New Agent	Old Agent	Subject ID No.	New Agent	Old Agent
106	20	29	994	18	26
135	22	28	871	24	21
321	24	28	401	27	26
348	24	23	458	24	28
652	23	29	310	27	30
78	27	25	990	27	29
342	19	28	505	22	33
628	25	28	177	24	22
240	23	23			

15. A poll was taken of a random sample of faculty and students concerning their opinion about a trimester academic year as opposed to the current semester system. The results of the survey are shown below. Is there evidence of a significant difference between the faculty and students?

Preference	Faculty	Students
Trimester	15	202
Semester	56	482

16. Use the Combined Z Transformation method to create standardized scores for the 8 groups in the following data set. Assume the data set contains all the observations of interest.

		VARIABLE	
GROUP	A	B	C
ONE	32	13	95
TWO	41	9	160
THREE	32	11	163
FOUR	40	7	144
FIVE	27	7	149
SIX	27	9	195
SEVEN	36	15	142
EIGHT	32	10	153

Statistical Methods for Researchers

Rank the groups from high to low based on the standardized score.

17. Your organization invested their pension fund in four stocks during the past year. Form a monthly index (starting with an index value of 100 for January) based on the stock prices given below:

Stock	Jan	Feb	Mar	Apr	May	June	July	Aug	Sept	Oct	Nov	Dec
Angel	18.0	18.5	16.5	17.0	18.0	18.5	19.5	19.0	21.0	22.5	21.0	24.0
Best Buy	33.0	34.5	32.0	30.0	28.5	26.0	26.5	24.0	27.0	28.5	30.0	31.0
Chess	12.5	12.5	12.5	12.5	12.0	12.5	13.0	13.5	13.0	14.0	13.5	13.0
Dog Hill	88.5	84.0	86.0	86.5	85.5	91.5	96.0	94.5	99.0	92.5	91.0	90.5

Excel Instruction

As noted in the text it is important to prepare a graph of the relationship between the two variables when doing simple linear regression. Chapter 2 explained how to prepare an XY graph, but it should be emphasized that Excel interprets the first variable in a data set as the X variable so when you use Excel to create the XY graph, be sure to arrange your data so that the X variable precedes the Y variable on the worksheet.

To conduct a regression analysis in Excel you will need the data for the X and Y variables on a worksheet. Excel carries out the needed procedures with a function appropriately called "REGRESSION". We access the REGRESSION function by selecting "Tools", "Data Analysis" and then choosing "REGRESSION" from the "Analysis Tools" listing. You then hit "OK" and the Regression dialog box appears on the screen (see below).

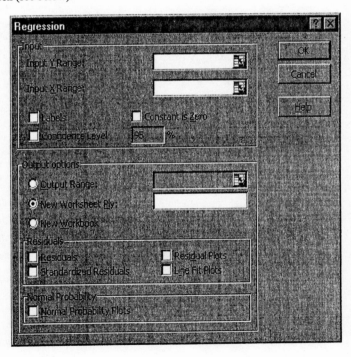

Give the location of the Y variable in the "Input Y Range" box - note that the Y variable is identified before the X variable in REGRESSION. Then give the location of the X variable in the "Input X Range" box. If your input range does not contain labels, leaves the "Labels" box empty. However, if the first row of the input range includes labels, click on the box so that a "✓" appears in the "Labels" box. We want our intercept to correspond to an X = 0, which is the default setting for "Constant is Zero", so we just leave this check box empty. Leave the box to the left of "Confidence Level" empty as well. You automatically will have a 95% confidence level for beta in your report. By activating this box you can have an additional confidence level. The box to the left of "Confidence level" and before "%" allows you to set the additional confidence level desired. Your selected confidence interval will appear in the summary table along with the 95% limits.

For the "Output options" section, note that the regression output is rather large so leave the circle for "New Worksheet Ply" with a dot in it. Remember to give the new worksheet a name you'll remember so you can refer to it when needed. We now only need to complete the bottom portions of the dialog box. We often want to examine the residuals so when that is the case, click on the "Residuals" box and a "✓" will appear in the box. All other boxes may be left blank. When you've entered all the information, hit "OK" and you should have a display of your regression analysis.

An example of the regression output plus an explanation of the output were provided in the text. Remember that the multiple R value in simple linear regression is always reported as a positive number even if the correlation is negative. Consequently, use the sign of the slope as the sign for the correlation coefficient. If the slope is positive then the correlation coefficient is positive. However if the slope is negative then you need to interpret the correlation as a negative relationship.

To do a multiple regression, you essentially perform the same steps described above. However, you will have more than one X variable and all the X variables must be adjacent to each other. Furthermore it is recommended that you include the labels for the variables. It is much easier to interpret the output if the names of the X variables are included.

To determine the value of the Partial F-test Criterion you need to conduct a simple regression analysis for each X variable independently. The ANOVA tables for a given X variable contain the SS(X) values. The $SS(X_1 \& X_2)$ and the $MS(X_1 \& X_2)$ are derived from the multiple regression output using all X variables. The calculated F value for the test is found by a simple division. To determine the critical F, you use a function called FINV. To access FINV, click on the Function icon, then "Statistical" from the Function Category list, followed by "FINV" from the Function Name listing. The FINV dialog box is shown on the next page.

The "Probability" entry is the Alpha Risk which is usually set at .05. "Deg freedom1" is the df value for the Regression row of the multiple regression ANOVA table. It is based on the following calculation: (the number of variables included in the multiple regression

−1). The "Deg freedom2" is the df value for the Residual row in the multiple regression ANOVA table. It is based on the following calculation (total number of observations) − (the number of independent variables + 1). In the multiple regression example we worked in the text the probability value would be .05, "Deg freedom1" would be 1 (2-1) and "Deg freedom2" would be 6 (9-3). After clicking on OK the critical F value is returned.

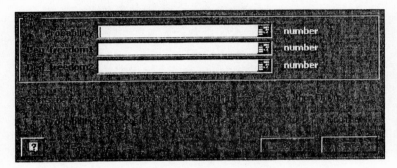

The problems listed below will allow you to practice using Excel.

1. Use your data set to examine the relationship between pay and age.
   a. What do you conclude?
   b. What is the 95% confidence interval for the slope?
2. Using your data base, conduct a regression to see the effect of years of service on pay.
   a. Create a graph showing the relationship
   b. How good is the regression?
   c. How much would a person with 15 years of service be paid?
   d. What is the residual value for the first person (i.e., observation 1) in the data set?
   e. What is the residual value for a person making $20,000?
3. Based on your data base, carry out a regression analysis for advancement potential using years of service and number of awards received as the independent variables.
   a. What is the likelihood someone with 5 years of experience and 5 awards will be said to have advancement potential
   b. Is there a significant relationship between the dependent and independent variables?
4. Use your data base to see the effect test scores and years of service have on the August proficiency scores.
   a. What is the correlation between the variables used in the regression?
   b. Is there a statically significant slope?

Chapter 14

Time Series and Forecasting

Introduction

There are many cases in which a researcher needs to make a forecast. In some instances it may involve a practical issue - what will costs and income be in the near future? Furthermore, after we gather a great deal of information about a phenomenon of interest we may want to use that data to make predictions.

Sometimes the forecast involves predicting whether a certain event will occur or it may entail a description of a future happening - e.g. forecasting the type of communication technology that will be available 5 years from now. These latter types of predictions do not primarily involve numerical information and one of the forecasting techniques for these non-numerical events will be reviewed in chapter 15. In this chapter, quantitative methods for forecasting will be explored. We will learn how to project the future value of a quantitative variable based on its past behavior.

Time Series

A time series is a set of observations of a quantitative variable, which are taken over a period of time. Perhaps one of the best known time series is the Dow Jones Industrial Average which tracks the stock market based on a sample of selected stocks. Most organizations keep track of important factors by developing time series. The variable may be the organization's payroll, the size of the work force or the amount of money spent on health care. From a time series one hopes to gain insight into the past behavior of the variable. What caused a rise or a fall in health care costs? Is there an upward or downward trend in the payroll? Are there times in the year when there are large jumps or falls in the work force? It is also common to use a time series to predict what the variable may do in the future. How much are we likely to spend on health care over the next two years?

The advent of the computer has made it much easier to explore time series. Many sophisticated methods exist, but some people still rely on intuition. In personal decision making, the individual is free to use intuition rather than a more mathematical approach. However, when working for an organization, decisions based on intuition may not be well received. Nevertheless, intuition does have a place in even complex time series analysis. If the answer doesn't "look right", you probably should go back and check your input and the assumptions made. If the answer seems wrong, then there is a reasonably good chance that it may be incorrect.

In a time series, the independent variable is always time and the dependent variable is the factor or phenomenon you are studying (e.g., payroll or health care costs). Time units can be days, weeks, years, etc. We have already learned how to create a graph and constructing a graph is frequently the best place to begin a time series analysis.

Fitting Trendlines

Usually a time series will exhibit a somewhat erratic pattern and it's helpful to identify what the underlying general trend is like. For example, is the variable increasing rapidly or decreasing slowly over time? There are a number of ways we can describe the trend. A very popular tool is a technique we have already studied. We can use linear regression analysis to identify the general behavior of a variable over time. However, it is important to remember that the phenomenon and time must be related in a linear manner when we use linear regression analysis. Linear regression forces a straight line through the data points and if the phenomenon is not linear you will get a misleading trend.

In addition to a linear pattern, a time series phenomenon can also follow other patterns. Two popular patterns are shown in Figure 14-A and Figure 14-B.

Figure 14-A   Figure 14-B

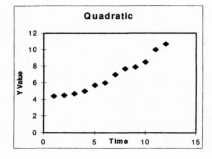
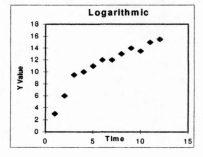

Note that the data points in these figures appear to follow curved patterns rather than linear ones. The term **trendline** is frequently used for the line or even a curve that best represents the dominant pattern and the process is called **line fitting**. Graphs that follow certain patterns are given a name based on the mathematical relationship between the X and Y variables. A popular relationship, called the **quadratic,** is displayed in figure 14-A. The quadratic is often felt to be the simplest curvilinear relationship. There is a generic equation for the curve associated with a quadratic relationship which is:

$$Y = a + b_1 * X + b_2 * X^2.$$

where a is the intercept and b_1 and b_2 are constants.

The specific trendline for the data shown in Figure 14-A is displayed in Figure 14-C (see next page) and is based on the following equation:

$$Y = 4.1 + 0.12 * X + .036 * X^2.$$

We relied on a computer program to generate the trendline equation and the mathematical derivation will not be discussed. Once we have the equation for the trendline (see above) we find a predicted Y value just as we did in linear regression. For example, when X is 2, the value of Y is predicted to be:

$$Y = 4.1 + (.12 * 2) + (.036 * 2^2) = 4.1 + .24 + .144 = 4.484$$

A **logarithmic** relationship (Figure 14-B) will typically rise rapidly at the beginning and then level off. The generic equation for the curve associated this relationship is:

$$Y = a + b * Ln(X)$$

where a and b are constants and Ln(X) is the natural log of X.

You can see that the logarithmic relationship gets its name from the presence of a logarithmic notation (i.e. Ln) in the formula. The logarithmic trendline, for the data shown in Figure 14-B, is displayed in Figure 14-D (see next page) and is based on the following equation:

$$Y = 3.16 + 4.83 * Ln(X)$$

Again we relied on a computer program to produce our equation. Using the above equation, we can now predict values for the logarithmic relationship. It should be noted that to use the equation, it is necessary to find the natural log of X (which is obtained from a natural log table). The predicted Y, when X is 2, is given below. The natural log of 2 (the value of X) is .693, so our equation becomes:

$$Y = 3.16 + (4.83 * .693) = 6.507$$

Figure 14-C

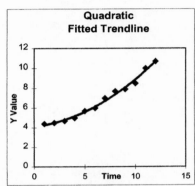

Figure 14-D

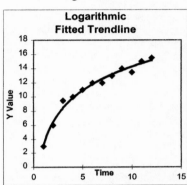

Two additional curvilinear relationships are shown below as Figure 14-E and 14-F illustrating a polynomial and exponential relationship respectively.

Figure 14-E

Figure 14-F

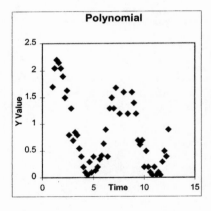

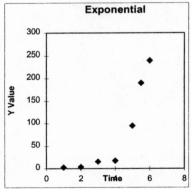

The most complex looking pattern is that shown in Figure 14-E. The oscillating relationship displayed here falls into a class of relationships called **polynomial**. The generic formula for a polynomial is:

$$Y = a + (b_1*X) + (b_2*X^2) + (b_3*X^3)...$$

where a, b_1, b_2, etc. are constants.

The polynomial trendline, for the data presented in Figure 14-E, is displayed in Figure 14-G (see below) and is based on the following equation:

$$Y = .05 + 3.77X - 2.27 X^2 + .489 X^3 - .0.04 X^4 + .001 X^5$$

Once we have the equation for the trendline (see above) we can again find a predicted Y. In this case, when X is 2, Y is predicted to be:

$$Y = .05 + (3.77*2) - (2.27*2^2) + (.489*2^3) - (.0.04*2^4) + (.001*2^5)$$

$$Y = .05 + 7.54 - 9.08 + 3.912 - 0.64 + 0.032 = 1.814$$

Technically the quadratic is a special form of a polynomial. However, because of its popularity we will treat the quadratic as a distinct and separate relationship.

Figure 14-G   Figure 14-H

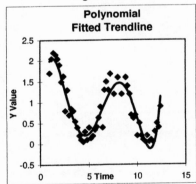

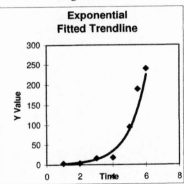

The **exponential** relationship (see Figure 14-F) is characterized by a slow start but then increasingly large values for Y as X becomes larger. The generic equation for an exponential is:

$$Y = c*e^{b*X}$$

where c and b and constants and e is the base of the natural logarithm (or roughly 2.718).

The exponential trendline, for the data shown in Figure 14-F, is displayed in Figure 14-H (see previous page) and is based on the following equation:

$$Y = .825 * e^{0.935X}$$

In this example, if X is 2, then the predicted Y is

$$Y = .825 * e^{0.935*2} = .825 * 2.718^{1.870} = .825 * 6.49 = 5.35$$

An analyst inspects a data set and sees how different trendlines fit the data, trying to find the best but simplest relationship. In chapter 13 we looked at different ways to determine how good a fit was. We used a statistic such as the R square value. The R Square values for the four relationships described above (i.e. the quadratic, logarithmic, polynomial and exponential) are given in Table 14-A. For comparison purposes the R Square value for a linear fit for each data set is also provided in Table 14-A.

Table 14-A
R Square Values for Different Fitted Relationships

Figure	Fitted Relationship	R Square
14-C	Quadratic	.99
	Linear	.96
14-D	Logarithmic	.98
	Linear	.87
14-G	Polynomial	.89
	Linear	.23
14-H	Exponential	.96
	Linear	.75

A linear relationship is the simplest, and therefore the preferred choice, unless one of the more sophisticated trendlines is clearly a better fit. For example, in the first two rows of Table 14-A, the quadratic relationship is marginally better than a linear fit (.99 versus .96). The difference is small and some analysts would be satisfied with the linear fit because it's easier to work with than a quadratic trendline. However, for the other comparisons in Table 14-A, the linear fit is clearly inferior. This is especially noticeable in the contrast between the polynomial and linear fit for Figure 14-G where the difference in R Square is very large (.89 versus .23).

Once you have settled on the type of relationship that provides the best fit, you can easily make a projection on what are the next most likely values for Y. For example, assume we want to predict the next value for the quadratic relationship shown in Figure 14-C. In order to predict the next Y value, we need to note that the time variable (X) is scaled from 1 to 12.

We would therefore need to substitute an X of 13 (the next time point) in the quadratic equation to arrive at our answer. The quadratic equation is reproduced below:

$$Y = 4.1 + 0.12*X + .036*X^2.$$

The calculation is straight forward:

$$Y = 4.1 + 0.12*13 + .036*13^2 = 4.1 + 1.56 + 6.08 = 11.74$$

In a similar fashion one can obtain predictions for each of the other relationship discussed above.

Moving Average Smoothing

In addition to line fitting, there are techniques called **smoothing** which reduce the irregular pattern of a time series and help an analyst identify long-term trends in the series. The most popular smoothing technique is the **moving average** method. When employing the moving average technique the analyst plots the mean for 2 or more consecutive points rather than each observed point, thereby eliminating some of the jagged appearance in a time series. By smoothing, the uneven appearance of a time series can be diminished. Refer to the data presented below (Table 14-B) and shown graphically in Figure 14-I (see next page).

Table 14-B
Raw Data for Smoothing

Time Point	Raw Data
1	24
2	35
3	51
4	56
5	68
6	53
7	59
8	52
9	38
10	62
11	44
12	50

To determine a moving average, based on a collection of 3 time points, we first calculate a mean value for moving average at time point 2. We use the term "moving average at time point 2" to denote that the time point we are calculating is for the moving average. Note that there is also an actual time point 2 - it is the value given in Table 14-B (i.e. 35).

Figure 14-I

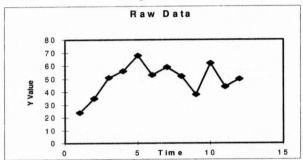

The mean for moving average at time point 2 is based on actual time points 1, 2 and 3. The mean of those time points is 36.7 ([24+35+51]/3=36.7). The value 36.7 is assigned to time point 2 of the moving average. Note that we can not calculate a 3 observation moving average for time point 1 since there is no observation prior to the first time point. For moving average time point 3, the mean is based on actual time points 2, 3 and 4 which has a mean of 47.3 ([35+51+56]/3 =47.3). For the next moving average time point (number 4) what value do you get for the moving average?

Your answer should be 58.3 ([51+56+68]/3). We continue calculating the means in this fashion through moving average time point 11. We can not do a mean for time point 12 since our data set does not have a time point 13. Table 14-C (see next page) shows the values for the moving average based on 3 time points.

If we wanted a moving average based on 4 time points we would take averages for each set of 4 adjacent time points. The first moving average at a time point that we can calculate utilizes actual time points 1, 2, 3 and 4. The average is 41.5 ([24+35+51+56]/4=41.5) and it is assigned to the moving average time point 2.5. Note that the moving time point 2,5 is the mean of the four actual time points ([1+2+3+4]/4=2.5). Furthermore, even though we did not begin with an actual time point of 2.5, there is no reason we can not use such a time point in a moving average. Whenever we use an even number of time points for our moving average, we will end up with moving average time points that are not identical to the original set of time points.

Table 14-C
Moving Average Values

Time Point	Raw Data	Moving Avg.
1	24	
2	35	36.7
3	51	47.3
4	56	58.3
5	68	59.0
6	53	60.0
7	59	54.7
8	52	49.7
9	38	50.7
10	62	48.0
11	44	52.0
12	50	

Figure 14-J graphs the original data set and a second set of points based on the moving average technique. It should be clear that the moving average presents the data in a flatter less erratic fashion.

Figure 14-J

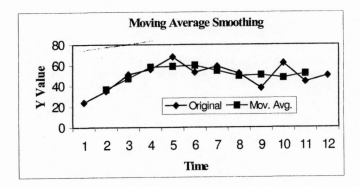

Exponential Smoothing

There is a second smoothing technique that is frequently used called exponential smoothing. Exponential smoothing provides a less erratic image of the movement of a time series by reducing the irregularities present in a data set. In addition, the exponential smoothing technique can be used to forecast a future outcome which is not possible with a moving average. The **exponential smoothing** technique is essentially a weighted average since each point of a smoothed curve is based on all the previously observed data points. The most recently used value receives the most weight, the one before that the next most weight, etc.

Exponential smoothing also requires a damping factor (D) which is a number between 0 and 1. The value of D is subjectively assigned and dependent on an analyst's goal. If the goal is to dampen the spikes in a data set, then the weighting coefficient should be close to 1. If the intention is to use the smoothed time series for prediction purposes, then D is set close to 0.

The formula for exponential smoothing is based on the following equation:

$$E_i = (1-D)*Y_i + D*E_{i-1}$$

where: E_i = the smoothed value at time point i
  D = the damping factor
  Y_i = the observed value at time point i
  E_{i-1} = the previous smoothed value

If we want to predict the Y value for time point 4, using a damping factor of .5, then i is 4 and the formula becomes:

$$E_4 = (1-.5)*Y_4 + .5*E_3$$

where: E_4 = the smoothed value at time point 4
  D = .5
  Y_4 = the observed value at time point 4
  E_3 = the previous smoothed value (i.e. the smoothed value at time point 3)

The exponential smoothing formula is applied to all time points except the first one. For the first time point, the observed value is used as the smoothed value. Our first smoothed value corresponds to the first time point value in Table 14-C and is therefore 24.

Using a damping factor of .50 for the second time point, the smoothed value (which we will call E_2) is:

$$E_2=(1-.5)*35 + (.5*24) = 17.5+12 = 29.5$$

The other smoothed values are calculated in a similar fashion. What answer do you get for the third time point, E_3?

Your answer should be 40.25 based on the following calculation: $(1-.5*51)+(.5*29.5)=40.25$. Table 14-D, shows the original and the values obtained by exponential smoothing.

Table 14-D
Exponential Values

Time Point	Raw Data	Exponential
1	24	24.0
2	35	29.5
3	51	40.3
4	56	48.1
5	68	58.1
6	53	55.5
7	59	57.3
8	52	54.6
9	38	46.3
10	62	54.2
11	44	49.1
12	50	49.5

To illustrate the effect of exponential smoothing, Figure 14-J will be amended to include the smoothed line based on exponential smoothing (see Figure 14-K). In Figure 14-K, Series 1 is the original data set, Series 2 is the moving average values and Series 3 refers to the values derived from exponential smoothing using a D of .5.

Figure 14-K

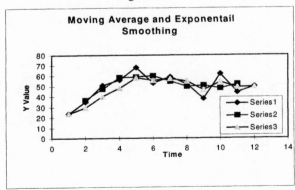

Decomposition

Even if we pick the best trendline we will still not have a perfect fit. There are inevitably discrepancies between the observed points and the fitted line. What we'd like to be able to do is to explain the cause behind the discrepancies. We'd like to identify the reason for any divergence. A common way to try and understand a time series is to break it down into component parts. This process is called **decomposition**. To simplify this section, we will only examine the components for a linear relationship.

Potentially there are four components of a linear time series. There can be an underlying trend, a cyclical component, a seasonal component or an irregular component. The long term direction of the data set is referred to as the **trend**. In Figure 14-L, we have a typical learning curve. At the beginning there is a relative large rise in knowledge gained, but over time the acquisition of knowledge tapers off.

Figure 14-L

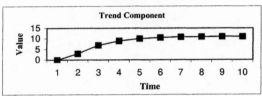

Figure 14-M, which plots entry fees collected in a state park system, illustrates a seasonal effect. In this figure the X-axis gives quarters 1-4 for the first year, quarters 5-8 for the second year, etc. A **seasonal component** means that the time series pattern is replicated and the repetitions occur on a yearly basis. The pattern in this time series is first seen in quarters 1-4 and then the pattern is repeated in quarters 5-8, 9-12, etc.

Figure 14-M

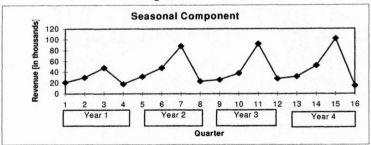

If patterns in a time series are repeated, but the repetitions are not based on a yearly sequence, the series is said to have a **cyclical** component. Figure 14-N, which plots the hypothetical infestation rate of the water beetle in a rural county over a number of years, has a positive trend but it dips and rises in six-year cycles. It would be useful to know what else follows a six-year pattern because that information could provide insight into the reasons for the infestation pattern of the water beetle.

Figure 14-N

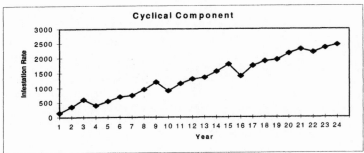

For any pattern there can also be an **irregular component**. A random or unpredictable event can suddenly distort any pattern that seems to be established. If we were monitoring absenteeism in an organization we might find a point, say the middle of the second week, that suddenly disrupts a fairly standard attendance pattern (see Figure 14-O).

Figure 14-O

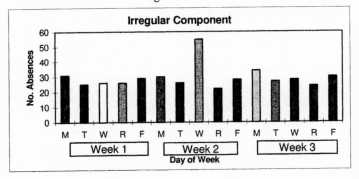

It would behoove the organization to determine what happened to account for the sudden pattern change on the Wednesday of week 2. Perhaps there was a snowstorm or maybe a special community event caused an unexpected rise in absenteeism.

Using decomposition, the analyst has a systematic way to review a time series, a means to identify the pattern the series follows and a way to come up with a rationale for why the pattern is followed. An explanation for any major deviations in the time series should also be sought.

Forecasting

There are alternative ways to make a forecast and many involve mathematical models that go beyond what can be presented in this text. However, in addition to a projection of a time series, based on substituting a future X value in the trendline equation, there are alternative methods worth mentioning. For instance, we could look for other variables, related to the variable that we are trying to forecast, and then incorporate those variables into a regression analysis to make the forecast. Assume we have five years of data on car registrations and our goal is to predict the number of registrations we will have in year 6. Our car registration data is given in Table 14-E and a graph of the data is provided as Figure 14-P.

Table 14-E Auto Registrations

Year	Number
1	1456
2	1409
3	1851
4	1701
5	1568

Figure 14-P

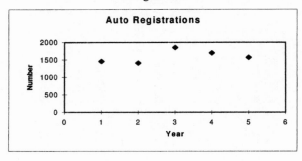

Our first forecast will be based on a simple projection of the time series for the auto registration data. Normally one would want more than 5 time points to construct a regression equation, but they will be suitable just to illustrate the forecasting technique.

A simple linear regression analysis could be run and the output is shown below.

OVA

	df	SS	MS	F	Significance F
gression	1	26625.6	26625.6	0.762	0.447
sidual	3	104772.4	34924.1		
al	4	131398			

	Coefficients	Standard Error	t Stat	P-value	Lower 95%	Upper 95%	Lower 95.000%
ercept	1442.2	196.0	7.36	0.0052	818.4	2066.0	818.436
ar	51.6	59.1	0.87	0.4468	-136.5	239.7	-136.47

From that output we can create the regression equation which is:

$$Y = 1442.2 + 51.6X$$

We can use the regression equation to predict the number of auto registrations for year 6. The year 6 projection is:

$$Y = 1442.2 + 51.5*6 = 1751.2$$

We expect there to be 1,751 registrations in year 6.

Another way we could make the forecast is to find a second variable that is related to auto registrations and use that variable to help predict the number of auto registrations in year 6. Identifying this type of variable is based on a person's knowledge of other variables that may be related to auto registrations. We will refer to the variable we are looking for as a **predictor variable**. There is one additional condition - the predictor variable must come from an antecedent event. In other words, the antecedent event must occur <u>before</u> the event we are trying to predict. Obviously if we are going to predict a stock market change based on economic news, the economic news must be available before the predicted stock market change occurs. In other words, the data needed for the prediction must be available at the time the prediction is made. If the useful information on the predictor variable is contemporaneous with the registration information, it comes too late to be useful in making the prediction.

Assume we find data on home building permits. The data set for home building permits is given in Table 14-F and is graphed in Figure 14-Q.

Table 14-F Home Building Permits

Year	Permits
1	28
2	115
3	67
4	34
5	45

Figure 14-Q

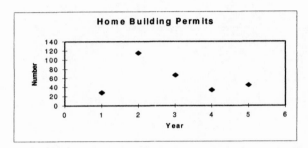

Next we match years 1-4 of the home building permits data to that of years 2-5 for the car registration data set. In other words, Year 2 for registrations is paired with Year 1 for permits, etc. When we do this we see the home building permits follow the pattern in car registrations (see Figure 14-R).

Figure 14-R

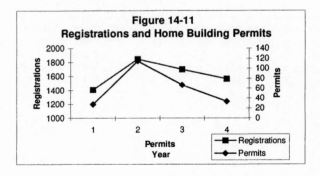

The similar pattern, after lagging car registrations one year, clearly shows that our search for a predictor variable has been successful. The connection between the two variables also makes sense - new homes bring in new residents who will then need to register their cars. Home building permits meets our requirement of having the information we need in advance of the prediction point.

To make the prediction, we will use regression analysis. As noted above, the data set we use is formed by pairing year 2 of auto registrations with year 1 of building permits, year 3 of auto registrations with year 2 of building permits, etc. The data set that results is shown in Table 14-G.

Table 14-G
Registration and Home Building Permits

Registration Year	Registrations	Permits	Permits Year
2	1409	28	1
3	1851	115	2
4	1701	67	3
5	1568	34	4
6		45	5

We do not have an auto registration value for the last row of the table because that is the value we are trying to predict - the number of auto registrations in year 6.

We next apply simple linear regression analysis to the first four rows of Table 14-G where the X variable is building permits and the Y variable is auto registrations. When this is done we obtain the following regression output:

SUMMARY OUTPUT

Regression Statistics	
Multiple R	0.946
R Square	0.895
Adjusted R Square	0.842
Standard Error	74.807
Observations	4

ANOVA

	df	SS	MS	F	Significance F
Regression	1	95354.49245	95354.492	17.039	0.0540
Residual	2	11192.25755	5596.1288		
Total	3	106546.75			

	Coefficients	Standard Error	t Stat	P-value	Lower 95%	Upper 95%
Intercept	1359.515	75.9241853	17.906215	0.0031	1032.839149	1686.19
Permits	4.471	1.083140164	4.1278768	0.054	-0.189310046	9.131448

Our last step is to derive the regression equation:

$$Y = 1359.5 + 4.47X$$

and solve for X = 45. The 45 represents the number of building permits issued in year 5 which is being used to predict the number of auto registrations in year 6. When we do this our new prediction becomes 1,561 registrations.

$$Y = 1359.5 + (4.47*45) = 1561$$

This prediction is less than the 1,752 registrations we obtained when we did the time series projection.

We can also add our permit information to the original analysis that just used year as the independent variable and conduct a multiple regression in an attempt to improve our prediction. In this analysis, registrations is the Y variable and there are two X variables - permits and registration year. Based on our earlier regression analysis, registration year did not appear particularly helpful when it was used by itself (see ANOVA Table). The t Test for the beta associated with the Year variable (t = 0.87) was not statistically significant in the analysis (p > .05). But, we also must remember that the sample size was quite small and there is a high risk of a type 2 error (i.e. claiming that there is no significance when, in fact, it is present).

After introducing both years and permits into a multiple regression analysis, the pertinent output from multiple regression is shown below.

	Coefficients	Standard Error	t Stat	P-value
Intercept	1234.2	33.8	36.5	0.017
Permits	4.6	0.3	14.8	0.043
Yr	46.6	9.6	4.8	0.130

The predicted value is now obtained from the following regression equation and indicates that we will have 1,674 car registrations in year 6.

$$Y = 1234.2 + (4.6*X_1) + (46.6*X_2) = 1234.2 + (4.6*45) + (46.6*6) = 1674$$

Mean Absolute Deviation

To determine the contribution of each of the independent variables (building permits and years) we can apply the partial F-test criterion which was described in chapter 13.

Which prediction is better, only time will tell. However, a popular method to identify the model that appears to fit the observed data best is based on the magnitude of the residual errors, which you may recall are the differences between the actual and predicted values. The method is called **Mean Absolute Deviation** or **MAD** and the formula is:

$$\text{MAD} = \frac{\Sigma \left| Y - \hat{Y} \right|}{N}$$

where Y is the observed value for a time point and $\hat{Y}$ is the predicted value for that time point.

Note that the lines before and after $Y\text{-}\hat{Y}$ mean you calculate the absolute value of the $Y\text{-}\hat{Y}$ difference. Remember the absolute value converts all negative values to positive values. As can be seen, MAD turns out to be the average of the absolute differences between the actual and the predicted values. If there were a perfect fit, MAD would be 0. Theoretically, the model with the smallest MAD is considered to be the best fit and, thus, the preferred model to use.

Based on the auto registration data given in Table 14-C, and the predictions from the simple regression model, the calculation of MAD is shown in Table 14-H.

Table 14-H MAD for Years Only Model

Year	Actual Number	Predicted Number	Absolute Difference
1	1456	1494	38
2	1409	1545	136
3	1851	1597	254
4	1701	1649	52
5	1568	1700	132
Sum			612
Mean			122.4 ◄── MAD

The MAD of 122.4 would then be compared to the MAD for the other models to see which one had the best fit - i.e. produced the smallest MAD. The MAD for the model using only Permits is 49.2 and the one using Years and Permits is 8.9. Therefore, the model that used both Years and Permits would be judged the best one.

Summary

In this chapter a time series was defined as a collection of observations, made on a quantitative variable, that are taken over a period of time. We learned that in addition to fitting a linear equation to a time series, there are other pattern such as the quadratic, exponential, logarithmic or polynomial which may provide a better fit. We also found that the irregular pattern of a time series can be reduced by a process called smoothing. Two popular techniques, moving average and exponential smoothing, were described. We saw that to help understand a time series, a process known as decomposition may be applied which identifies the components of a series - trend, seasonal, cyclical and irregular. The last issue we explored had to do with forecasts for a time series. We found that finding variables related to the principal variable, and incorporating the related variables into regression analysis, is a reasonable way to improve a forecast. We also learned that one way to decide which model provides the best prediction is to use the Mean Absolute Deviation (MAD) method.

PROBLEMS
Unless otherwise specified do all statistical tests at
the .05 level of significance and use two tail testing.

1. For the following data set, what kind of trend is there and what other component(s)
(cyclical, seasonal and irregular) appears to be present?

Observation	Value
1	80
2	83
3	68
4	77
5	69
6	70
7	26
8	68
9	66

2. Match the following graphs to the four relationships: linear, logarithmic, polynomial or
quadratic.

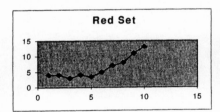

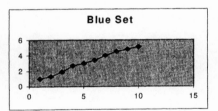

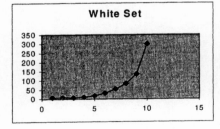

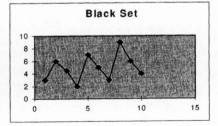

3. Based on the following trendline equations, what type of relationship is evident?
   a $Y = 14.3 + 2.4*Ln(X)$
   b $Y = .87 + .033*X$
   c $Y = 1.5 + (53.2*X) - (27.0*X^2) + (4.55 *X^3) - (.88*X^4) + (.06*X^5)$
   d $Y = 125*e^{1.42*X}$
   e $Y = 245 + (44.7*X) + (20.2* X^2)$

4. If you fit two trendlines to the same data set and the first trendline has an R Square of .60 and the second trendline an R Square of .39, what trendline should you use?

5. If a linear trendline had an R Square of .55 and a polynomial an R Square of .57 for the same data set, what trendline would an analyst most likely use?

6. For each of the following data sets, which of the relationships discussed in this chapter (linear, quadratic, logarithmic, or polynomial) seem to be the best fit?

Year	Alpha	Beta	Gamma	Delta
1	19	4	30	55
2	16	6	120	30
3	13	9	180	20
4	16	10	220	30
5	14	14	240	50
6	19	14	270	35
7	28	16	280	25
8	42	17.5	285	45
9	44	18	290	70
10	55	21	290	30

7. For the Alpha data set in problem 6, calculate a 3-year moving average

8. For the Beta data set in problem 6 calculate the exponential smoothed values using a damping factor of .25.

9. In an agricultural region the tax receipts have been monitored in hopes of being able to predict future governmental revenues. A regression analysis was run on the results for the past 12 years in which tax revenues for the fall were recorded in millions of dollars (i.e. 5.5 means 5.5 million dollars). The following information was extracted from the results of the analysis.

	Coefficients	Standard Error	t Stat	P-value
Intercept	16.57	2.751949116	6.02	0.00013
X Variable 1	2.79	0.373916254	7.45	0.00002

a   What are the predicted tax revenues for the fall of next year (i.e. year 13)?
An analyst also was able to record the rainfall (in inches) during the Spring planting season for each of the years. His input was added to the analysis and the following data were taken from the output of a multiple regression.

Regression Statistics	
Multiple R	0.98
R Square	0.96
Adjusted R Square	0.95
Standard Error	2.49
Observations	12

ANOVA

	df	SS	MS	F	Significance F
Regression	2	1253.730026	626.86501	100.98	6.8433E-07
Residual	9	55.87247446	6.2080527		
Total	11	1309.6025			

	Coefficients	Standard Error	t Stat	P-value
Intercept	10.70	1.9589144	5.46	0.00040
year	2.44	0.2200621	11.11	0.00000
rainfall	0.82	0.1694753	4.82	0.00095

b   What is the predicted tax revenue for the fall of next year if in the spring there were 10.1 inches of rain?
c   How much of the variation in tax revenues is accounted for by the 2 independent variables (year and rainfall)?
d   Is there a statistically significant regression?
e   How much do the tax revenues increase for each 1 inch of rain?

10. For the past 6 years, the local Department of Health has provided vaccinations for 1 year old children. Statistic on the program are given below.

Year	No. of Vaccinations
1	1245
2	1258
3	1345
4	1462
5	1044
6	1568

Starting in year 2, the department predicted the number of vaccinations to be given the next year by adding 100 to the current year count. This system worked reasonable well until year 5 when there was a sharp decrease in the number of 1 year olds requiring a vaccination. To improve the process the department has started to do their estimates based on linear regression using year as the independent variable. The latest regression analysis gave the following result:

Regression Statistics	
Multiple R	0.318
R Square	0.101
Adjusted R Square	-0.124
Standard Error	194.271
Observations	6

ANOVA

	df	SS	MS	F	Significance F
Regression	1	16972.86	16972.86	0.450	0.539
Residual	4	150964.48	37741.12		
Total	5	167937.33			

	Coefficients	Standard Error	t Stat	P-value
Intercept	1211.3	180.9	6.698	0.003
Year	31.1	46.4	0.671	0.539

a. What is the predicted number of vaccinations based on the earlier (i.e. original) system?
b. What is the predicted number of vaccinations for year 7 based on the regression analysis?
c. Is there a statistically significant slope?
d. Into what interval will about two thirds of the predicted number of vaccinations fall relative to the actual number of vaccinations?
e. What is the correlation coefficient between year and tax revenue?
f. What model (the original plan of adding 100 or the regression analysis) provides the best fit based on MAD?

The analyst also obtained the birth rate in the community for the 6 year period and those findings are shown below:

Year	Birth Rate
1	8.6
2	8.4
3	8.8
4	7.2
5	8.5
6	8.2

A regression analysis, after lagging the vaccination frequency data by 1 year, gave the following result.

Regression Statistics	
Multiple R	0.985
R Square	0.969
Adjusted R Square	0.939
Standard Error	49.728
Observations	5

ANOVA

	df	SS	MS	F	Significance F
Regression	1	104091.0063	104091.006	5.4749975	0.101233574
Residual	3	57036.19375	19012.0646		
Total	4	161127.2			

	Coefficients	Standard Error	t Stat	P-value
Intercept	-781.6	906.858	-0.862	0.45
Birth Rate	255.1	109.007	2.340	0.10

g. What is the predicted number of vaccinations needed for year 7?

The 2 variables (year and birth rate) were introduced into a multiple regression and the results are shown below and continue on to the next page.

Regression Statistics	
Multiple R	0.985
R Square	0.969
Adjusted R Square	0.939
Standard Error	49.728
Observations	5

ANOVA

	df	SS	MS	F	Significance F
Regression	2	156181.5	78090.8	31.58	0.031
Residual	2	4945.7	2472.8		
Total	4	161127.2			

	Coefficients	Standard Error	t Stat	P-value
Intercept	-1649.36	377.77	-4.37	0.049
Year	77.05	16.79	4.59	0.044
Birth Rate	322.48	41.97	7.68	0.017

    h. Based on the above analysis , what is the predicted number of vaccinations needed for year 7?

11. Refer to problem 10.
If you used the partial F-test criterion what variables would you retain or drop? The critical value for F can be found in the F table of the Appendix (i.e. Table 3).

12. What are three types of controls that can be applied to a research project?

13. A review of the records for 4 obstetricians produced the following results:

Birth	Dr. North	Dr. South	Dr. East	Dr. West	Total
Normal	457	262	241	678	1638
Complications	12	9	15	21	57

    a. Is there a statistically significant difference among the physicians?
    b. Cite the probability for obtaining this distribution if there is no difference among the obstetricians.

14. Samples were taken of the tonnage of recycling material collected at 18 different sites in the city's north side and compared to another set of 18 different sites taken in the city's south side. The results are shown below and the recycling administrator examined the mean (North side mean = 22 tons and South side mean = 24 tons) and decided the means showed little difference. However, the administrator asks you to examine the results using an appropriate statistical test. What test would you use and what would you tell her after completing the test?

	North Side			South Side	
25	24	23	24	26	24
21	21	24	25	25	23
17	20	22	20	26	25
19	27	20	27	22	26
25	24	21	23	26	27
20	23	20	22	21	20

15. A sample of 780 college graduates showed that 470 admitted to driving after they had three or more drinks at least once in their college career. What is the 80% confidence interval for the reported incidence?

Excel Instruction

In Excel you can see how a data set may fit some of the most popular type of trends. Data sets that will be useful for our demonstration of different trends were discussed in chapter 14 include: quadratic, logarithmic, polynomial and exponential. To fit different trend lines to a data set you need to have the data sets in a worksheet. You will first need to prepare an XY graph. The creation of an XY graph was done in chapter 2. Trendlines should then be fitted to the XY graph.

To fit a trendline you need to activate the XY chart by clicking on the chart (avoid the plot area). Next select "Chart" from the Main Menu and from the drop down menu select "Add trendline". A dialog box like the one shown below with the "Type" tab displayed will appear. You now have a choice of different trendlines.

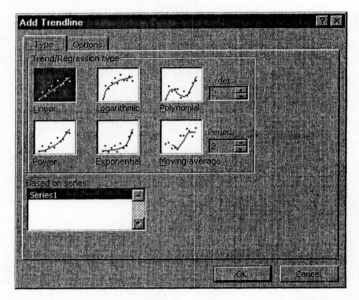

One of the trendline choices is LINEAR which is the type of relationship we studied in chapter 13. Note that besides LINEAR, there are options such as LOGARITHMIC, POLYNOMIAL, POWER and EXPONENTIAL. Each of these options creates a curve through the data points. You select the fit you think is best and Excel will fit the line/curve for you. To obtain relevant statistics on the fitted line you need to click on the "Options" tab of the Trendline dialog box. When you select "Options" the following screen appears:

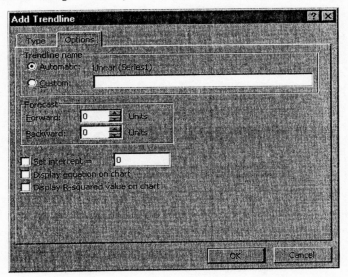

To forecast a time point you use the Forecast section of the dialog box. In the area described as "Forward:" click on the upper arrow of the "Units" box until the number in the box represents the time point you want to predict. A "1" indicates you want the prediction for the first time point beyond your time series. If you wanted to know the predicted value for a time point prior to the first time point of your data set, you use the units arrow in the "Backward:" section of the Forecast area.

If you want the equation of the regression line, you click on the box by "Display equation on chart" so a ✔ appears in the box. For the R Square value, click on the box by "Display R-squared value on chart" box so a ✔ appears in that box. When you click on "OK" you should have a figure with a fitted line, an equation and an R square value. To see all the data on the chart, you may need to do some rearranging. Click on the item you want moved (e.g. R Square value), and use the handles on the R square box to reposition it.

We will work an example based on the data set for the logarithmic figure (Figure 14-D of the text). The data set for the figure is given below.

Time	1	2	3	4	5	6	7	8	9	10	11	12
Value	3	6	9.5	10	11	12	12	13	14	13.5	15	15.5

You should be able to create the graph, fit a logarithmic trendline, provide the regression equation and R square value, and predict the value for time point 13 by following the above steps. Your final graph should look like this:

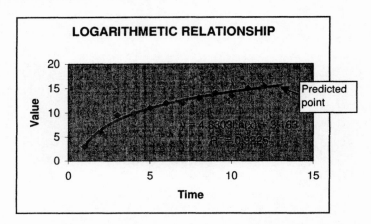

Note that the regression line is extended to represent where the predicted point would be on the curve. The line is extended one time point because the number of units in the Forecast Section in the Trendline dialog box contained a 1. You can obtain the actual numeric value of this predicted time point using the regression equation and solving it for an X of 13.

To create a trendline for a quadratic relationship you need to highlight the XY chart, select "Chart" from the Menu Bar, and then click on "Add Trendline" from the drop down menu. Select the POLYNOMIAL option in the Add Trendline dialog box for Type. To the right of the Polynomial illustration is a box labeled "Order". For the quadratic, the order is always 2 so be sure that that number is showing in the "Order" box. If you want a higher order polynomial, increase the number in the "Order" box. Figure 14-E uses an order of 5. The maximum order available is 6. For an exponential relationship (see figure 14-F) you select the EXPONENTIAL option. A trendline type, not discussed in the text, but available in Excel is represented by the power relationship. The trendline type is appropriately called POWER and the general equation for a power relationship is $Y = a*(X^b)$. The steps to produce the equation and the R square value for these last three relationships are the same as those described above.

Although there is a Moving Average option in the Trendline dialog box, it is best to execute this procedure by another Excel routine. To smooth a curve by the Moving Average method, select "Data Analysis" from the "Tools" menu. From the list of tools presented you chose "Moving Average". After clicking on "OK" you will be presented with the Moving Average dialog box (see below).

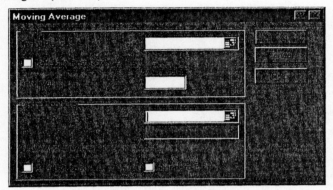

The "Input Range" refers to the location of the raw data for your moving average and you only include the location of the Y variable. If you included a cell for the name of the variable, check the "Labels in First Row" box. As mentioned in the text, Moving Average smoothes the line by forming averages for consecutive observations. You specify the number of data points you want included in the averages by inserting the number desired in the "Interval" box. The location of the moving average data set that will be produced is entered in the "Output Range" box. Only the first cell of the range needs to be cited.

We'll work an example. Copy the raw data from table 14-C onto a worksheet and then use the instructions provided above to find the values for the moving average data set. Remember to enter 3 in the "Interval" box because our example used a 3-year moving average. The output you obtain should be essentially the same as that shown below:

#N/A
#N/A
36.66667
47.33333
58.33333
59
60
54.66667
49.66667
50.66667
48
52

Unfortunately the Excel output does not match the correct time sequence. As stated in the text, with a 3 year moving average you would not get a moving average value for the first and last time points. However, Excel lists an "#N/A" entry, which means not applicable, for the first two years and gives a value for the last year. What needs to be done is to move the moving average data set up one cell. This can be done easily by highlighting the first cell (i.e. the cell at the first "#NA" notation, clicking on "Edit" and then picking the "Delete" option from the drop-down menu. The delete dialog box appears and you select the "Shift Cells Up" option. The Excel output will now be in the correct order. In the Moving Average dialog box, Excel has an option for "Chart Output", but because of the problem noted above it is not recommended that this option be used.

To do exponential smoothing you again start with "Tools" from the Menu Bar and then select the "Data Analysis" option. From the Analysis tools list, select "Exponential Smoothing". After clicking on "OK" you will be at the Exponential Smoothing screen (see below).

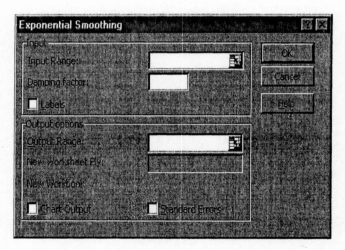

You complete this screen in a similar fashion as the Moving Average screen but you need to specify a "Damping factor" instead of an "Interval". In the text we pointed out that the damping factor could be any value between 0 and 1.

We can again use the Table 14-C data set to illustrate how we do an exponential smoothing operation. Follow the steps described above using a damping factor of .5 and you should produce the exponential output values shown on the next page.

#N/A
24
29.5
40.3
48.1
58.1
55.5
57.3
54.6
46.3
54.2
49.1

This output also needs to be adjusted so it matches the original time points. It is again necessary to delete the first data point and move all the others up one unit. This is done by highlighting the first cell (i.e. the cell at the top with the "#NA" notation), clicking on "Edit" and then picking the "Delete" option from the drop-down menu. The delete dialog box appears and you select the "Shift Cells Up" option. The Excel output will now be in the correct order, but there is no value for the last observation. You can fill in the last observation by making a forecast. The calculations are straightforward and the necessary formula is given in the text.

All other calculations done in this chapter use methods previously reviewed, such as linear regression, or they require the use of basic mathematical operations, which also have been discussed in previous lessons.

Here are some problems you can practice on using Excel.

1. Refer to the Alpha data set in problem 6. For that data set find the best fitting relationship and prepare a graph that shows the trendline, equation, R square value and next predicted value on the graph.
2. For the following data set, smooth the data using a] a 5 year moving average and b] the exponential smoothing technique with D = .70. Show on a single graph the original data set and the two smoothed data sets.

Month	Accidents	Month	Accidents
1	340	7	289
2	412	8	279
3	367	9	295
4	354	10	352
5	311	11	341
6	340		

Chapter 15

Qualitative Methods and Ethics

Introduction

Thus far in this text the focus has been on quantitative methods for the interpretation, testing and presentation of data. In fact, it was emphasized in chapter 6 that when you conduct a study, and the dependent variable is a concept rather than a quantitative measurement, you must transform the concept into a quantitative measurement (the term operationalizing was used for this step). However, in research, out of necessity or preference, there are many useful qualitative techniques that can be employed. This chapter reviews many of these methods including content analysis, focus groups and the Delphi technique. A significant portion of the chapter is devoted to questionnaires. Recommendations on activities in the preparation stage as well as guidelines for the wording of specific questions are covered. The last part of the chapter presents an overview of research ethics. Past and more contemporary studies that have stimulated debate about what should or should not be done in the name of research are presented.

Definitions

It is not always possible to confine one's research projects to studies that begin with an independent variable and end up with a quantitative dependent variable. What if we want to know the degree and type of alienation teenagers feel? To assess this phenomenon, a different approach than introducing an independent variable and observing its effect on a quantitative dependent variable must be followed. To do this latter type of investigation, an approach referred to as qualitative research is often required.

Qualitative research may be defined as research in which the results are events or themes that are described in words or pictures rather than numerical values. **Quantitative research** involves collecting and analyzing numeric data whereas qualitative research refers to the use of methods that gather and study non-numeric information. Hence, a **qualitative measurement** may be defined as a measurement that is non-numeric. The form of a qualitative measurement begins as words or images that can not be directly ranked or compared without first requiring a major transformation effort on the part of the researcher. We will define **quantitative measurements** as measurements that have numeric properties so that ranking or calculating the mean for the variables results in a meaningful answer.

In making assessments both quantitative and qualitative measurements are commonly employed. For example, in a college admittance process certain information is obtained to determine if a prospective student is eligible to attend the school. Typically the information gathered consists of SAT score, grade point average (GPA) in High School, an essay, letters of recommendation, etc. It is clear that in the determination of acceptability both quantitative (SAT and GPA) and qualitative (essay and letters of recommendation) are used. To confine the evaluation process to only quantitative or qualitative measurements would result in an incomplete assessment - it would lack content validity (see chapter 6).

Qualitative Versus Quantitative Research

There are several major differences between how quantitative and qualitative research is conducted. The usual role of the researcher in qualitative research is to become personally involved in the data collection exercise. In fact, it would be fair to say that the researcher and the information collected are linked and one can not be easily separated from the other. In contrast the researcher in a quantitative study tries to disassociate himself or herself from the data gathering phase. The researcher wants the data collection to be impersonal so that the same quantitative data would be collected no matter who the researcher was. In qualitative work the researcher tends to constantly interpret the data coming in, but in a quantitative study the researcher seeks to obtain the data without the need for interpretation on his or her part. In a qualitative study the researcher becomes a part of the measurement device, but the goal of a quantitative researcher is to be personally independent of the data acquisition phase. It is also fair to note that quantitative methods often involve a finite amount of data - a set of blood pressures or performance ratings. Qualitative material however, often begins as a large amount of information that needs to be deciphered and organized before any meaning can be attached to it.

It should be clear that quantitative and qualitative approaches use different investigative methods. A quantitative researcher would prefer to work with measurement devices that can be easily reduced to numbers and have structured formats. In contrast a qualitative researcher would be content with devices that start out as descriptive material which is unstructured.

In terms of conducting a research study, each approach has its advantages and disadvantages. A quantitative research project must be run in a very rigid fashion. The experimenter would resist making changes in his/her design or measurements, but a qualitative researcher would be more open to the possibility of revising these features. However, the lack of structure means external validity is relatively low for the qualitative approach compared to the quantitative method. The success seen with a program evaluated using a qualitative method may not be replicated when transferred to a new environment. On the other hand, due to the more structured approach in quantitative research, there is greater reliance on, and need for, background information to design the study and identify relevant measurements. But it is also true that when doing contemporaneous research, the time commitment of subjects in qualitative studies will generally be more demanding than those who participate in a quantitative investigation.

Qualitative Research Methods

A good way to distinguish between quantitative and qualitative approaches is to review methods used in qualitative investigations that would not be followed in a quantitative program. Qualitative research often favors making direct observations in natural settings. Two techniques, the participant observer and non-participant observer, are typical examples of the qualitative approach.

Participant Observer Method

A **participant observer**, as the name implies, is a researcher who interacts with the research subjects in an intrinsic (i.e. intimate) fashion. The researcher (as a study participant) observes, records and acquires experiences in the "real world" to better understand and interpret pertinent information sought in the study. The research role of the participant observer may or may not be known by the subjects being observed. If the participant observer's role would not distort the behavior of the subjects, and there is no ethical problem with posing as a participant, then the participant observer's identity would not be made known. If the aim of a research project is to determine whether a new training program is better than the current program, then a participant observer could pose as a recipient of the service and have sessions with trainers teaching the current and new method.

Non-Participant Observer Method

This technique differs from the one just discussed in that the researcher is only an observer of the research subject's behavior. The spectator role of the **non-participant observer** is usually clearly evident. The researcher could be present at a training session, but his or her presence and role would be obvious. An apparent concern with the non-participant observer method is that the subjects may not act naturally. Knowing they are being "watched" may cause nervousness and prompt them to perform at a subpar level. On the other hand, knowing they are being observed may make them perform at a heightened level of performance, but in either situation there is a distortion in their behavior.

Reliance on Records

Other qualitative techniques include the use of **records** that contain written as opposed to arithmetic material. The material may be in many forms - letters, meeting minutes, reports, graffiti on walls, photographs illustrating damage to buildings, etc. It is self evident that in most organizations there exist past records to aid a researcher. Some records are clearly quantitative - past blood pressure readings or grades on an exam. However, other pertinent records may be qualitative - a written evaluation report or a memo covering a person's past history. These written documents may be very revealing and can provide valuable information for an analysis of a past or baseline condition.

Interviewing

For many, the most useful qualitative tool is the interview. An **interview** typically involves an oral exchange between the researcher and a study subject in which open ended questions, that are germane to the research question, are asked and answered. There is an art to good interviewing which can be learned and perfected. Recommended approaches include developing a friendly/trusting relationship with the interviewee and using a setting that is conducive to a frank and open exchange. It is also felt desirable to ask questions that require the interviewee to articulate a position or provide an explanation rather than just giving single word responses (e.g. "Can you tell me about what you do in your job" rather than "Are you a teacher?"). One should also be sure that the interviewee understands the purpose of the interview and gives his or her consent. In addition it is wise to interject confirmatory statements throughout a session to be sure the interviewee is understanding what is being stated (e.g. "As I understand it you believe ...."") and to sum up the major conclusions before concluding the interview (e.g. "Based on what I heard I conclude that ....").

How to record the information from an interview opens up several possibilities. One may simply take notes during the interview and quickly review them for completeness and accuracy shortly after terminating the session. Another option is to tape record the interview. Taping requires advanced permission by the interviewee, but does leave a physical record that can be most helpful to confirm initial impressions and assessments. The downside to the tape recording is the imposing effect it may have on the person being interviewed. More guarded and misleading responses may be the result.

A researcher also may have a choice on whether to use a personal interview, a telephone interview or a mail interview (i.e. the questions are mailed to the interviewee to complete and return). The main advantage of a personal interview lies in the opportunity to use visual aides and to read "body language". Some concepts are best seen and understood by means of graphics or pictures. Furthermore, the appearance and personality of the interviewer may also be transmitted in a personal interview, but curtailed in a phone interview and absent in a mail interview. It is usually felt that the personal interaction in a face to face session results in a greater commitment on the part of the interviewee and it is also possible to hold the person's attention longer.

The personal and phone interview also provide an opportunity to clarify questions and explain concepts which give these methods a major advantage over the mail interview. The persons being questioned are in a more "captive" setting and, as a result, they may be more willing to respond to questions when an interviewer is physically present. The scourge of telemarketers may translate to a lower response rate for phone interviews since it's harder to turn down a request in person than on the phone. Furthermore, the mail interview has by far the lowest response rate.

There are disadvantage to the personal interview as well. The cost is usually highest for personal interviews and lowest for mail interviews. If a question involves a personal or sensitive area, the person may be more reluctant to give the information to an interviewer as opposed to submitting the information as part of a questionnaire completed in the home. There can be more bias in the personal interview if the respondent senses a lack of respect or sensitivity on the part of the interviewer. An interviewer may also have personal beliefs about the "right" answers to some questions and may lead the interviewee rather than allow the person to respond truthfully. Another drawback of the personal interview can be the reluctance of the interviewer to go to assigned homes or locations which the interviewer believes are unsafe. It also needs to be stressed that if multiple interviewers are used, it is highly desirable to hold training session for them to enhance consistency.

Content Analysis

Another technique used in qualitative research is content analysis. Content analysis uses written documents, film or other kinds of physical records as source material. The most popular items to which content analysis may be applied are newspaper articles, books, TV programs, movies, transcripts as well as letters and journals. **Content analysis** reviews the source material looking for terms, phrases or ideas that were pre-determined to be relevant to the research question. The frequency with which these elements are encountered are noted. Many times categories are developed for elements that connote the same thought (e.g. terms that represent a positive attitude such as "good", "worthy" or "valuable" could be grouped into one category. Percentage distributions for the categories are prepared and can then be subjected to statistical testing using methods such as a test of proportions that we discussed in an earlier chapter.

In terms of an example, one could imagine an investigation into a politician's attitude toward women and men by reviewing the politician's speeches and counting the number of favorable references to each gender. Strict rules on what terms or expressions constitute a favorable or unfavorable comment must be defined in advance and clear guidelines on what represents a redundant comment must also be established.

There are some clear cut advantages to content analysis. It allows researchers to investigate issues for which subjects need not be physically present. You may conduct the research retrospectively, as long as useable record are available. For example, the attitude of a deceased politician could be researched providing a representative sample of his or her pronouncements could be found. Compared to quantitative research where the data must be generated, content analysis is relatively inexpensive since there is usually a low cost in procuring old records. The time to do content analysis may be relatively short - for example to see how a politician's attitude may have changed during a 20 year pubic service career does not require a 20 year longitudinal study - all the researcher needs is 20 years of speeches or other public domain records.

There are also negative aspect associated with content analysis. There is the possibility of bias in selecting and interpreting statements or taking statement or words out of context. The threats to internal validity may apply as well. Are the documents assembled complete or has there been a selection bias? If one is working with historical documents, concurrent events could play a role in any attitudinal changes discovered, but it may be all but impossible to be aware of this due to the time lag involved. One also needs to be wary of terminology differences over time that may cause classification and interpretation problems for the individual(s) making the assessments. To avoid bias associated with a single rater, multiple judges are often used, but then instrumentation becomes an issue.

Focus Groups

Another technique classified as qualitative research that has grown in popularity is the focus group. The **focus group** usually consists of a relatively small number of people (e.g. 6-10) that are assembled to discuss a specific subject in which the group has particular knowledge or experience. Frequently, the members are unknown to each other, but they share a common experience or body of knowledge. There is a group leader who represents the research team's interest and is not a member of the group. The job of the group leader is to guide but not dominate the session. The leader comes prepared with an open ended set of questions that are designed to elicit information pertinent to the research goal. The opinions and conclusions of the group are recorded and become part of the permanent record.

Marketing research has been particularly interested in using focus groups to understand consumer attitudes toward new or future products. What must be kept in mind when relying on focus groups as part of a research strategy is that the group is almost never randomly chosen. Great care is, therefore, called for in terms of what one can conclude from the comments and deliberations of a focus group. One should not treat focus group results as having tested and proven a research hypothesis. The focus group may be quite helpful in generating hypotheses or ideas on cause and effect relationships, but the non-random nature of the group selection process requires healthy skepticism about the generalizability of the findings.

Case Study

The **case study**, which was discussed in chapter 12, may be viewed as a type of qualitative research design. The investigator takes an in-depth and comprehensive look at one or a few experimental units. An experimental unit is referred to as a case and can be a person, a group or an entire organization. Since a case study includes only one or a very few cases it can afford to thoroughly assess a variety of factors germane to the research question being explored. For example, in medicine, the case study methodology is used to describe in detail a specific patient who has an unusual illness or disease course and goes into detail about the pertinent variables and other factors associated with the patient and his/her illness. In organizational research, the case study approach is often used to describe how and why an organization succeeded or failed in carrying out a specific task such as implementing a policy.

Note that the case study stands in sharp contrast to research which uses the survey type of approach. In a survey the data are gathered on a large number of people, but the amount of information collected is restricted and limited. In comparison, the case study has the potential to give a fine grained picture of an issue, capturing subtleties and amplifying details that may be missed by the more general survey approach. Unfortunately the strength of the case study method is also its weakness. The "truth" revealed by a case study is highly dependent on the skill and integrity of the researchers. Unintentional bias, lack of knowledge, non-random observations, flawed interviewing and too little or too much empathy can result in misleading conclusions. Perhaps the most serious disadvantage of a case study is that it shares the drawback of other qualitative methods – the difficulty of generalizing the results to other people, groups or organizations. However, if consumers of the research use the results more as descriptive information rather than for predictive purposes, the richness and documented portrayal of people and group dynamics can be highly rewarding.

Delphi Technique

In chapter 14 we used a few different quantitative methods to do forecasts, but mentioned that there were also qualitative techniques available. The qualitative forecasting system we will look at is called the Delphi Technique. Essentially the **Delphi Technique** uses a group of experts who operate in an anonymous fashion and through recursive reviews generate a forecast. Generally, in a Delphi exercise a group of experts on a certain subject are asked to make an anonymous prediction on how a future event might occur. Their answers include their rationale and are given to a central administrative group. The group summarizes the information in such a way that the areas of agreement and disagreement are clearly identified and the information fed back to the experts for a second round. The experts see what others thought and are asked to resubmit their forecast and its rationale a second time revising it as they see fit. Since the exercise is carried out anonymously, the belief is that even experts will feel free to change positions without embarrassment or concern for their egos. The process is repeated until the central administrative group believes that an acceptable forecast has emerged or that there is little likelihood that a consensus will emerge.

Credibility of Qualitative Methods

In the end, due to the greater subjectivity inherent in qualitative techniques it is important to make the findings as credible as possible. **Credibility** can be strengthened by documenting the expertise of the research team and being sure all operational steps, such as conducting thorough interviews and not omitting any individuals selected for an interview, are carefully observed. The records and documentation used should also be well organized and accessible to others who have a "need to know". Good research practice dictates that enough information is present so that others using the same data and records could do their own evaluation and hopefully come to a similar conclusion. For qualitative research this means documenting all interviews, maintaining well organized files of the records used in the research etc. In this day and age a computerized data base offers the most efficient and practical way to identify the information collected in the investigation. There are, in fact, software tools devoted to qualitative research and a researcher interested in using this technique should become familiar with the current software.

In doing qualitative research it is obvious that mistakes (e.g. misinterpretation of the information) can be made. Therefore, it is often recommended that researchers use multiple methods to verify an interpretation based on a qualitative evaluation. The replication could be in the form of two independent documents which result in the same assessment, but it is preferable that different sources be used - for example, a non-participant observer finding collaborated by the same conclusion obtained from written records.

Questionnaires and Surveys

Whether a questionnaire is a qualitative or quantitative method has no simple answer. The truth is that it can be considered as a type of either approach because the individual items in a questionnaire are frequently a mix of quantitative and qualitative questions. In any event, the questionnaire is the backbone of surveys and deserves our attention.

A **questionnaire** is simply a list of questions that seeks information about people's opinions, behaviors, attitudes, etc. The questionnaire is the main ingredient in most surveys. Although, sometimes the terms questionnaire and survey are used interchangeably, a **survey** is an examination or assessment of a phenomena which almost always uses a sample to give us information about the underlying population. A questionnaire is frequently the instrument used to collect the data for the survey.

Criticisms of Surveys

Surveys are criticized because they tend not to establish causal relationships. For example a survey could find that most college graduates voted Republican but that does not mean a college education results in more conservative political beliefs. There is also a concern that a survey may be misleading since the wording of questions can alter the way people will perceive an issue. Here is an example of two different versions of the same proposition. The proposition was to go before Houston, Texas residents in 1997 and called for an end to affirmative action in awarding city contracts and in hiring.

> Version A. The City of Houston shall not discriminate against or grant preferential treatment to any individual or group on the basis of race, sex, ethnicity or national origin in the operation of public employment and public contracting.
> Version B. Shall the Charter of the City of Houston be amended to end the use of preferential treatment (affirmative action) in the operation of the city of Houston employment and contracting?

Sample surveys found that 68% of the residents favored proposition A but only 47% favored proposition B. Hence, the same proposition would either pass or be defeated depending on the wording of the question. Surveys also require a level of reading skill and a vocabulary to properly comprehend the individual questions and ignoring those conditions can cause sampling bias. Proponents of questionnaires like to point out that although all the above criticisms are true they may apply to other research methods as well.

Questionnaire Preparation

When one is asked to do a survey it is important to be sure the concept to be investigated is clearly defined. For example, assume the assignment is to determine the public's opinion of inequality in a certain region of the U.S. Note that there is more than one dimension of inequality that could be examined. Does the inequality refer to race, gender or age? Are we interested in the extent of the inequality (e.g. the proportion facing discrimination), the causes of the inequality (e.g. did it come about from economic or educational reasons) or the effects (e.g. does it lead to increased unemployment or less advancement). The point is that one can not start preparing a questionnaire until the research problem is fully understood.

It is also not necessary to start questionnaire preparation from scratch. There may be previous surveys that can help identify the appropriate questions to ask. Thus a careful literature search should be conducted. This is especially critical since if certain questionnaires have been used in the past, they may be considered the established way to investigate the issue. If you do not use the same forms you may be open to criticism. In addition the components of the questionnaire may have useful reliability statistics that you can use to bolster the credibility of your research.

It sometimes is also useful to interview potential respondents to find out what data they feel is important to collect. The interviews may suggest vocabulary to avoid or use, the educational level at which the questions should be asked, etc. Finally discussions with others who have researched the same topic may also produce useful ideas that you can use in designing your questionnaire.

The sample size for your project needs to be determined and that number can be obtained using the approach described in chapter 7. Remember to build in a factor for the number of no-returns you may experience. Not all who receive a questionnaire complete it and you need a good estimate of the response rate to avoid ending up with too few responses.

The length of the questionnaire is also a consideration. There is no simple rule to determine the length of a questionnaire. You need only enough questions to cover your topic, but that advise isn't very helpful in knowing what questions may be cut. A pragmatic approach to rejecting useless questions is for the researcher to develop a hypothetical, but reasonable distribution of the results for each question. Then study the hypothetical results and see what useful information was obtained. Some questions may make a negligible contribution and are candidates for elimination. As a further incentive for researchers to reduce unnecessary question, they should realize that respondents, who are asked to answer questions that do not seem relevant, suspect that the survey is not being conducted by a knowledgeable team and they are more likely to not take the time to fill it out.

To encourage completion of the questionnaire there are some obvious things you can do. Incentives include stressing to the respondent pool the value and importance of the research in a cover letter. You can also encourage their cooperation by promising to send them a tabulation of the results. Some organizations conducting a survey also offer monetary rewards usually in the form of a meal at a gourmet restaurant, donation to a charity or chance to win a major prize.

A low response rate is indeed cause for alarm. The bias that may be introduced is almost always indeterminable. We are then left to ask - to whom do these results apply? Sometimes there may be demographic characteristics available for the population you are sampling (e.g. the sex and age distribution of an organization). You can use the population data to see if your returned sample surveys have similar characteristics to that of the population. Similar profiles do not guarantee that you have a representative sample, but a dissimilar profile is very good evidence that you have a non-representative sample and as heart breaking as that may be, you should know the truth. The lack of compatibility between a sample and population should also be included in any report you prepare of your study so that those who read the report can determine how useful the results may be for them.

We usually think of questionnaires as documents that are mailed to prospective respondents, but you can also follow a questionnaire format in a phone call or give the questionnaire to a group in a face to face session. The differences between personal, phone and mail interviews previously discussed also apply to questionnaires.

Guidelines for Questions

In respect to the development of questions, guidelines can be provided. As with any set of guidelines, if there are good reasons to ignore them, then do so.

1. USE SIMPLE LANGUAGE. Avoid using jargon, acronyms or technical terms that may not be fully understood. You do not have to be condescending in your language but, if in doubt, use the simpler rather than the more complex term or expression.

2. THE SHORTER THE QUESTION LENGTH THE BETTER. Long questions are harder to comprehend and the reader can even loose interest in a lengthy question.

3. REVISE AMBIGUOUS QUESTIONS. It is easy to write an ambiguous question and not realize it. For example, examine the following question:

   How often do you visit your parents?

Its flaws may not be immediately obvious. However, after a moments thought it should occur to you that the parents may be deceased or not live together and a respondent would be confused as to how to answer the question in that situation.

4. AVOID LEADING QUESTIONS. Some people who do a survey want a given answer and therefore compose questions so that their desired position wins support. In the two versions of the affirmative action proposition presented earlier, those who were against affirmative action developed version A and those in favor of retaining it developed version B. Each side was rewarded with the response they wanted. A prime example of a leading question is:

   Do you favor cutting defense spending and allowing communism to rise again and become a national threat?

Of course, most leading questions are more subtle than this example, but that only makes them harder to detect.

5. AVOID QUESTIONS THAT INCLUDE NEGATIVE WORDS SUCH AS "NOT". Take a moment and decide what is meant by the following question:

   Marijuana should not be illegal.

This question has two negative terms (i.e. not and illegal) which makes it difficult to interpret. A preferable way to ask the question is:

Should marijuana be legal?

6. AVOID QUESTIONS WHICH FORCE A RESPONDENT TO GUESS AT AN ANSWER. One option to reduce guessing is to include a "Don't Know" option. Another choice is to precede the question with a screening question. A typical screening question could be " Are you familiar with ...." and insert the subject in the last portion of the question. For example, before asking the lay public whether they support the President's trade policy proposal, it is wise to first ask:

Are you familiar with the President's trade policy proposal?
Yes (Answer the next question)
No (Go to question --)

7. PROVIDE DIRECTIONS FOR RELATED QUESTIONS. Inform the respondent what to do after completing a question that has sub-questions. This was done in the preceding example. Nevertheless, here is a second example which solicits information about smoking habits:

4. Do you smoke cigarettes?
    Yes [Answer question 4A]   No [Go to question 5]

4A. How many cigarettes do you smoke per day?   ☐

5. Do you smoke a pipe?
    Yes [Answer question 5A]   No [Go to question 6]

8. AVOID REPETITIVE OR CONFIRMATORY QUESTIONS. Asking the same question twice will only demonstrate that some people give a different answer when asked the same question twice. Your research doesn't need to reveal this fact. In survey work it is just accepted that there will be some inconsistency by people when answering a question. If correctness is critical, be sure the question is asked in a clear and understandable fashion rather than asking it twice.

9. PROVIDE A CLEAR FRAME OF REFERENCE. As the next example shows, time units need to be identified:

How often do you visit your mother?

First note that the question assumes that the mother is alive (a "not applicable" should be one of the response options). Still, you also need to include a time unit such as "number of times per year" in the question.

10. BE SURE THERE IS FAIRNESS IN THE OPTIONS PROVIDED FOR A QUESTION. For example if you ask:

> The cost of car insurance in New York City is:
> Much too high   Too high   About right

you should balance the options with "Too low" and "Much too low" options.

11. ELIMINATE IRRELEVANT QUESTIONS. It is tempting to add questions that you think may be useful. However, questionnaire length is critical and if you can't justify a clear value to a question consider dropping it. As noted earlier developing hypothetical, but realistic results for the questions should help you appreciate the potential value of each question.

12. BE SURE YOUR QUESTION OPTIONS ARE EXHAUSTIVE AND MUTUALLY EXCLUSIVE. The options should cover the full range of answers (i.e. exhaustive) and they should not overlap (i.e. mutually exclusive). In way of an example, note the set of options to the following question:

> How many hours per day do you watch TV?
> 1-6   6-12   12-18

These options fail both standards. The following revision is called for:

> How many hours per day do you watch TV?
> Less than 1   1-6   7-12   13-18   More than 18

13. BE SURE THE QUESTION HAS OPTIONS THAT CAN DETECT IMPORTANT DIFFERENCES. The question identified in the last guideline (number 12) may have categories that are so broad that they will miss identifying an important difference. Most people may fall into the 1-6 hours per day category and to recognize differences between your comparison groups, a greater breakdown of that category would be advisable. Of course, you could just have the person enter the number of hours of TV watched, thereby eliminating the need to pre-determine the appropriate categories. This may be a workable solution, but when asking for confidential information it may be necessary to create categories rather than asking for a specify number. For instance, if you wish salary data it is best to list ranges rather than ask for a specific amount.

Scales for Questions

There are different ways to present response options for a question. For quantitative information a respondent may be requested to just enter a number (e.g. age) or select from a pre-set list of numerical ranges (e.g. 10-19, 20-29, etc.). For qualitative information the request may be to select the correct answer (e.g. circle male or female), but qualitative information can also be obtained based on a scale. There are a number of options regarding the scale one can use. Five widely used scales are Likert, Thermometer, Checklists, Rankings and Ratings.

A Likert scale is also sometimes referred to as an Intensity scale. For a **Likert Scale** the items in the questionnaire are asked in terms of the degree to which a person agrees or disagrees with a statement. For example:

President Jones is doing a good job handling the economy.
   Strongly Agree   Agree   No Opinion   Disagree   Strongly Disagree

The respondent circles the option that is consistent with his or her belief. The middle position, "No Opinion", is sometimes omitted or it may contain other expressions such as "Undecided" or "Neutral". For analysis purposes, the options are assigned a ranking - in the above case the options can be ranked from 1 to 5 since there are five options. Statistical procedures suitable to ordinal (i.e. ranked measurements) can then be applied.

A **thermometer scale** typically just defines the extreme ends of a scale but lets the person pick any level at or between those extremes. An example is shown below:

How often do you understand the instructor's assignments?
Never   Always
   1   2   3   4   5   6   7   8   9   10

Thermometer scales can use other ranges (e.g. 1 to 100) and some may have no numbers at all - the respondent simples places an X on a line to represent their position. There also are many choices for the anchor definitions (e.g., Absolutely True and Absolutely False or Critically Essential and Not Important At All). Typically the researcher records the numeric value circled or measures the linear distance from the extreme low end of the scale to the marked spot with un-numbered scales. In either event the value derived is then used in a statistical analysis suitable for interval scale measurements.

A **Checklist,** also known as a Cafeteria scale, has the respondent check off items from a listing that apply to a statement or concept. For example a listing of important job characteristics could be listed on a form in which the respondent circles all those that apply to his or her job. The responses are treated as nominal measurements and the statistical procedures suitable for that level of measurement applied. See below for an example.

Circle all the terms that apply to skills needed in your job.
         Considerate   Competent
         Creative   Accurate

Ranking scales are just what the term implies. A list is given and the respondent assigns a rank to the items with 1 indicating his or her first choice, a 2 the second choice etc. We could use the example above as our listing but in this case the respondents would be asked to place a rank number by the characteristic.

Rank the characteristics required in your job by placing a 1 in the box by the most important term, a 2 in the box for the next most important term, etc. until all the terms have been ranked from 1 to 4.

Considerate ☐   Competent ☐

Creative ☐   Accurate ☐

The last format is the **Rating scale** where a respondent assigns a grade to the elements in a list. The grades or ratings need to be specified in advance. A typical example is shown below.

Rate your job in respect to each of the following traits (check one box per trait)

Job Trait	Very Important	Important	Unimportant	Very Unimportant
Considerate	☐	☐	☐	☐
Competent	☐	☐	☐	☐

The terms "Very Important" to "Very Unimportant" represent an ordinal measurement scale and statistics appropriate for ordinal scales can be applied.

Questionnaire Formats

Another choice one has when preparing a questionnaire is whether to use open or closed formats. In a **closed format** the respondent selects an answer from a pre-set number of options (the examples cited above for different types of scales were all closed format question) or enters a numeric value (e.g. age or weight). In the **open format** the person is free to write a narrative answer. For instance one could include this item in a questionnaire:

Describe the kind of job you think President Jones is doing

and then leave space for the answer.

The closed style for questions requires careful planning because if you leave out an important option some of the respondents will have to select an option that fails to reflect their belief or they will have to leave the question blank. The disadvantage of the open format is that one has to devote time to classifying the response using an appropriate qualitative technique. In addition, it is frequently necessary to first interpret a handwritten answer which is time consuming and sometimes unsuccessful. In many questionnaires there is a place for both types of formats.

It is highly recommended to always include a last open ended question for "Comments". A respondent has been cooperative by filling out the questionnaire and should have the opportunity to provide his or her final observations. Furthermore, the information provided may contain important clarification or explanation for the research team.

Questionnaire Development

Too often people who first become engaged in developing a questionnaire may overlook a few planning steps that can cause problems later. Listed below are two of the issues one needs to address.

CODING. For all but the very smallest surveys, it is more efficient to store and process the responses by computer. Therefore a computer coding plan should be in place. You may wish to code the actual form to indicate where each item of information will appear on the computerized file or develop a template to be placed over a completed form showing where the data should be recorded on the computer file. For example, the data may be entered onto a spreadsheet and the identity of the spreadsheet cell(s) (e.g. "C4") that match each answer would be added to a question.

APPEARANCE.   The aesthetics of a questionnaire are important. If the form looks professional the chances of it being completed are enhanced and, in addition, the quality of the form represents you and your organization.

Questionnaire Sections

There are three major parts to a questionnaire. The parts are a cover letter, instructions on how to complete the questionnaire and finally the actual set of questions. In the cover letter it is wise to include the purpose of the research and explain why it is important for a respondent to complete the form. The cover letter is you best opportunity to motivate the recipient to complete the questionnaire in a prompt and complete manner.

In the second section, instruction on how to answer questions should be given - does the respondent circle, check or fill-in the answer? It may be necessary to have short instructions on the actual questionnaire, especially if different directions are needed (e.g. how to complete a thermometer scale). It is also critical to explain how to return the questionnaire after it is completed. Do not expect the respondent to pay the cost of mailing back the form - a self addressed postage paid return envelope is expected.

The third part, the questions themselves, was covered earlier. However, it is worthwhile to add that the questions need to follow a logical sequence (e.g. historical and demographic information usually comes first).

Quality Control Guidelines

Pre test your questionnaire. Logical choices for the pre test are 1] colleagues, 2] potential respondents, 3] people on the research staff who may have to code terms or interpret responses to open ended questions or 4] anyone else who can give you feedback. Pilot testing of the whole process - sending, receiving, coding, entering results into a computer file, etc. is also advantageous if the budget and time permit. In the computerization step consider a computer editing function - i.e. designing specifications that would flag questionable responses as they are being entered into a computer file. For instance, if the survey was for high school students an age range might be set so that any responses below or above a certain age would be flagged.

All statistical analyses will almost assuredly be done by computer and the package that can do the proposed analyses should be secured and run on previously analysed data to be certain that the program functions properly. This step goes beyond verifying the integrity of the program. It allows the researcher to test to be sure that the user is inputting the information correctly and entering the proper computer instruction/command for the desired statistical applications.

One final comment has to do with the revisions of questions. The rule of thumb is not to revise questions in the course of a study unless it is absolutely necessary. Any revision can create validity and reliability problems. If a question must be revised it is wise to keep the original question in the questionnaire so the responses to the original and the revised question can be compared. It may be possible, based on the comparison, to make an adjustment/correction to the revised question so the data from the original question can be retained in an analysis.

ETHICAL ISSUES

Although science is said to be a search for the truth this does not mean researchers and statistician are always truthful and never engage in dishonest or irresponsible acts. We all have probably heard the observation attributed to Benjamin Disraeli that there are three kinds of lies - lies, damned lies and statistics.

The Research Consumer

The concern for responsible research practices effects not only the research community, but it also extends to those who depend on the conclusions and decisions made by the scientists. Even though you may not intend to become a full-time research worker or statistician, you will still be exposed to the work of these professionals and as a consumer you should be aware of the ethical problems that can occur. Perhaps the best example of how consumers are affected is in the health care field, including pharmaceutical research. Decisions about acceptable health practices and drugs are based on research methods and statistical practices described in this text. For example, the Food and Drug Administration requires at least two adequate and well controlled studies before they can approve a new drug for the marketplace.

Furthermore, you may also find that you are a direct consumer of research findings - i.e. your position calls on you to rely on the research results of others. In this latter situation you should feel free to approach your task in a skeptical fashion. By now you should have a good understanding of all the problems that can occur in a research endeavor and it should not come as a surprise that even with the best methodology and intentions, erroneous results occur. Researchers can never be 100% sure that they picked the most beneficial treatment schedule, used the right measurements, or took the measurements at the right time. Maybe they did not include enough subjects or perhaps there were too many. The study may have been run too long or not long enough, and then we have all those threats to internal validity.

As a recipient of research done by others, be watchful of a biased sample, especially if the sample is not a random sample for a definable population. Be concerned if the response rate is low or there are a large number of drop outs unless there is a convincing argument as to why these problems can be ignored. If the sample size or response rate is omitted beware - it is unacceptable to omit these elements when reporting research findings. Did the researchers push statistical significance as the only way to decide if the conclusions were sound or did they also demonstrate that there was practical significance? If you can, obtain access to the original reporting forms (e.g. questionnaire) and assure yourself that serious validity or reliability problems are not present. Also carefully review any displays to see if the distortions discussed in chapter 2 exist.

If we examine ethics from the viewpoint of those doing the research, it's interesting to note what some scholars consider unethical. The list includes a poorly designed study, a poorly executed study or a faulty analysis even if the parties involved had no intention to deceive or misrepresent the results. The point is that it is unethical to engage in research if you are not an informed and competent researcher.

If one is not competent in the work they do then they should abstain from doing the work. However, many scientists are motivated to have their research accepted and published even if it is not of high quality. Unfortunately what drives much of research is the importance of having publications in scientific journals. For some, publishing is the most important way to have a successful and rewarding academic or professional career. In addition, there is a much greater chance of having a research study accepted for publication if it has a statistically significant result. Consequently, there is a temptation to go to lengths to find a statistically significant difference even if it may require some dubious decisions along the way. Some actions are easily recognized, such as doing a one tail test when a two tail test is more reasonable or applying interval scale testing to data that are ordinal scale measurements.

Ethical Problems

Although the great majority of researchers may shun the temptation to engage in unethical practices, there are exceptions. Examples of some of the types of problems that have been reported are cited below.

FALSIFYING RESULTS. A Yale researcher painted mice with a black stripe running down their back to mimic a genetic change he had predicted. In other trials, researchers entered non existent or ineligible patients into studies since they were compensated by the number of subjects they entered into a trial. In one large project, consisting of many research sites, one of the investigational teams committed this faulty practice. When the project's sponsors found out they eliminated the questionable data, but never communicated the problem to the public because they feared the adverse publicity it would generate about their project.

THE TUSKEEGEE STUDY. There are some classic cases in which the rights of subjects in studies were abridged. Perhaps the most frequently cited case is the infamous Tuskeegee study. In the 1930s medical people wanted to understand syphilis better, but there was very little information on the course of the disease. There was, at the inception of the study, only one treatment available and its efficacy was unknown. Tuskeegee was a rural community in Alabama with a high prevalence of syphilis. The US Public Health agency, which set up and conducted the trial, believed that Tuskeegee would be a good area to study syphilis. The study was not done in secret and in 1936 an article in the *Journal of the American Medical Association* included reference to it. Enrolled in the study were 399 blacks who had never been treated for syphilis. A control group, identified as blacks without syphilis, was also created. The unseemly features of this trial can be summarized below.

A. The participants with syphilis were never told they had the disease. In fact the experimental and control group were offered free burials, not as a gesture to somehow repay their heirs for the subject's participation, but because this was a way to perform an autopsy and determine the destruction caused by the disease.

B. At one point the subjects were given spinal cord taps (a potentially dangerous and painful procedure) to determine the disease status. However, the subjects were told that the rationale for the taps was because they had "bad blood".

C. Penicillin, discovered and used during World War II, was found to be highly effective for the treatment of syphilis. It became available for general use shortly after the war ended. However, the Tuskeegee subjects were never offered the drug because it would obviously interfere with the purpose of the trial - to study the natural course of the disease.

D. In 1969 the health agency convened a group of experts to decide whether to continue or terminate the Tuskeegee study. They voted to continue.

E. The local medical society in Tuskeegee county agreed not to treat subjects with antibiotics for any disease because such an act would cause confounding.

The nature of the Tuskeegee trial was exposed by the media in late 1969. Congress held hearings on the trial shortly thereafter and subsequently the trial was terminated. The most egregious ethical flaws with the trial involve the use of deception - the subjects did not know they were participants in an experimental study. There was also harm done to the subjects - the risk from the spinal taps. There was harm done to the subject's family - the risk that the subject's illness could be spread to them, including their children who could acquire congenital syphilis.

THE US RADIATION EXPERIMENTS. Other U.S. government sponsored studies may also be cited - for example, the Radiation Experiments performed by our military establishment in the post World War II period. In 1994 news of the study became public and exposed the fact that people were given radiation, sometimes repeated doses as well as very high doses, without their permission and even without their knowledge.

The argument to perform both the Tuskeegee and the Radiation studies was based on the usefulness of the results for society. Syphilis was a dreaded disease and knowing how it progressed had valuable scientific implications. The US military felt it was important to know the effects of radiation and indeed it was. But, in both cases, the cost of the information was borne by a few who weren't given an option about their participation in the research endeavor, nor were their best interests considered a value worth protecting.

WORLD WAR II RESEARCH PRACTICES AND THE NUREMBERG CODE. Perhaps the most infamous studies in modern times were conducted during World War II when German and Japanese physicians and other "scientists" performed inhuman studies on military and political prisoners. Surgery without anesthesia, immersion in freezing tubs of water and spraying deadly biological products on unsuspecting enemy civilians are only a short list of the crimes committed. From the experiences in Nazi Germany the research community developed a set of principles to guide experimentation, frequently identified as the **Nuremberg Code**. At the core are two requirements - a subject's right to participate in a study only after giving informed consent and the need to protect the confidentiality of the subjects.

Informed Consent. Informed subjects are volunteers who has been told that they are participating in a research study, what the objectives are of the study, what the benefits and risks are of their participation, that they have a right to confidentiality and that they may drop out of the study at any time without repercussions. The consent portion of the requirement means the subject's agreement to participate was done voluntarily. The notion of giving voluntary consent raises the spectrum of coercion in some settings such as prisons, hospitals and nursing homes. In these custodial environments the total dependence of the subjects on their institutions may mean they will participate out of fear. They fear that their decision not to participate could be resented by their care takers.

We also need to be sensitive to the immature and incompetent subject (e.g. children and the mentally retarded) when it comes to obtaining informed consent. The researcher has, in these settings, an extra burden to be sure the subjects' parents/guardians are in fact freely agreeing to the subjects' participation and that there are benefits of that participation that outweigh the risks.

Confidentiality.   Confidentiality requires the researchers to safeguard any personal information acquired directly or indirectly about subjects. In many trials a great deal of personal information can be collected bout a subject and the research team must protect their subject's privacy rights.

In spite of the Nuremberg code, examples of poor ethical practices are still reported. In one study senile patients were injected with live cancer cells. The patients' guardians were told that the patients would be given injections to test the patient's immune system without any mention of the live cancer cells. In another trial, the goal was to prevent breast cancer in women by giving them prophylactic treatment with a potent drug. The drug also was associated with a high risk of producing other types of cancers and causing blood clots that could be fatal. A review of the informed consent documents used in the study showed that the form was missing risks, and especially a serious omission of the risk associated with contracting other forms of cancer.

THE MILGRAM STUDY.   Before we leave the subject of ethics a few more experiments merit our attention. The Milgram experiment is often cited as an example of unethical research but defended on the grounds that no harm accrued to the subjects. The trial was carried out in the 1960s and subjects were told that the study was investigating the learning process and, in particular, the effect of punishment on the ability to memorize information.

Two people participated in a role play scenario in which one participant was called the teacher and the other participant the learner. The two met before the "experiment" began and the experiment consisted of the teacher reading a list of words along with matching terms. The learner was then tested to see if he could recall the matching term. If the learner failed to remember what the matching term was, an electrical shock was given to the learner by the teacher. Different physical arrangements were used, but the most popular version had the teacher sitting on one side of a screen with switches for different levels of shock and the learner seated on the other side. The two could hear, but not see each other in this scenario. The shock was increased with each missed answer and some of the shocks were marked with multiple Xs implying the strength of the shock was very strong.

Unbeknown to the teacher the learner was an actor who received no shock whatsoever. However, the learner, when "shocked" would scream out in pain, claimed he had a heart condition, asked for the trial to end as well as making other pronouncements of great distress. If the teacher stated he did not want to continue with the experiment he was told he "must" proceed by the research team. Some teachers refused to continue the experiment, but many shocked the learner through the entire set of shocks possible.

The issues that are debated in this case include the ironic situation of whether the goal, which was based on the need to gain insight into why so many individuals followed inhumane orders during the Nazi reign, justified the means. Furthermore, the study was criticized because it forced participants to learn that they possessed the type of personality that could hurt a fellow human being. Opponents claim people should not be tricked into revealing an undesirable personal trait about themselves.

The researchers of the study had participants examined for adverse effects from the study and reported that there were no serious after effects from their participation. Even if it is assumed that this finding is true, critics point out that this step was taken after the study was finished - how did the researchers know in advance that there would be no harm to the unsuspecting subjects?

AIDS RESEARCH. In 1997 another trial that produced controversy was an AIDS trial sponsored in part by the U.S. government that involved, as subjects, pregnant women infected with HIV (human immunodeficiency virus). The women lived in Africa, Thailand and the Dominican Republic. Half of the women were given doses of drugs that could prevent the transmission of HIV to their unborn babies, but the other half received only inert pills (i.e. placebo). Due to the high cost of drug treatment, which was beyond what citizens of third world countries could afford, the trial set out to find if lower doses of medications would be effective, thereby reducing the cost of treatment to a level third world countries could afford. It has been estimated that 1,000 children, whose mothers were in the control group, would come down with HIV even though the transmission was preventable if they received the active treatment. Proponents of the study argued that the use of the inert pills was the only way to reach a quick reliable result. It was also noted that the sponsors weren't denying the women anything since the women wouldn't be able to afford the effective treatment, even if they were not part of the study.

Summary

In this chapter we saw that there are a large and rich number of qualitative methods available for use in research projects. In qualitative research the investigator becomes much more personally involved in the data collection process. For example, experimenters may be an active part of the research setting when they play the role of participant observers. He or she is clearly instrumental in shaping information gathered in a study when content analysis or interviews are part of the research methodology. Due to the greater subjectivity associated with qualitative methods, experimenters need to take extra care in documenting their work.

Although questionnaires are used in both quantitative and qualitative research their preparation must be done with care. Thirteen recommendations on how to prepare individual questions are offered. The advantages and disadvantages of different ways to record responses to questions are presented for the Likert, thermometer, checklist, ranking and rating scales.

The last section on ethics points out that there are strong motivational factors that can tempt an investigator to engage in questionable research practices. The horrible studies conducted in World War II led to the adoption of the Nuremberg Code that continues to serve as a standard to protect the rights of research subjects. However, research continues to be done that raises issues of ethical and socially irresponsible behavior. In some cases unethical performance by a researcher is clearly evident, but who is right when it comes to ethical behavior is sometimes hard to tell. And so on this uncertain note we finally come to an end.

It is the wish of the author that you enjoyed our exploration and found that statistics and research methods aren't so bad after all.

PROBLEMS

1. What type of research (quantitative or qualitative) would be more accepting of the following practices.
   a. Modifying or adding variables during the course of the project.
   b. Using subjects who have limited time to devote to the project.
   c. Recording free form comments and observations by the subjects
   d. Expecting the research results from one study to be applicable in other situations.
   e. Finding little information is available about the phenomenon to be researched.
   f. A trial involving variables that have ideational rather than numeric properties.

2. What is the difference between a participant observer and a non-participant observer?

3. What type of records are used in qualitative research projects?

4. Name the 6 steps that an interviewer can take to enhance the likelihood of a successful interview session.

5. What are the advantages and disadvantages of a personal interview and a phone interview.

6. What are major differences between the case study approach and a survey type of research design?

7. Refer to the survey request letter and the questionnaire starting on page 453. Note those elements that should be brought to the researcher's attention.

8. What are the five features from the Tuskeegee study that represent unethical practices.

9. What is meant by informed consent?

10. What arguments did the sponsors of the AIDS study in pregnant women use to defend their trial plan?

11a. For the following data set on electrical power usage, calculate the 4 year moving average
   b. Calculate the exponential smoothed values for the set using a D of .80.

Electrical Power Usage									
Year	1988	1989	1990	1991	1992	1993	1994	1995	1996	1997
Usage	458.2	465.6	467.3	497.1	523.0	489.5	514.3	524.6	518.2	548.7

12. The following computer output was based on a study of the relationship between the average ACT score for entering freshmen at local colleges and that group's graduation rate.

SUMMARY OUTPUT

Regression Statistics
Multiple R	0.839
R Square	0.703
Adjusted R Square	0.661
Standard Error	5.799
Observations	9

ANOVA

	df	SS	MS	F	Significance F
Regression	1	558.15	558.15	16.60	0.005
Residual	7	235.41	33.629		
Total	8	793.56			

	Coefficients	Standard Error	t Stat	P-value	Lower 95%	Upper 95%
Intercept	-0.98	18.07	-0.05	0.96	-43.71	41.75
X Variable 1	3.05	0.75	4.07	0.00	1.28	4.82

a. What was the independent and what was the dependent variable?
b. How many colleges were included in the analysis?
c. What is the correlation between the independent and dependent variables?
d. Write a summary paragraph for the analysis using the terms ACT score and graduation rate - not X and Y to refer to the variables.

13. What is a double blind study and what does it try to achieve?

14. Are the following statistical procedures parametric or non-parametric tests?
　Chi Square　　　ANOVA　　　Wilcoxson Rank Sum
　Spearman r　　　Pearson r　　t Test

15. A manager believes that to change equipment there must be at least a 2.5% increase in performance. The current supplier has a performance index score of 34.8. Materials were submitted by 4 competitors who are seeking to replace the current supplier. The performance index score for their products are shown below. In addition, based on the tests done, a probability of the performance score being significantly better than the current supplier was determined.

Supplier	Universal	Statewide	National	Federal
Performance Score	35.6	35.9	36.1	35.2
P-value	.12	.02	.09	.03

a. Based on the grid below, place the 4 competitors in their proper cell. Assume the manager uses the conventional .05 level for statistical significance.

Practical	Statistical Significance	
Significance	Yes	NO
Yes		
No		

b. Which competitor, if any, should the manager select?

There are no instructions in Excel for this chapter.

State College
Office of   March 13, 1999
Student Affairs

Dear State College Faculty and Staff,

We need your help. The College is conducting a campus wide survey of the use of alcohol and other drugs. Your responses will help us get a more accurate picture of alcohol and drug use and opinions about alcohol and other drug use by the faculty, staff and administrative contingents of the College. A similar survey has been distributed to all undergraduate students.

The study is being conducted by the Task Force on Alcohol and Other Drug Education. The survey findings will play an important role in the development of programs and delivery of support services by the College. This is your opportunity to guarantee that future programming will be relevant to the needs and concerns of the campus community.

The enclosed survey is designed to guarantee anomymity. Your name and identity cannot be associated with your answers. Do not put your name on the instrument. The survey will take about 15 minutes to fill out. When you have completed the survey, please return it in the enclosed self-addressed envelope.

After you have mailed the survey, also sign and mail **separately** the enclosed post card. This will let us know that you have completed and returned the survey. Return of the post card will also automatically enter you into a drawing for a **complete dinner for two at the Le Canard Restaurant or one of two $50 gift certificates at the store of your choice in the College Mall.** Surveys and post cards should be returned no later than Thursday, March 28.

We appreciate your participation. If you have any questions, please call Deborah Koed, extension 2517 or Bob Bmoc, extension 2201.

    Appreciatively,

    Deborah Koed,
    Chair, Task Force on Alcohol and Drug Education.

CORE INSTRUMENT
(Modified for use by State College)
(Faculty and Staff Version)

Brookfield Group - Drug Prevention Program
State University System
State Capital, 123455.

1. Classification:
Faculty   O
Staff   O

4. Marital status:
  Single   O
  Married   O
  Separated   O
  Divorced   O
  Widowed   O

2. Age:
O O   O O
O 1   O 1
O 2   O 2
O 3   O 3
O 4   O 4
O 5   O 5
O 6   O 6
O 7   O 7
O 8   O 8
O 9   O 9

3. Ethnic origin:
American Indian/Native   O
American/Alaskan Native   O
Hispanic   O
Asian/Pacific Islander   O
White (non-Hispanic)   O
Black (non-Hispanic)   O
Other   O

5. Gender:
Male ....O
Female .....O

6. Are you working?
  Full time   O
  Part time   O

7. Highest degree
   Earned:
   High school   O
   Bachelor of Arts. O
   Master of Arts   O
   Ph.D.   O

8. Living
   Arrangement:
   (mark all that apply)
   With roommate(s).   O
   With significant other   O
   Alone   O
   With parent(s)   O
   With spouse   O
   With children   O

9. Length of Employment
   at State College:
Less than 5 yrs   O
Between 5 and 10 yrs   O
   Between 11 and 20 yrs   O
   More than 20 yrs   O

10. Campus situation on alcohol and drugs:	For Students			For Employees		
	yes	no	don't know	yes	no	don't know
a. Does State College have drug and alcohol policies?	O	O	O	O	O	O
b. If so are they enforced	O	O	O	O	O	O
c. Does State College have a drug and alcohol prevention program?	O	O	O	O	O	O
d. Do you believe State College is concerned about the prevention of drug and alcohol use?	O	O	O	O	O	O
e. Are you actively involved in efforts to prevent drug and alcohol use problems at State College	O	O	O	O	O	O

11. Think over the last two weeks. How many times have you had five or more drinks at a sitting?

None	O
Once	O
Twice	O
3 to 5 times	O
6 to 9 times	O
10 or more times	O

12. Average # of drinks you consume a week

O....O	O....O
O....1	O....1
O....2	O....2
O....3	O....3
O....4	O....4
O....5	O....5
O....6	O....6
O....7	O....7
O....8	O....8
O....9	O....9

* a drink is a bottle of beer, a glass of wine, a
wine cooler, a shot glass of liquor or a mixed drink.

13. At what age did you use (mark one for each line)	never	<10	10-11	12-13	14-15	16-17	18-25	26+
a. Tobacco (smoke, chew, snuff)	O	O	O	O	O	O	O	O
b. Alcohol (beer, wine, liquor)	O	O	O	O	O	O	O	O
c. Marijuana (pot, hash, hash oil)	O	O	O	O	O	O	O	O
d. Cocaine (crack, rock, freebase)	O	O	O	O	O	O	O	O
e. Amphetamines (uppers, speed)	O	O	O	O	O	O	O	O
f. Sedatives (downers, ludes)	O	O	O	O	O	O	O	O
g. Hallucinogens (LSD, PCP)	O	O	O	O	O	O	O·	O
h. Opiates (heroin, smack, horse)	O	O	O	O	O	O	O	O
i. Inhalants (glue, solvents, gas)	O	O	O	O	O	O	O	O
j. Designer drugs (estacy, MDMA)	O	O	O	O	O	O	O	O
k. Steroids	O	O	O	O	O	O	O	O
l. Other drugs	O	O	O	O	O	O	O	O

14. Within the last year about   never 1x/yr 6x/yr 1x/mo 2x/mo 1x/wk 3x/wk 5x/wk daily
    how often have you used: (mark one for each line)

	never	1x/yr	6x/yr	1x/mo	2x/mo	1x/wk	3x/wk	5x/wk	daily
a. Tobacco(smoke, chew, snuff)	O	O	O	O	O	O	O	O	O
b. Alcohol (beer, wine, liquor)	O	O	O	O	O	O	O	O	O
c. Marijuana (pot, hash, hash oil)	O	O	O	O	O	O	O	O	O
d. Cocaine (crack, rock, freebase)	O	O	O	O	O	O	O	O	O
e. Amphetamines (uppers, speed)	O	O	O	O	O	O	O	O	O
f. Sedatives (downers, ludes)	O	O	O	O	O	O	O	O	O
g. Hallucinogens (LSD, PCP)	O	O	O	O	O	O	O	O	O
h. Opiates (heroin, smack, horse)	O	O	O	O	O	O	O	O	O
i. Inhalants (glue, solvents, gas)	O	O	O	O	O	O	O	O	O
j. Designer drugs (estacy, MDMA)	O	O	O	O	O	O	O	O	O
k. Steroids	O	O	O	O	O	O	O	O	O
l. Other drugs	O	O	O	O	O	O	O	O	O

15. Have any of your family had alcohol or other drug problems: (mark all that apply)

O Mother	O Mother's parents	O Great grandparents
O Father	O Father's parents	O Aunts/Uncles
O Stepmother	O Spouse	O Children
O Stepfather	O Brothers/Sisters	

16. In the past year where have   never   on   off   where   Friend's in a   other
    you used: (mark one for each line)   campus campus you live house   car

	never	on campus	off campus	where you live	Friend's house	in a car	other
a. Tobacco(smoke, chew, snuff)	O	O	O	O	O	O	O
b. Alcohol (beer, wine, liquor)	O	O	O	O	O	O	O
c. Marijuana (pot, hash, hash oil)	O	O	O	O	O	O	O
d. Cocaine (crack, rock, freebase)	O	O	O	O	O	O	O
e. Amphetamines (uppers, speed)	O	O	O	O	O	O	O
f. Sedatives (downers, ludes)	O	O	O	O	O	O	O
g. Hallucinogens (LSD, PCP)	O	O	O	O	O	O	O
h. Opiates (heroin, smack, horse)	O	O	O	O	O	O	O
i. Inhalants (glue, solvents, gas)	O	O	O	O	O	O	O
j. Designer drugs (estacy, MDMA)	O	O	O	O	O	O	O
k. Steroids	O	O	O	O	O	O	O
l. Other drugs	O	O	O	O	O	O	O

17. Please indicate how often you have experienced the following due to your drinking or drug use during the last year: (mark one for each line)

	never	once	twice	3-5 times	6-9 times	10+ times
a. Had a hangover	O	O	O	O	O	O
b. Performed poorly on an important project	O	O	O	O	O	O
c. Been in trouble with police or other authorities	O	O	O	O	O	O
d. Damaged property	O	O	O	O	O	O
e. Got into an argument or fight	O	O	O	O	O	O
f. Got nauseated or vomited	O	O	O	O	O	O
g Driven a car while under the influence	O	O	O	O	O	O
h. Missed work	O	O	O	O	O	O
i. Been criticized by someone I know	O	O	O	O	O	O
j .Thought I might have a drinking or other drug problem	O	O	O	O	O	O
k. Had a memory loss	O	O	O	O	O	O
l. Done something I later regretted	O	O	O	O	O	O
m. Been arrested for DWI/DUI	O	O	O	O	O	O
n. Have been taken advantage of sexually or have taken advantage of another sexually	O	O	O	O	O	O
o. Tried unsuccessfully to stop using	O	O	O	O	O	O
p. Thought about or tried to commit suicide	O	O	O	O	O	O
q. Been hurt or injured	O	O	O	O	O	O

18. How many of the students

	None	A Few	Several	Many	Most	All
at State College do you think use:						
(mark one for each line)						
a. Tobacco (smoke. chew, snuff)	O	O	O	O	O	O
b. Alcohol (beer, wine, liquor	O	O	O	O	O	O
c. Marijuana (pot, hash, hash oil)	O	O	O	O	O	O
d. Cocaine (crack, rock, freebase	O	O	O	O	O	O
e. Amphetamines (uppers, speed)	O	O	O	O	O	O
f. Sedatives (downers, ludes)	O	O	O	O	O	O
g. Hallucinogens (LSD, PCP)	O	O	O	O	O	O
h. Opiates (heroin, smack. horse)	O	O	O	O	O	O
i. Inhalants (glue, solvents, gas)	O	O	O	O	O	O
j. Designer drugs (ecstasy, MDMA)	O	O	O	O	O	O
k. Steroids	O	O	O	O	O	O
1. Other illegal drugs	O	O	O	O	O	O

19. How many of the employees

	None	A Few	Several	Many	Most	All
at State College do you think use:						
(mark one for each line)						
a. Tobacco (smoke. chew, snuff)	O	O	O	O	O	O
b. Alcohol (beer, wine, liquor	O	O	O	O	O	O
c. Marijuana (pot,hash,hashoii)	O	O	O	O	O	O
d Cocaine (crack, rock, freebase	O	O	O	O	O	O
e. Amphetamines (uppers, speed)	O	O	O	O	O	O
f. Sedatives (downers, ludes)	O	O	O	O	O	O
g. Hallucinogens (LSD, PCP)	O	O	O	O	O	O
h. Opiates (heroin, smack. horse)	O	O	O	O	O	O
i. Inhalants (glue, solvents, gas)	O	O	O	O	O	O
j. Designer drugs (ecstasy, MDMA)	O	O	O	O	O	O
k. Steroids	O	O	O	O	O	O
1. Other illegal drugs	O	O	O	O	O	O

20. Do you believe the programming and services relevant to alcohol and other drug use which are currently available at State College are:

O Adequate
O Inadequate

Answers to Selected Problems

Chapter 1
1. Mean: 76.3
   Median: 76
   Mode: 74
   Avg. Deviation: 9.8
   Stand. Deviation: 11.6
   Variance: 135.7
2. Mean: Increase
   Median: Increase
   Mode: No change
   Standard Deviation: Can't tell is an acceptable answer, but statisticians would probably point out that you can be pretty certain the standard deviation will increase because the new values represent a larger mean deviation than the standard deviation (12.7 v. 11.6) and therefore the new standard deviation will increase a little.
5. Mean: 76.6
   Medeian: 76
   Mode: 74
   Standard Deviation: 11.77

Chapter 2
4. Line - time trends
   Bar - comparison of quantities
   Pie -subparts of a larger unit
   XY - relationship between 2 variables
7a. Mean = 77.2, Median = 79, Standard deviation = 8.83
 b. 78
8a. Range = 18 (70-88), Mode = 72,78,83 Variance = 29.4
 b. 77

Chapter 3
1a. 1/3
 b. ½
3a. .51
 b. .85
 c. .33
5a. .68
 b. .09
 c. .19
7a. .50

  b. .50
  c. .50
9a $2.90
  b. Yes
13. 1/18
15a. Standard Deviation = Decrease
  b. Range = Unchanged
  c. Median = Can't tell
  d. Mode = Can't tell
  e. Variance = Decrease

Chapter 4
1a. 20
3a. 840
5.   15
7.   10
9.   435
11. 252
13. 015625
17a. .015
19a. 3/8

Chapter 5
2a. $Z \geq +1.00$ or from $Z=+1.00$ to $+\infty$
  b. Z between -1.41 and +1.41
  c. $Z \leq -0.84$ from $Z=-0.84$ to $-\infty$
  d. Z between -.67 and + .67
4a. -.52
  b. 0
  c. +1.64
  d. -1.04
6a. .11
  b   .62
  c. .50
  d. .01
8a. .006
  b. .067
  d. .494
10a. 18/38 or .47
  b. 1044/1444 or .72
  c. 324/1444 or .22

  d.   324/1444 or .22
12a. 3,628,800
  b.   210
  c.   0.2880
  e.   0.8208

Chapter 6
1a.   Military = Ordinal
2.1a. Reliability = Baker
2.2a. Damingo = Stability
2.3a. Hadad = Face
3.1 – 3.3 for Event 1

Event	3.1 Threat	3.2 Variable(s) Affected	3.3 Group Bias
1. Competitor Report	Statistical Regression	The Side Effect variable because the incidence of side effects during the trial is probably exaggerated. Only subjects with no symptoms were allowed into the trial. The absence of side effects possible occurred because some of the women were having a "good" day.	We cannot tell which group's values would be biased because we do not know the identify of the women whose side effects were under reported.

5a.   .09
 c.   .547
 e.   .498
7a.   .083
9a.   .2743
 b.   61

Chapter 7
1a.   Stratified
2a.   Mean = No change; Standard deviation = Small increase
3a.   27.2 – 44.4
3c.   35.5 - 36.1
5.   N=27
7.   A mean equal to or smaller than this year's average has a Probability = .078
9.   N =208
10.   Need a mean contribution of $433
13a. 0625
 b.   .3750
15a. .211
 b.   <.0014

Chapter 8
1a. 56-74
  b. 23-29
4a. +1.65
  b. -1.28
  c. -2.58 and +2.58
  d. Yes there is a statistically significant difference
6a.   -2.33 and + 2.33
  b. +1.75
  c. The difference is not statistically significant
9a. The null hypothesis is that there is no change in weight. The mean weight change is
     0 under the null hypothesis.
10a. .24/69 or 35
  b. 120/314,364 or <.001
  c. 300/4,692 or .06
13. 21,924
15. 273
17b. .077

Chapter 9
1a. No
  b. Yes
3a. Two tail
  c. .05
  d. Yes
5a. Two tail
  b. .05
  c. No
7a. 18
  b1. Yes
8a. 40
  b1. No
9a. 20
  b1. No
10a. 52.2-67.8

Chapter 10
1a. Yes, there is statistical significance
2d. The standard deviation is the square root of the variance or 5.2.
4a. .21 - .32
9b. The analysis of variance

12a. There is no statistical significance

Chapter 11
1. The difference between Republicans and Democrats is statistically significant - Republicans are more likely to vote than Democrats.
3. The difference between urban and suburban residents is statistically significant. Urban residents tend to give lower ratings to their Congressional representative than suburban residents.
5. There is not sufficient evidence to declare that the difference, in respect to the appeal of MPA and MBA students to health care organizations, is statistically significant.
6a. .86
11a. Group B

Chapter 12
1a. A Two Way Analysis of Variance.
 b. Because there are two factors - a treatment factor and a site factor.
9b. When there is no recognized treatment or when the use of a placebo is contraindicated.
11. Appelton Standardized Score = 229
    You get to figure the others out yourself
12. Year   1   2   3   4
    Index   100   105   (Oh Oh - the last 2 are for you to do)
13. Atomic Standardized Score = 4.0
15.   The difference is not statistically significant.
17a. The difference is not statistically significant.
19a. The difference is statistically significant and action should be taken.

Chapter 13
1a. 212
     b.   Loss of 55 units
     c.   Positive because the slope (b) is positive
     d.   -8
2a. Y = 474.7 + 0.011X
     d. $55
5a. No
6b. 20,797
7a. 6.8, the predicted value is smaller
 b. 10.2
12. p = .142625
13. r_s = .82
14, There is a statistically significant difference.

Chapter 14

1. Linear trend with irregular component
3a. Logarithmic
    b. Linear
5. Most analysts would use the linear trendline because it is simpler to use and it's lower
    R Square value is a negligible difference.
7. The values for the first two observations are:

Obs	Value
1	NA
2	16.0

9a. 52.86 million
    c. 96%
10a. 1668 vaccinations
    d. ±194.271
    h. 1534
13a. The difference is not statistically significant.

Chapter 15
1a. Qualitative
    b. Quantitative
6. In a survey design the data are gathered on a large number of people, but the amount of
    information collected is restricted and limited compared to a typical case study.
    The case study has the potential to give a fine grained picture of an issue, capturing
    subtleties and amplifying details that may be missed by the more general survey
    approach.
    The case study is highly dependent on the skill and integrity of the researchers and
    there is a greater risk of misleading conclusions.
    There is a danger if the results of a case study are generalized to other people, groups or
    organizations.
    The in depth and documented portrayal of people and group dynamics from a case
    study can be of great use when used for descriptive purposes.
11a.

Year	1989.5	1990.5	1991.5	1992.5	1993.5	1994.5	1995.5
Usage	472.1						526.5

    b.

Year	1988	1989	1990	1991	1992	1993	1994	1995	1996	1997
Usage				468.4						

15a.

Practical	Statistical Significance	
Significance	Yes	No
Yes		
No		Universal

  b. Statewide

Appendix — List of Tables

Statistical Methods for Researchers

Table 1 Areas Under the Normal curve

Z	Area in left tail	Area in right tail	Z	Area in left tail	Area in right tail	Z	Area in left tail	Area in right tail
-3.0	0.0013	0.9987	-2.64	0.0041	0.9959	-2.28	0.0113	0.9887
-2.99	0.0014	0.9986	-2.63	0.0043	0.9957	-2.27	0.0116	0.9884
-2.98	0.0014	0.9986	-2.62	0.0044	0.9956	-2.26	0.0119	0.9881
-2.97	0.0015	0.9985	-2.61	0.0045	0.9955	-2.25	0.0122	0.9878
-2.96	0.0015	0.9985	-2.6	0.0047	0.9953	-2.24	0.0125	0.9875
-2.95	0.0016	0.9984	-2.59	0.0048	0.9952	-2.23	0.0129	0.9871
-2.94	0.0016	0.9984	-2.58	0.0049	0.9951	-2.22	0.0132	0.9868
-2.93	0.0017	0.9983	-2.57	0.0051	0.9949	-2.21	0.0136	0.9864
-2.92	0.0018	0.9982	-2.56	0.0052	0.9948	-2.2	0.0139	0.9861
-2.91	0.0018	0.9982	-2.55	0.0054	0.9946	-2.19	0.0143	0.9857
-2.9	0.0019	0.9981	-2.54	0.0055	0.9945	-2.18	0.0146	0.9854
-2.89	0.0019	0.9981	-2.53	0.0057	0.9943	-2.17	0.0150	0.9850
-2.88	0.0020	0.9980	-2.52	0.0059	0.9941	-2.16	0.0154	0.9846
-2.87	0.0021	0.9979	-2.51	0.0060	0.9940	-2.15	0.0158	0.9842
-2.86	0.0021	0.9979	-2.5	0.0062	0.9938	-2.14	0.0162	0.9838
-2.85	0.0022	0.9978	-2.49	0.0064	0.9936	-2.13	0.0166	0.9834
-2.84	0.0023	0.9977	-2.48	0.0066	0.9934	-2.12	0.0170	0.9830
-2.83	0.0023	0.9977	-2.47	0.0068	0.9932	-2.11	0.0174	0.9826
-2.82	0.0024	0.9976	-2.46	0.0069	0.9931	-2.1	0.0179	0.9821
-2.81	0.0025	0.9975	-2.45	0.0071	0.9929	-2.09	0.0183	0.9817
-2.8	0.0026	0.9974	-2.44	0.0073	0.9927	-2.08	0.0188	0.9812
-2.79	0.0026	0.9974	-2.43	0.0075	0.9925	-2.07	0.0192	0.9808
-2.78	0.0027	0.9973	-2.42	0.0078	0.9922	-2.06	0.0197	0.9803
-2.77	0.0028	0.9972	-2.41	0.0080	0.9920	-2.05	0.0202	0.9798
-2.76	0.0029	0.9971	-2.4	0.0082	0.9918	-2.04	0.0207	0.9793
-2.75	0.0030	0.9970	-2.39	0.0084	0.9916	-2.03	0.0212	0.9788
-2.74	0.0031	0.9969	-2.38	0.0087	0.9913	-2.02	0.0217	0.9783
-2.73	0.0032	0.9968	-2.37	0.0089	0.9911	-2.01	0.0222	0.9778
-2.72	0.0033	0.9967	-2.36	0.0091	0.9909	-2.0	0.0228	0.9772
-2.71	0.0034	0.9966	-2.35	0.0094	0.9906	-1.99	0.0233	0.9767
-2.7	0.0035	0.9965	-2.34	0.0096	0.9904	-1.98	0.0239	0.9761
-2.69	0.0036	0.9964	-2.33	0.0099	0.9901	-1.97	0.0244	0.9756
-2.68	0.0037	0.9963	-2.32	0.0102	0.9898	-1.96	0.0250	0.9750
-2.67	0.0038	0.9962	-2.31	0.0104	0.9896	-1.95	0.0256	0.9744
-2.66	0.0039	0.9961	-2.3	0.0107	0.9893	-1.94	0.0262	0.9738
-2.65	0.0040	0.9960	-2.29	0.0110	0.9890	-1.93	0.0268	0.9732

Table 1 Areas Under the Normal curve

Z	Area in left tail	Area in right tail	Z	Area in left tail	Area in right tail	Z	Area in left tail	Area in right tail
-1.92	0.0274	0.9726	-1.56	0.0594	0.9406	-1.2	0.1151	0.8849
-1.91	0.0281	0.9719	-1.55	0.0606	0.9394	-1.19	0.1170	0.8830
-1.9	0.0287	0.9713	-1.54	0.0618	0.9382	-1.18	0.1190	0.8810
-1.89	0.0294	0.9706	-1.53	0.0630	0.9370	-1.17	0.1210	0.8790
-1.88	0.0301	0.9699	-1.52	0.0643	0.9357	-1.16	0.1230	0.8770
-1.87	0.0307	0.9693	-1.51	0.0655	0.9345	-1.15	0.1251	0.8749
-1.86	0.0314	0.9686	-1.5	0.0668	0.9332	-1.14	0.1271	0.8729
-1.85	0.0322	0.9678	-1.49	0.0681	0.9319	-1.13	0.1292	0.8708
-1.84	0.0329	0.9671	-1.48	0.0694	0.9306	-1.12	0.1314	0.8686
-1.83	0.0336	0.9664	-1.47	0.0708	0.9292	-1.11	0.1335	0.8665
-1.82	0.0344	0.9656	-1.46	0.0721	0.9279	-1.1	0.1357	0.8643
-1.81	0.0351	0.9649	-1.45	0.0735	0.9265	-1.09	0.1379	0.8621
-1.8	0.0359	0.9641	-1.44	0.0749	0.9251	-1.08	0.1401	0.8599
-1.79	0.0367	0.9633	-1.43	0.0764	0.9236	-1.07	0.1423	0.8577
-1.78	0.0375	0.9625	-1.42	0.0778	0.9222	-1.06	0.1446	0.8554
-1.77	0.0384	0.9616	-1.41	0.0793	0.9207	-1.05	0.1469	0.8531
-1.76	0.0392	0.9608	-1.4	0.0808	0.9192	-1.04	0.1492	0.8508
-1.75	0.0401	0.9599	-1.39	0.0823	0.9177	-1.03	0.1515	0.8485
-1.74	0.0409	0.9591	-1.38	0.0838	0.9162	-1.02	0.1539	0.8461
-1.73	0.0418	0.9582	-1.37	0.0853	0.9147	-1.01	0.1562	0.8438
-1.72	0.0427	0.9573	-1.36	0.0869	0.9131	-1	0.1587	0.8413
-1.71	0.0436	0.9564	-1.35	0.0885	0.9115	-0.99	0.1611	0.8389
-1.7	0.0446	0.9554	-1.34	0.0901	0.9099	-0.98	0.1635	0.8365
-1.69	0.0455	0.9545	-1.33	0.0918	0.9082	-0.97	0.1660	0.8340
-1.68	0.0465	0.9535	-1.32	0.0934	0.9066	-0.96	0.1685	0.8315
-1.67	0.0475	0.9525	-1.31	0.0951	0.9049	-0.95	0.1711	0.8289
-1.66	0.0485	0.9515	-1.3	0.0968	0.9032	-0.94	0.1736	0.8264
-1.65	0.0495	0.9505	-1.29	0.0985	0.9015	-0.93	0.1762	0.8238
-1.64	0.0505	0.9495	-1.28	0.1003	0.8997	-0.92	0.1788	0.8212
-1.63	0.0516	0.9484	-1.27	0.1020	0.8980	-0.91	0.1814	0.8186
-1.62	0.0526	0.9474	-1.26	0.1038	0.8962	-0.9	0.1841	0.8159
-1.61	0.0537	0.9463	-1.25	0.1056	0.8944	-0.89	0.1867	0.8133
-1.6	0.0548	0.9452	-1.24	0.1075	0.8925	-0.88	0.1894	0.8106
-1.59	0.0559	0.9441	-1.23	0.1093	0.8907	-0.87	0.1922	0.8078
-1.58	0.0571	0.9429	-1.22	0.1112	0.8888	-0.86	0.1949	0.8051
-1.57	0.0582	0.9418	-1.21	0.1131	0.8869	-0.85	0.1977	0.8023

Statistical Methods for Researchers

Table 1 Areas Under the Normal curve

Z	Area in left tail	Area in right tail	Z	Area in left tail	Area in right tail	Z	Area in left tail	Area in right tail
-0.84	0.2005	0.7995	-0.48	0.3156	0.6844	-0.12	0.4522	0.5478
-0.83	0.2033	0.7967	-0.47	0.3192	0.6808	-0.11	0.4562	0.5438
-0.82	0.2061	0.7939	-0.46	0.3228	0.6772	-0.1	0.4602	0.5398
-0.81	0.2090	0.7910	-0.45	0.3264	0.6736	-0.09	0.4641	0.5359
-0.8	0.2119	0.7881	-0.44	0.3300	0.6700	-0.08	0.4681	0.5319
-0.79	0.2148	0.7852	-0.43	0.3336	0.6664	-0.07	0.4721	0.5279
-0.78	0.2177	0.7823	-0.42	0.3372	0.6628	-0.06	0.4761	0.5239
-0.77	0.2206	0.7794	-0.41	0.3409	0.6591	-0.05	0.4801	0.5199
-0.76	0.2236	0.7764	-0.4	0.3446	0.6554	-0.04	0.4840	0.5160
-0.75	0.2266	0.7734	-0.39	0.3483	0.6517	-0.03	0.4880	0.5120
-0.74	0.2296	0.7704	-0.38	0.3520	0.6480	-0.02	0.4920	0.5080
-0.73	0.2327	0.7673	-0.37	0.3557	0.6443	-0.01	0.4960	0.5040
-0.72	0.2358	0.7642	-0.36	0.3594	0.6406	0.0	0.5000	0.5000
-0.71	0.2389	0.7611	-0.35	0.3632	0.6368	0.01	0.5040	0.4960
-0.7	0.2420	0.7580	-0.34	0.3669	0.6331	0.02	0.5080	0.4920
-0.69	0.2451	0.7549	-0.33	0.3707	0.6293	0.03	0.5120	0.4880
-0.68	0.2483	0.7517	-0.32	0.3745	0.6255	0.04	0.5160	0.4840
-0.67	0.2514	0.7486	-0.31	0.3783	0.6217	0.05	0.5199	0.4801
-0.66	0.2546	0.7454	-0.3	0.3821	0.6179	0.06	0.5239	0.4761
-0.65	0.2578	0.7422	-0.29	0.3859	0.6141	0.07	0.5279	0.4721
-0.64	0.2611	0.7389	-0.28	0.3897	0.6103	0.08	0.5319	0.4681
-0.63	0.2643	0.7357	-0.27	0.3936	0.6064	0.09	0.5359	0.4641
-0.62	0.2676	0.7324	-0.26	0.3974	0.6026	0.1	0.5398	0.4602
-0.61	0.2709	0.7291	-0.25	0.4013	0.5987	0.11	0.5438	0.4562
-0.6	0.2743	0.7257	-0.24	0.4052	0.5948	0.12	0.5478	0.4522
-0.59	0.2776	0.7224	-0.23	0.4090	0.5910	0.13	0.5517	0.4483
-0.58	0.2810	0.7190	-0.22	0.4129	0.5871	0.14	0.5557	0.4443
-0.57	0.2843	0.7157	-0.21	0.4168	0.5832	0.15	0.5596	0.4404
-0.56	0.2877	0.7123	-0.2	0.4207	0.5793	0.16	0.5636	0.4364
-0.55	0.2912	0.7088	-0.19	0.4247	0.5753	0.17	0.5675	0.4325
-0.54	0.2946	0.7054	-0.18	0.4286	0.5714	0.18	0.5714	0.4286
-0.53	0.2981	0.7019	-0.17	0.4325	0.5675	0.19	0.5753	0.4247
-0.52	0.3015	0.6985	-0.16	0.4364	0.5636	0.2	0.5793	0.4207
-0.51	0.3050	0.6950	-0.15	0.4404	0.5596	0.21	0.5832	0.4168
-0.5	0.3085	0.6915	-0.14	0.4443	0.5557	0.22	0.5871	0.4129
-0.49	0.3121	0.6879	-0.13	0.4483	0.5517	0.23	0.5910	0.4090

Table 1 Areas Under the Normal curve

Z	Area in left tail	Area in right tail	Z	Area in left tail	Area in right tail	Z	Area in left tail	Area in right tail
0.24	0.5948	0.4052	0.6	0.7257	0.2743	0.96	0.8315	0.1685
0.25	0.5987	0.4013	0.61	0.7291	0.2709	0.97	0.8340	0.1660
0.26	0.6026	0.3974	0.62	0.7324	0.2676	0.98	0.8365	0.1635
0.27	0.6064	0.3936	0.63	0.7357	0.2643	0.99	0.8389	0.1611
0.28	0.6103	0.3897	0.64	0.7389	0.2611	1.0	0.8413	0.1587
0.29	0.6141	0.3859	0.65	0.7422	0.2578	1.01	0.8438	0.1562
0.3	0.6179	0.3821	0.66	0.7454	0.2546	1.02	0.8461	0.1539
0.31	0.6217	0.3783	0.67	0.7486	0.2514	1.03	0.8485	0.1515
0.32	0.6255	0.3745	0.68	0.7517	0.2483	1.04	0.8508	0.1492
0.33	0.6293	0.3707	0.69	0.7549	0.2451	1.05	0.8531	0.1469
0.34	0.6331	0.3669	0.7	0.7580	0.2420	1.06	0.8554	0.1446
0.35	0.6368	0.3632	0.71	0.7611	0.2389	1.07	0.8577	0.1423
0.36	0.6406	0.3594	0.72	0.7642	0.2358	1.08	0.8599	0.1401
0.37	0.6443	0.3557	0.73	0.7673	0.2327	1.09	0.8621	0.1379
0.38	0.6480	0.3520	0.74	0.7704	0.2296	1.1	0.8643	0.1357
0.39	0.6517	0.3483	0.75	0.7734	0.2266	1.11	0.8665	0.1335
0.4	0.6554	0.3446	0.76	0.7764	0.2236	1.12	0.8686	0.1314
0.41	0.6591	0.3409	0.77	0.7794	0.2206	1.13	0.8708	0.1292
0.42	0.6628	0.3372	0.78	0.7823	0.2177	1.14	0.8729	0.1271
0.43	0.6664	0.3336	0.79	0.7852	0.2148	1.15	0.8749	0.1251
0.44	0.6700	0.3300	0.8	0.7881	0.2119	1.16	0.8770	0.1230
0.45	0.6736	0.3264	0.81	0.7910	0.2090	1.17	0.8790	0.1210
0.46	0.6772	0.3228	0.82	0.7939	0.2061	1.18	0.8810	0.1190
0.47	0.6808	0.3192	0.83	0.7967	0.2033	1.19	0.8830	0.1170
0.48	0.6844	0.3156	0.84	0.7995	0.2005	1.2	0.8849	0.1151
0.49	0.6879	0.3121	0.85	0.8023	0.1977	1.21	0.8869	0.1131
0.5	0.6915	0.3085	0.86	0.8051	0.1949	1.22	0.8888	0.1112
0.51	0.6950	0.3050	0.87	0.8078	0.1922	1.23	0.8907	0.1093
0.52	0.6985	0.3015	0.88	0.8106	0.1894	1.24	0.8925	0.1075
0.53	0.7019	0.2981	0.89	0.8133	0.1867	1.25	0.8944	0.1056
0.54	0.7054	0.2946	0.9	0.8159	0.1841	1.26	0.8962	0.1038
0.55	0.7088	0.2912	0.91	0.8186	0.1814	1.27	0.8980	0.1020
0.56	0.7123	0.2877	0.92	0.8212	0.1788	1.28	0.8997	0.1003
0.57	0.7157	0.2843	0.93	0.8238	0.1762	1.29	0.9015	0.0985
0.58	0.7190	0.2810	0.94	0.8264	0.1736	1.3	0.9032	0.0968
0.59	0.7224	0.2776	0.95	0.8289	0.1711	1.31	0.9049	0.0951

Statistical Methods for Researchers

Table 1 Areas Under the Normal curve

Z	Area in left tail	Area in right tail	Z	Area in left tail	Area in right tail	Z	Area in left tail	Area in right tail
1.32	0.9066	0.0934	1.68	0.9535	0.0465	2.04	0.9793	0.0207
1.33	0.9082	0.0918	1.69	0.9545	0.0455	2.05	0.9798	0.0202
1.34	0.9099	0.0901	1.7	0.9554	0.0446	2.06	0.9803	0.0197
1.35	0.9115	0.0885	1.71	0.9564	0.0436	2.07	0.9808	0.0192
1.36	0.9131	0.0869	1.72	0.9573	0.0427	2.08	0.9812	0.0188
1.37	0.9147	0.0853	1.73	0.9582	0.0418	2.09	0.9817	0.0183
1.38	0.9162	0.0838	1.74	0.9591	0.0409	2.1	0.9821	0.0179
1.39	0.9177	0.0823	1.75	0.9599	0.0401	2.11	0.9826	0.0174
1.4	0.9192	0.0808	1.76	0.9608	0.0392	2.12	0.9830	0.0170
1.41	0.9207	0.0793	1.77	0.9616	0.0384	2.13	0.9834	0.0166
1.42	0.9222	0.0778	1.78	0.9625	0.0375	2.14	0.9838	0.0162
1.43	0.9236	0.0764	1.79	0.9633	0.0367	2.15	0.9842	0.0158
1.44	0.9251	0.0749	1.8	0.9641	0.0359	2.16	0.9846	0.0154
1.45	0.9265	0.0735	1.81	0.9649	0.0351	2.17	0.9850	0.0150
1.46	0.9279	0.0721	1.82	0.9656	0.0344	2.18	0.9854	0.0146
1.47	0.9292	0.0708	1.83	0.9664	0.0336	2.19	0.9857	0.0143
1.48	0.9306	0.0694	1.84	0.9671	0.0329	2.2	0.9861	0.0139
1.49	0.9319	0.0681	1.85	0.9678	0.0322	2.21	0.9864	0.0136
1.5	0.9332	0.0668	1.86	0.9686	0.0314	2.22	0.9868	0.0132
1.51	0.9345	0.0655	1.87	0.9693	0.0307	2.23	0.9871	0.0129
1.52	0.9357	0.0643	1.88	0.9699	0.0301	2.24	0.9875	0.0125
1.53	0.9370	0.0630	1.89	0.9706	0.0294	2.25	0.9878	0.0122
1.54	0.9382	0.0618	1.9	0.9713	0.0287	2.26	0.9881	0.0119
1.55	0.9394	0.0606	1.91	0.9719	0.0281	2.27	0.9884	0.0116
1.56	0.9406	0.0594	1.92	0.9726	0.0274	2.28	0.9887	0.0113
1.57	0.9418	0.0582	1.93	0.9732	0.0268	2.29	0.9890	0.0110
1.58	0.9429	0.0571	1.94	0.9738	0.0262	2.3	0.9893	0.0107
1.59	0.9441	0.0559	1.95	0.9744	0.0256	2.31	0.9896	0.0104
1.6	0.9452	0.0548	1.96	0.9750	0.0250	2.32	0.9898	0.0102
1.61	0.9463	0.0537	1.97	0.9756	0.0244	2.33	0.9901	0.0099
1.62	0.9474	0.0526	1.98	0.9761	0.0239	2.34	0.9904	0.0096
1.63	0.9484	0.0516	1.99	0.9767	0.0233	2.35	0.9906	0.0094
1.64	0.9495	0.0505	2.0	0.9772	0.0228	2.36	0.9909	0.0091
1.65	0.9505	0.0495	2.01	0.9778	0.0222	2.37	0.9911	0.0089
1.66	0.9515	0.0485	2.02	0.9783	0.0217	2.38	0.9913	0.0087
1.67	0.9525	0.0475	2.03	0.9788	0.0212	2.39	0.9916	0.0084

Table 1 Areas Under the Normal curve

Z	Area in left tail	Area in right tail	Z	Area in left tail	Area in right tail	Z	Area in left tail	Area in right tail
2.4	0.9918	0.0082	2.6	0.9953	0.0047	2.8	0.9974	0.0026
2.41	0.9920	0.0080	2.61	0.9955	0.0045	2.81	0.9975	0.0025
2.42	0.9922	0.0078	2.62	0.9956	0.0044	2.82	0.9976	0.0024
2.43	0.9925	0.0075	2.63	0.9957	0.0043	2.83	0.9977	0.0023
2.44	0.9927	0.0073	2.64	0.9959	0.0041	2.84	0.9977	0.0023
2.45	0.9929	0.0071	2.65	0.9960	0.0040	2.85	0.9978	0.0022
2.46	0.9931	0.0069	2.66	0.9961	0.0039	2.86	0.9979	0.0021
2.47	0.9932	0.0068	2.67	0.9962	0.0038	2.87	0.9979	0.0021
2.48	0.9934	.0.0066	2.68	0.9963	0.0037	2.88	0.9980	0.0020
2.49	0.9936	.0.0064	2.69	0.9964	0.0036	2.89	0.9981	0.0019
2.5	0.9938	0.0062	2.7	0.9965	0.0035	2.9	0.9981	0.0019
2.51	0.9940	0.0060	2.71	0.9966	0.0034	2.91	0.9982	0.0018
2.52	0.9941	0.0059	2.72	0.9967	0.0033	2.92	0.9982	0.0018
2.53	0.9943	0.0057	2.73	0.9968	0.0032	2.93	0.9983	0.0017
2.54	0.9945	0.0055	2.74	0.9969	0.0031	2.94	0.9984	0.0016
2.55	0.9946	0.0054	2.75	0.9970	0.0030	2.95	0.9984	0.0016
2.56	0.9948	0.0052	2.76	0.9971	0.0029	2.96	0.9985	0.0015
2.57	0.9949	0.0051	2.77	0.9972	0.0028	2.97	0.9985	0.0015
2.58	0.9951	0.0049	2.78	0.9973	0.0027	2.98	0.9986	0.0014
2.59	0.9952	0.0048	2.79	0.9974	0.0026	3.00	0.9987	0.0013

Adapted from *Statistical Tables for Biological, Agricultural and Medical Research* © 1963 R. A. Fisher and F. Yates. Reprinted by permission of Pearson Education Limited.

Table 2 t Table - Critical t Values for Various Alpha Risks

	One-Tail Test						
df	.10	.05	.025	.01	.005	.0005	← One-Tail Alpha Risk
	Two-Tail Test						
df	.20	.10	.05	.02	.01	.001	← Two-Tail Alpha Risk
1	3.078	6.314	12.71	31.82	63.66	636.6	
2	1.886	2.920	4.303	6.965	9.925	31.60	
3	1.638	2.353	3.182	4.541	5.841	12.94	
4	1.533	2.132	2.776	3.747	4.604	8.610	
5	1.476	2 015	2.571	3.365	4.032	6.859	
6	1.440	1.943	2.447	3.143	3.707	5.959	
7	1.415	1.895	2.365	2.998	3.499	5.405	
8	1.397	1.860	2.306	2.896	3.355	5.041	
9	1.383	1.833	2.262	2.821	3.250	4.781	
10	1.372	1.812	2.228	2.764	3.169	4.587	
11	1.363	1.796	2.201	2.718	3.106	4.437	
12	1.356	1.782	2.179	2.681	3.055	4.318	
13	1.350	1.771	2.160	2.650	3.012	4.221	
14	1.345	1.761	2.145	2.624	2.977	4.140	
15	1.341	1.753	2.131	2.602	2.947	4.073	
16	1.337	1.746	2.120	2.583	2.921	4.015	
17	1.333	1.740	2.110	2.567	2.898	3.965	
18	1.330	1.734	2.101	2.552	2.878	3.922	
19	1.328	1.729	2.093	2.539	2.861	3.883	
20	1.325	1.725	2.086	2.528	2.845	3.850	
21	1.323	1.721	2.080	2.518	2.831	3.819	
22	1.321	1.717	2.074	2.508	2.819	3.792	
23	1.319	1.714	2.069	2.500	2.807	3.767	
24	1.318	1.711	2.064	2.492	2.797	3.745	
25	1.316	1.708	2.060	2.485	2.787	3.725	
26	1.315	1.706	2.056	2.479	2.779	3.707	
27	1.314	1.703	2.052	2.473	2.771	3.690	
28	1.313	1.701	2.048	2.467	2.763.	3.674	
29	1.311	1.699	2.045	2.462	2.756	3.659	
30	1.310	1.697	2.042	2.457	2.750	3.646	
35	1.306	1.690	2.030	2.438	2.724	3.591	
40	1.303	1.684	2.021	2.423	2.704	3.551	
50	1.299	1.676	2.009	2.405	2.680	3.500	
60	1.296	1.671	2.000	2.390	2.660	3.460	
90	1.291	1.662	1.990	2.368	2.632	3.402	
120	1.289	1.658	1.980	2.358	2.617	3.373	
∞	1.282	1.645	1.960	2.326	2.576	3.291	

Adapted from *Statistical Tables for Biological, Agricultural and Medical Research*
© 1963 R. A. Fisher and F. Yates. Reprinted by permission of Pearson Education Limited.

Table 3.   F Table - Critical F Values for the .05 Level of Significance

Denominator	Numerator Degreesof Freedom								
Degrees of Freedom	1	2	3	4	5	6	7	8	9
1	161.4	199.5	215.7	224.6	230.2	234.0	236.8	238.9	240.5
2	18.51	19.00	19.16	19.25	19.30	19.33	19.35	19.37	19.38
3	10.13	9.55	9.28	9.12	9.01	8.94	8.89	8.85	8.81
4	7.71	6.94	6.59	6.39	6.26	6.16	6.09	6.04	6.00
5	6.61	5.79	5.41	5.19	5.05	4.95	4.88	4.82	4.77
6	5.99	5.14	4.76	4.53	4.39	4.28	4.21	4.15	4.10
7	5.59	4 74	4.35	4.12	3.97	3.87	3.79	3.73	3.68
8	5.32	4.46	4.07	3.84	3.69	3.58	3.50	3.44	3.39
9	5.12	4.26	3.86	3.63	3.48	3.37	3.29	3.23	3.18
10	4.96	4.10	3.71	3.48	3.33	3.22	3.14	3.07	3.02
11	4.84	3.98	3.59	3.36	3.20	3.09	3.01	2.95	2.90
12	4.75	3.89	3.49	3.26	3.11	3.00	2.91	2.85	2.80
13	4.67	3.81	3.41	3.18	3.03	2.52	2.83	2.77	2.71
14	4.60	3.74	3.34	3.11	2.96	2.85	2.76	2.70	2.65
15	4.54	3.68	3.29	3.06	2.90	2.79	2.71	2.64	2.59
16	4.49	3.63	3.24	3.01	2.85	2.74	2.66	2.59	2.54
17	4.45	3.59	3.20	2.96	2.B1	2.70	2.61	2.55	2.49
18	4.41	3.55	3.16	2.93	2.77	2.66	2.58	2.51	2.46
19	4.38	3.52	3.13	2.90	2.74	2.63	2.54	2.48	2.42
20	4.35	3.49	3.10	2.87	2.71	2.60	2.51	2.45	2.39
21	4.32	3.47	3.07	2.84	2.68	2.57	2.49	2.42	2.37
22	4.30	3.44	3.05	2.82	2.66	2.55	2.46	2.40	2.34
23	4.28	3.42	3.03	2.80	2.64	2.53	2.44	2.37	2.32
24	4.26	3.40	3.01	2.78	2.62	2.51	2.42	2.36	2.30
25	4.24	3.39	2.99	2.76	2.60	2.49	2.40	2.34	2.28
26	4.23	3.37	2.98	2.74	2.59	2.47	2.39	2.32	2.27
27	4.21	3.35	2.96	2.73	2.57	2.46	2.37	2.31	2.25
28	4.20	3.34	2.95	2.71	2.56	2.45	2.36	2.29	2.24
29	4.18	3.33	2.93	2.70	2.55	2.43	2.35	2.28	2.22
30	4.17	3.32	2.92	2.69	2.53	2.42	2.33	2.27	2.21
40	4.08	3.23	2.84	2.61	2.45	2.34	2.25	2.18	2.12
60	4.00	3.15	2.76	2.53	2.37	2.25	2.17	2.10	2.04
120	3.92	3.07	2.68	2.45	2.29	2.17	2.09	2.02	1.96
∞	3.84	3.00	2.60	2.37	2.21	2.10	2.01	1.94	1.88

Adapted from E. S. Pearson and H. O. Hartley, eds., *Biometrika Tables for Statisticians*, by permission of Oxford University Press.

Table 4 Studentized Range (Q) Table - Critical Q Values for the .05 Level of Significance

Demoninator Degrees of Freedom	Numerator Degrees of freedom								
	2	3	4	5	6	7	8	9	10
1	18.0	27.0	32.8	37.1	40.4	43.1	45.4	47.4	49.1
2	6.09	8.3	9.8	10.9	11.7	12.4	13.0	I3.5	14.0
3	4.50	5.91	6.82	7.50	8.04	8.48	8.85	9.18	9.46
4	3.93	5.04	5.76	6.29	6.71	7.05	7.35	7.60	7.83
5	3.64	4.60	5.22	5.67	6.03	6.33	6.58	6.83	6.99
6	3.46	4.34	4.90	5.31	5.63	5.89	6.12	6.32	6.49
7	3.34	4.16	4.68	5.06	5.36	5.61	5.82	6.00	6.16
8	3.26	4.04	4.53	4.89	5.17	5.40	5.60	5.77	5.92
9	3.20	3.95	4.42	4.76	5.02	5.24	5.43	5.60	5.74
10	3.15	3.88	4.33	4.65	4.91	5.12	5.30	5.46	5.60
11	3.11	3.82	4.26	4.57	4.82	5.03	5.20	5.35	5.49
12	3.08	3.77	4.20	4.51	4.75	4.95	5.12	5.27	5.40
13	3.06	3.73	4.15	4.45	4.69	4.88	5.05	5.19	5.32
14	3.03	3.70	4.11	4.41	4.64	4.83	4.99	5.13	5.25
15	3.01	3.67	4.08	4.37	4.60	4.78	4.94	5.08	5.20
16	3.00	3.65	4.05	4.33	4.56	4.74	4.90	5.03	5.15
17	2.98	3.63	4.02	4.30	4.52	4.71	4.86	4.99	5.11
18	2.97	3.61	4.00	4.28	4.49	4.67	4.82	4.96	5.07
19	2.96	3.59	3.98	4.25	4.47	4.65	4.79	4.92	5.04
20	2.95	3.58	3.96	4.23	4.45	4.62	4.77	4.90	5.01
24	2.92	3.53	3.90	4.17	4.37	4.54	4.68	4.81	4.92
30	2.89	3.49	3.84	4.10	4.30	4.46	4.60	4.72	4.83
40	2.86	3.44	3.79	4.04	4.23	4.39	4.52	4.63	4.74
60	2.83	3.40	3.74	3.98	4.16	4.31	4.44	4.55	4.65
120	2.80	3.36	3.69	3.92	4.10	4.24	4.36	4.48	4.56
∞	2.77	3.31	3.63	3.86	4.03	4.17	4.29	4.39	4.47

Adapted from E. S. Pearson and H. O. Hartley, eds., *Biometrika Tables for Statisticians*, by permission of Oxford University Press.

Table 5 Chi Square (X^2) Table - Critical Values for the 0.10 to 0.01 Level of Significance

Degrees of Freedom	Level of Significance			
	0.10	0.05	0.02	0.01
1	2.706	3.841	5.412	6.635
2	4.605	5.991	7.824	9.210
3	6.251	7.815	9.837	11.341
4	7.779	9.488	11.668	13.277
5	9.236	11.070	13.388	15.086
6	10.645	12.592	15.033	16.812
7	12.017	14.067	16.622	18.475
8	13.362	15.507	18.168	20.090
9	14.684	16.919	19.679	21.666
10	15.987	18.307	21.161	23.209
11	17.275	19.675	22.618	24.725
12	18 549	21.026	24.054	26.217
13	19.812	22.362	25.472	27.688
14	21.064	23.685	26.8/3	29.141
15	22.307	24.996	28.209	30.578
16	23.542	26.296	29.633	32.000
17	24.769	27.587	30.995	33.409
18	25.989	28.869	32.346	34.805
19	27.204	30.144	33.687	36.191
20	28.412	31.410	35.020	37.566
21	29.615	32.671	36.343	38.932
22	30.813	36.415	40.270	42.980
25	34.382	37.652	41.566	44.314
26	35.563	36.885	42 856	45.642
27	36.741	40.113	44.140	46.963
28	37.916	41.337	45.419	48.278
29	39.087	42.557	46.693	49.588
30	40.256	43.773	47.962	50.892

Adapted from *Statistical Tables for Biological, Agricultural and Medical Research* © 1963 R. A. Fisher and F. Yates. Reprinted by permission of Pearson Education Limited

Formulae

Name	Formula	Page Reference
Descriptive Statistics		
Mean	$\overline{X} = \dfrac{\Sigma X_i}{N} = \dfrac{X_1 + X_2 + X_3 + \ldots\ldots X_n}{N}$	7
Mean Expected Value, Binomial Variable	$\mu = N * P$	134
Mean, Proportion	$p = \dfrac{\text{Number with the characteristic of interest}}{\text{Total Number}}$	211
Variance, Population	$\sigma^2 = \dfrac{\Sigma(X_i - \mu)^2}{N}$	16
Variance, Sample	$S^2 = \dfrac{\Sigma(X_i - \overline{X})^2}{N-1}$	174
Standard Deviation, Population	$\sigma = \sqrt{\dfrac{\Sigma(X_i - \mu)^2}{N}}$	16
Standard Deviation, Sample	$S = \sqrt{\dfrac{\Sigma(X_i - \overline{X})^2}{N-1}}$	175
Standard Deviation, Binomial Mean	$\sigma = \sqrt{N * P *(1 - P)}$	134
Standard Deviation, Proportion	$\sigma = \sqrt{p *(1 - p)}$	187

| Standard Deviation, Estimation for Population | $\sigma \cong \dfrac{\text{Range}}{6}$ | 186 |

| Convert Population to Sample Variance | $S^2 = \sigma^2 * \dfrac{N}{N-1}$ | 176 |

| Convert Population to Sample Standard Deviation | $S = \sigma * \sqrt{\dfrac{N}{N-1}}$ | 177 |

Probability Formulae

| Probability | $\dfrac{\text{Number of times the event A occcurs}}{\text{Number of times the trial is repeated}}$ | 59 |

| Probability | $\Sigma \dfrac{\text{Number of Times an Event of Interest Occurs}}{\text{Total Number of Possible Events}}$ | 95 |

Simple Addition Probability	$P(A \text{ or } B) = P(A) + P(B)$	62
General Addition Probability	$P(A \text{ or } B) = P(A) + P(B) - [P(A)* P(B)]$	63
Simple Multiplication Probability	$P(A \text{ and } B) = P(A)* P(B)$	66

| Size, Probability Tree | n^r | 67 |

General Multiplication Probability	$P(A \text{ and } B) = P(A)* P(B\int A)$	69
Outcome Value (OV)	$OV = \text{Probability} * \text{Payoff}$	76
Expected Value (EV)	$EV = \Sigma(\text{Outcome Values})$	76

| Binomial Probability Formula | $\dfrac{n!}{r! *(n-r)!} * p^r * (1-p)^{(n-r)}$ | 102 |

Permutation and Combination

Permutation, Complete	$w = n!$	89
Permutation, Partial	$w = \dfrac{n!}{(n-r)!}$	92
Combination	$w = \dfrac{n!}{r! *(n-r)!}$	94

Z Formulae

Z Score
$$Z = \frac{X - \mu}{\sigma}$$
121

Z Test, One Sample
$$Z = \frac{\overline{X} - \mu}{SE} \text{ where } SE = \frac{\sigma}{\sqrt{N}}$$
178

Delineation of a proportion of a distribution

    Mean $\overline{X} = (Z * SE) + \mu$ 182

    Lower Limit $LL = (Z_l * SE) + \mu$ 183

    Upper Limit $UL = (Z_u * SE) + \mu$ 183

    Condensed UL & LL $LIMITS = \pm(Z * SE) + \mu$ 183

Sample Size for Continuous Variable
$$N = \left(\frac{Z * \sigma}{E} \right)^2$$
184

Sample Size, Binomial Variable
$$N = \left(\frac{Z * \sigma}{E} \right)^2$$
187

Confidence Interval

    Lower Limit $LL = (Z_l * SE) + \overline{X}$ 205

    Upper Limit $UL = (Z_u * SE) + \overline{X}$ 205

t Formulae

t Test, One Sample
$$t = \frac{\overline{X} - \mu}{SE} \text{ where } SE = \frac{S}{\sqrt{N}}$$
236

t Test, General Formula
$$t = \frac{Difference\ between\ Group\ Means}{Standard\ Error}$$
240

t Test, Unequal Standard Deviations

241

$$t = \frac{(\overline{X}_1 - \overline{X}_2)}{SEu}$$
$$SE_u = \sqrt{(SE_1)^2 + (SE_2)^2}$$
$$SE_1 = S_1 / \sqrt{N_1} \text{ and } SE_2 = S_2 / \sqrt{N_2}$$

Non-Parametric Formulae

Regression and Forecasting Formulae

Residual	Actual Value − Predicted Value	365
Multiple Regression	$Y = a + b_1X_1 + b_2X_2 + \ldots$	371
Partial F-Test Criterion Sum of Squares	$SS(X_1|X_2) = SS(X_1 \& X_2) - SS(X_2)$	375
F formula	$F = \dfrac{SS(X_1 \mid X_2)}{MS(X_1 \& X_2)}$	375
Linear Relationship	$Y = a + bX$	360
Quadratic Relationship	$Y = a + b1*X + b2*x^2$	395
Logarithmic Relationship	$Y = a+b*Ln(X)$	395
Polynomial Relationship	$Y = a+(b1*X)+(b2*X^2)+(b3*X^3)\ldots$	397
Exponential Relationship	$Y=c*e^{b*x}$	397
Exponential Smoothing	$E_i + (1-D) * Y_i + D*E_{i-1}$	402
Mean Absolute Deviation	$MAD = \dfrac{\Sigma|Y - \overline{Y}\|}{N}$	411

Other Formulae

Covariance	$\sigma_{xy} = (X_1 - \overline{X})*(Y_1 - \overline{Y}) + (X_n - \overline{X})*(Y_n - \overline{Y})$	306
Critical Range	$CriiticalRange = Q*\sqrt{\dfrac{MS_{within}}{2}*\left(\dfrac{1}{N_1}+\dfrac{1}{N_2}\right)}$	275
F Ratio	$F = \dfrac{MS_{Between}}{MS_{Within}}$	272
Pearson Correlation Coefficient	$r = \dfrac{\sigma_{xy}}{\sigma_x * \sigma_y}$	306
Percentizing	$X_a = 100*\dfrac{Y_a - Y_w}{Y_b - Y_w}$	348
Standard Error	$SE = \dfrac{\sigma}{\sqrt{N}}$	180

Formulae

Statistical Index $Index = 100 * \dfrac{Current\ Year}{Base\ Year}$ 347

Systematic Sampling n $n = \dfrac{Total\ No.\ Available}{Desired\ Sample\ Size}$ 170

Weighted Index $Weighted\ Index = \Sigma(W_1 * V_1)$ 349

Excel Commands and Functions

Bibliography

Adams, Gerald R. and Schvaneveldt. 1985. *Understanding Research Methods.* New York: Longman.

Dixon, W. and Massey, F. 1969. *Introductoion to Statistical Analysis.* New York: McGraw Hill.

Fallik, Fred and Bruce Brown. 1983. *Statistics for Behavioral Sciences.* Homewood, IL: Dorsey.

Hintz, J., 1992. *Number Cruncher Statistical System.* Kaysville, Utah: Jerry Hintz.

Holcomb, Zealure. 1998. *Fundamentals of Descriptive Statistics.* Los Angles, Ca: Pyrczak Publishing.

Hintze, Jerry L. *Number Cruncher Statistical System.* Kaysville, Utah.

Huff, Darrell. 1954. *How to Lie with Statistics.* New York: W.W. Norton.

Jaeger, Richard M. 1990. *Statistics: A Spectator Sport.* New York: Sage, 2nd ed..

Katz, Jay. 1972. *Experimentation with Human Beings.* New York: Russell Sage Foundation.

Levine, D., M. Berenson, and D.Stephan. 1997. *Statistics for Managers.* Upper Saddle River, NJ: Prentice Hall.

Meier, Kenneth J. and Jeffery Brudney. 1993. *Applied Statistics for Public Administration.* Belmont, CA: Wadsworth.

O'Sullivan, Elizabethann and Gary R. Rassel. 1995. *Research Methods for Public Administrators*. New York: Longman.

Patten, M. 1997. *Understanding Research Methods*. Los Angles, Ca: Pyrczak Publishing.

Pyrczak, F. 1995. *Making Sense of Statistics*. Los Angles, Ca: Pyrczak Publishing.

Person R. 1993. *Using Excel*. Indianapolis, IN: Que Corporation.

Siegel, Sidney and Catellan, N. John. 1988. *Nonparametric Statistics for the Behavioral Sciences*. New York: McGraw-Hill.

Szafran, R.1994. *Social Science Research*. Los Angles, Ca: Pyrczak Publishing.

Sylvia, Ronald D., Meier, Kenneth J. and Gunn, Elizabeth M. 1985. *Program Planning and Evaluation for the Public Manager*. Monterey, CA: Brooks/Cole.

Triola, Mario. 1998. *Elementary Statistics*. Reading MA: Addison-Wesley.

Tufte, Edward R. 1974. Data Analysis for Politics and Policy. Englewood Cliffs, NJ: Prentice-Hall.

Welch, Susan and John Comer. 1985. *Quantitative Methods for Public Administration.* Homewood, IL: Dorsey.

Winer, B. J. 1971. *Statistical Principles in Experimental Design*. New York, McGraw-Hill.

Index and Glossary

a priori **probability** The probability of an event that is known in advance because of the structure of the event. 60

Adjusted R square The modified value of R square based on sample size and the number of independent variables used in the regression model. 377

Alpha risk The chance of rejecting the null hypothesis when the null hypothesis is true i.e. concluding there is a difference between treatments when no difference actually exists. 217

Alternative hypothesis The assumption that there is a difference between the comparison groups. 213

Analysis of variance The parametric test that should be used when comparing the means of two or more groups. Abbreviated as ANOVA. 269

Analysis of covariance The parametric test that is used when the comparison groups are not represented in an equal fashion for an important variable. Abbreviated as ANCOVA. 334

AND probability The probability that all events specified occur. For example if there are two events of interest, event A and event B, then it is the probability that both events A and B occur. (See simple multiplication rule and general multiplication rule). 65

Association There is a correlation between two variables, but not necessarily in a cause and effect manner. 380

Attrition threat A threat to internal validity because subjects discontinue a study before its completion and their discontinuation or lost information may bias the results. (Synonym = experimental mortality). 157

Average deviation A method to measure variation which is based on absolute deviations from the mean. 15

Bar graph A graph in which the length of a bar denotes the numerical value for a variable. The bar graph is often used to show comparisons between quantities. (Synonym = histogram) 36

Bernoulli process A process in which there can be only two outcomes, the outcomes are independent and time does not influence the outcome probabilities. 99

Beta The value in a regression equation for b (i.e. the slope of the regression line). 360

Beta risk When planning a study, the chance of accepting the null hypothesis when the null hypothesis is false - i.e. the chance of concluding that there is no difference between treatments when a difference actually exists. 217

Between groups sum of squares A statistic calculated in an ANOVA that reflects the variation among the means of the treatment groups. 162

Bias The introduction of an inaccurate value because of a problem in sampling , data collection or study conduct. 162

Binomial probability distribution The distribution that results from a Bernoulli process. (See Bernoulli process) 99

Bivariate table A presentation of the interrelationship between two variables in a table. (Synonyms = Cross-tabs table and Two way table). 34

Break even point The amount of investment in a probability event that in the long run results in neither a financial gain nor loss. 76

Case control method A research method, popular in epidemiology, where the researcher starts with an outcome and looks retrospectively for a cause. 337

Case study A type of qualitative research design in which the investigators study a phenomenon at one or a few experimental units/locations, but examine those sites in-depth. 337, 443

Central limit theorem The tendency for means to distribute themselves normally as the sample size increases, even if the underlying distribution from which the elements are taken is not normal. 180

Checklist A type of question in which the respondent checks off items from a list that match a specified condition, statement, etc. (Synonym = cafeteria scale). 440

Chi square test A non-parametric test that compares the frequency distributions of nominal variables. 308

Classical randomized design A design in which subjects are randomly allocated to either an active treatment or a control treatment group, and observation are made both before and after the treatments are administered. 343

Classical randomized with experimental model control design A design in which subjects are randomly allocated to one of 3 groups - active treatment, established treatment or a no treatment control group. Observation are made both before and after the treatments are administered. 344

Closed format A type of question in which the respondent selects an answer from a pre set number of options or enters a numeric value. 441

Cluster sampling A sampling plan that begins with a general population and randomly selects a series of subgroups at progressively more specific levels within the population. 172

Coefficient of determination A measure of how well a regression model fits the sample data by specifying the amount of variation in Y that can be explained by X (Synonym = R square). 366

Combination The number of ways items can be selected from a group when the order of selection doesn't matter. 88, 93

Combined Z transformation A method used to form an index in which all the variables are converted to Z values and then the Z values are added together. 346

Complete permutation The number of ways items can be selected from a group when the order of selection is important and all items in the group are to be selected. 88

Concept An abstract or general idea that needs to be converted to a quantitative measurement for research purposes. 152

Concurrent event threat A threat to internal validity because external events may have an affect on the subjects while they are participating in a study that influence the result. 155

Conditional probability The probability of the outcome of one event which is dependent on the result of another event. 69

Confidence interval A range of values within which there is a specific degree of certainty that the range includes the population value. 204

Confidentiality An obligation on the part of a researcher to respect and protect information about a subject that is considered private. 447

Confounding A situation in which an extraneous variable is disproportionately represented in one of the treatment group. Thus, we can't be sure if the treatment effect is due to the treatment given, the extraneous variable or a combination of the two. 338

Content analysis A qualitative research method that reviews source material and records the frequency of terms, phrases or ideas that were previously determined to be relevant to the research question. 431

Content validity The type of validity that is concerned with how completely the measurements used encompass the underlying concept. 154

Continuous distribution The number of items that are classified/assigned to the specific categories of a continuous variable. Is in contrast to a discrete variable. 116

Continuous variable A variable that can take on greater and greater specificity depending on the precision of the measuring devise. Is in contrast to a discrete variable. 4, 116

Control group A group that participates in a study but is not given the new or experimental treatment. 333

Control variable A variable that has the potential to affect the response of the dependent variable and is used to examine comparability between the treatment groups. 334

Controlled experimental setting Incorporating a variety of techniques in a study in order to reduce the effects of extraneous variables and .increase the likelihood that the result is due only to the independent variable. 334

Convenience sampling A non probability sampling technique in which items are selected by the researcher on the basis of what's available. 173

Correlation coefficient A measure of the direction and degree of relationship between two variables. The symbol r is usually used to represent a correlation coefficient. 298

Countered balanced design A design in which all treatments are given to all subjects. 341

Covariance A measure of the degree to which two variables move in a similar (e.g. they both increase) or dissimilar (e.g. one increases while the other decreases) fashion. 306

Covariance of X and Y A measure of the degree to which variables X and Y move in the same or opposite direction. 306

Criterion validity The type of validity that involves a comparison between a new measurement and one that is traditionally used to measure a concept. 154

Critical value The value of a test statistic which is at the point that divides the rejection and non-rejection regions in a hypothesis testing situation. 218

Critical Value Method A method to determine statistical significance based on the Alpha Risk, the critical value(s) of the test statistic and the observed value of the test statistic. If the observed test statistic value is greater than or equal to the critical value, there is statistical significance. 217

Cross sectional design A research design in which all the data are collected at essentially one time. 336

Cross tabs table A presentation of the interrelationship between two variables in a table. (Synonyms = Bivariate table and Two way table). 34

Cumulative frequency A count of all items in a given category plus all the items in prior categories. 6

Cyclical component The pattern in a time series which is repeated, but the repetitions are not based on the seasons of a year. 405

Data mining A search for variables that appear to affect the dependent variable, but there was no *a priori* intention to include them as independent variables in the study (Synonym = Fishing exercise). 370

Data set A collection of observations about a phenomenon in which we have an interest. 1

Decomposition The process of breaking down the components of a time series into trend, seasonal, cyclical and irregular components. 404

Degrees of freedom The number of values that are allowed to vary when calculating a statistic such as the mean. For the mean, the degrees of freedom are equal to N - 1. 236

Delphi technique The use a group of experts who operate in an anonymous fashion and, through recursive reviews, generate a forecast. 433

Dependent variable The variable that is affected by the independent variable(s). In a cause and effect analogy, it is the effect. (Synonym = Outcome variable) 3

Descriptive Statisitcs The branch of statistics that involves the organization and display of data, primarily for illustrative rather than analytical purposes. 3

Discrete distribution The number of items that are classified/assigned to the specific categories of a discrete variable. 116

Discrete variable A variable that can only take on values that are whole numbers. Is in contrast to a continuous variable. 4, 116

Disproportionate stratified sample A stratified sampling plan in which the groups making up the population are selected in a proportion not equal to their representation in the population. 172

Distribution An enumeration of the elements in a data set that are classified/assigned to specific categories. 3

Double blind study A study in which both the evaluator and the subject are unaware of the treatments the subjects are receiving. 334

Dummy variable The term applied to nominal variables introduced in multiple linear regression. 379

Equivalence A form of reliability which examines the consistency of a measurement between different devices and operators. 151

Error The term used to refer to the random inconsistent result observed with many phenomenon. 11

Estimation The use of a statistic from a sample to deduce what is the most likely population value for the statistic. 203

Expected value The mean value of an event when there are a great many replications. It is the sum of the outcome values for a given problem (See outcome value). 76

Experimental mortality threat A threat to internal validity because subjects discontinue a study before its completion and their discontinuation or lost information may bias the results. (Synonym = attrition) 156

Experimentwide error The chance of a type 1 error for all the treatment comparison made in a study. 267

Exponential relationship The type of association between the X and Y variable in a time series that is characterized by a very sharp rise in Y as X becomes larger. 397

Exponential smoothing A technique to reduce the irregular pattern in a time series which can also be used to forecast a future outcome. 402

External validity The ability to generalize the results and conclusions of a study to other populations and conditions than those used in the study. 153

Extraneous variable A variable present in a study, other than the independent or a control variable, that can influence the dependent variable. 338

Figure A pictorial representation of important features associated with a data set. See line, bar, scatter and pie graph. (Synonyms = Chart and Graph) 36

Face validity An opinion that the measurements look as if they are appropriate for a study. 154

Fisher Exact Test A non-parametirc test for proportions that is used when the chi square expected frequencies are small. 315

Fishing exercise A search for variables that appear to affect the dependent variable, but there was no *a priori* intention to include them as independent variables in the study. (Synonym = Data mining). 370

Focus group Usually a relatively small number of people (e.g. 6-10) that are assembled to discuss a specific subject in which the group has particular knowledge or experience. 432

Frame The portion of the population that is accessible to the experimenter. 166

Frequency table A table that contains one or more categories and for each category a count of the number of items in the category is provided. 4, 33

General addition rule The probability calculation for an OR probability in which the events are not mutually exclusive. 63

General multiplication rule The probability calculation for an AND probability in which the events lack independence. 69

Hawthorne effect A behavioral change by subjects because they know they are participating in a study. 335

Hypothesis generating research Research that is designed to look to see what variables may be related to or affect other variables. Variables identified should then be used in hypothesis testing research to confirm their effect. 369

Hypothesis testing A formal process used in research studies to see if one treatment is different from another treatment. A pre determined criterion is established and used to reach a conclusion on whether to accept or reject the null hypothesis. 213

Independent group design A design in which the treatments are given to separate and distinct groups of subjects. (Synonym = Unpaired design). 242

Independence The condition that exists when the occurrence of one event is not influenced by the occurrence of a second event. 65

Independent variable The variable that produces an affect on another variable. In a cause and effect analogy, it is the cause. (Synonym = Treatment) 3

Inferential statistics The branch of statistics that uses the results from a sample to provide information about a population. 166, 203

Informed consent The process of explaining the nature of a study to a potential subject and gaining the person's acceptance to participate in the study. 446

Instrumentation threat A threat to internal validity because of a change in, or differences between, measuring devices which can be mistaken for a treatment effect. 156

Intercept The term used in regression to refer to the value of Y when X = 0. (Mathematical symbol = a). 360

Internal consistency A form of reliability which examines the uniformity of the multiple elements that make up a measuring devise. 151

Mean Absolute Deviation (MAD) The average of the absolute differences between the actual and the predicted values of a forecast. 411

Measurement The rules that are used to assign numeric values to a variables. 145

Median The middle value in a data set after the set is rearranged in sequential order. 8

Milgram study An example of an unethical study because of the deception used, but defended on the grounds that no harm accrued to the subjects. 447

Mode The most frequently occurring value in a data set. 8

Moving average A technique to reduce the irregular pattern in a time series which plots the mean for 2 or more consecutive points rather than each observed point. 399

Moving scale Alteration of the Y axis so it does not begin at Y = 0 but at a different point that exaggerates changes or differences. 37

Multi-stage sampling A sampling approach that uses at least two phases. In the initial phase a sample is taken and then from that first sample a second sample is drawn. There can be additional replications of this process. 172

Multiple R In simple linear regression it is the Pearson correlation coefficient without the sign. In multiple regression it is simply the square root of R square. 366, 377

Multiple range test A supplementary test to an ANOVA (when there is statistical significance between the groups) that allows identification of what particular groups are different from the others. 275

Multiple regression Regression analysis in which more than one independent variable is included in the regression equation and the relationship between the independent and dependent variables is linear (i.e. a straight line). 370

Mutually exclusive Events that can not co-exist i.e., they can't happen together. 61

Nominal scale A scale that can merely place objects, persons, characteristics, etc. into categories. The categories can not be distinguished in terms of how much of a difference there is between them. 146

Non-parametric tests Tests that make no assumptions about the nature of the underlying population distributions and are appropriate for data measured by either the nominal or ordinal scale. 297

Non participant observer method A technique in which the researcher observes the research subject's behavior and the spectator role of the researcher is clearly evident. 429

Non-probability sample Sampling plans that can be described as subjective or arbitrary. (See Convenience and Judgmental sampling) 170

Non-rejection region The set of values for a test statistic that cause acceptance of the null hypothesis. 218

Normal curve A bell shaped symmetrical curve that is peaked in the middle with long tapering tails. The mean, median and mode are all at the center of a normal curve. (Synonym = Gaussian curve) 116

Normal distribution A theoretical distribution that, when graphed, takes on the appearance of a bell. The units of the variable are used as the X axis scale. 116

Null hypothesis A statement that there is no difference between the groups being compared. In practice, the null hypothesis is a strawman - a proposition set up with the intention of disproving it. 213

Nuremberg code A set of principles to guide ethical experimentation in humans. A subject's right to participate in a study only after giving informed consent and the need to protect the confidentiality of the subjects are core elements of the code. 446

One sample t test The t test used to compare a mean from a single sample with a population or established mean. 36, 239

One tail test A test that specifies a direction for the group differences. The comparison tested is either whether one treatment is better than another treatment or whether one treatment is worse than another treatment, but both of these comparisons are not tested in one analysis. The entire Alpha Risk is located in one tail of the test statistic. 222

One way analysis of variance An ANOVA test in which there is only a single independent variable which is the treatment factor (Synonym = Single factor ANOVA). 269

One way table A presentation of a single variable in a table. (Synonym = Univariate table) 34

Open format A type of question in which the respondent is free to write a narrative response. 441

Operational definition A description of how the dependent variable(s) will be measured after operationalizing the underlying concept.. 152

Operationalizing The process of transforming a concept into a surrogate notion that can be measured in a quantitative manner. 152

OR probability The probability that either of several events will occur. For example if there are two events of interest, event A and event B, then it is the probability that event A or B occurs. (See simple addition rule and general addition rule). 61

Ordinal scale A scale that can place items in rank order, but the units of the scale are not uniform. 146

Outcome Value The probability an event will occur times the financial payoff for the event. 76

Outlier An atypical element in a data set. 128

P-value The probability for the occurrence of a test statistic. Statistical significance is present for p-values that are equal to or smaller than the probability specified in the Alpha Risk and absent for p-values that are greater than the probability specified in the Alpha Risk. 225

Paired design A design in which the treatments are given to all subjects or to sets of subjects who have been matched. 247

Paired t test A t test used for paired designs in which there are only two treatments. 247

Parameter The mean, variance and other descriptive statistics when they refer to a population. 175

Parametric tests Tests that are appropriate for data measured by the interval scale. 297

Partial F-test criterion A method used to determine the contribution of each independent variable in a multiple regression model. 374

Partial permutation The number of ways items can be selected from a group when the order of selection is important and only a portion of the total group are to be selected. 92

Partial regression coefficient The beta values for the slope of the independent variable in multiple linear regression (Synonym = Partial slope). 373

Participant observer method The technique in which the researcher plays the role of a study participant and observes/records information relevant to the study's purpose. 429

Pearson correlation coefficient A parametric procedure using interval measurements which shows the strength and direction of the relationship between two variables. (Mathematical symbol = r) 305

Percentizing technique A way to transform variables to make them unit free in which the relative position (i.e. percentile rank) of an item is determined. 348

Permutation The number of ways items can be selected from a group when the order of selection is important. 87

Pie graph A graph that shows the proportional components of an item. The pie is often used to compare the size of the subparts of different items. (Synonym = Circle graph). 36

Point estimate Use of a single sample statistic to estimate a value of the population. 203

Polynomial relationship The type of association between the X and Y variable in a time series that is characterized by an oscillating relationship between X and Y. 396

Population The collection of all items for which the researcher has an interest. 165

Population standard deviation The standard deviation of all items for which the researcher has an interest. The mathematical symbol is σ and the calculation uses N as the denominator. (In contrast see Sample standard deviation). 16

Population variance The variance of all items for which the researcher has an interest. The mathematical symbol is σ^2 and the calculation uses N as the denominator. (In contrast see Sample variance). 16

posterior **probability**. The probability of an event that is known because of observations that were made concerning its past behavior. 60

Post test design A design in which the subjects are only measured following the completion of treatment and no pre treatment measurement is taken. 339

Post test with inactive control group design A design in which the subjects are only measured following the completion of treatment and a non-randomized control group is included. 340

Power The probability of correctly rejecting the null hypothesis because it is false. 221

Practical significance A difference between treatments that is meaningful/important to the researcher. 283

Pre - post test design A design in which the subjects are measured prior to the initiation of treatment as well as following completion of treatment. 340

Predictor variable A variable that is used to make a forecast which is derived from an antecedent event (i.e. the antecedent event occurs before the event being predicted occurs). 407

Probability The proportion of times an event will occur in a very large series of identical repeated trials. 59, 95

Probability distribution A distribution in which all the possible outcomes of a trial are plotted against the probability of each outcome occurring. 97

Probability sample Sampling plan in which the likelihood of selecting an item is based on chance. 170

Probability Value Method. A method to determine statistical significance based on the Alpha Risk and the p-value associated with the test statistic. If the p-value is less than or equal to the Alpha Risk probability, there is statistical significance. 224

Proportionate stratified sample A stratified sampling plan in which the groups making up the population are selected in a proportion equal to their representation in the population. 172

Quadratic relationship The type of association between the X and Y variable in a time series that is characterized by a curve in which the increase in Y becomes progressively larger for each unit increase in X. 395

Qualitative measurement A measurement that is non-numeric - i.e. it is in the form of words or images. 428

Qualitative research Research in which the information collected consists of events or themes that are described in words or pictures rather than numerical values. 428

Quantitative measurement A measurement that has numeric properties so that finding the sum or calculating the mean for the measurement results in a meaningful answer. 428

Quantitative research Research in which the information collected consists of numeric data. Data are usually analysed and summarized using standard statistical methods. 428

Quasi experimental design A design in which good design principles are only partially satisfied because the researcher is often limited by practicalities. 338

Questionnaire A list of questions that seeks information about people's opinions, behaviors, attitudes, etc. 434

R square A measure of how well a regression model fits the sample data by specifying the amount of variation in Y that can be explained by X (Synonym = Coefficient of determination). 366

Range The interval from the lowest to the highest value in a data set. 13

Random variation Unexplained variation that is inherent in the behavior of any variable. 244

Randomized post test with active control group A design in which subjects are randomly allocated to an active treatment or a control treatment group, but observation are made only after the treatments are administered. 342

Random sampling A process in which every member of the frame has an equal chance of being included in the sample. 168

Ranking scale A type of question in which a respondent is asked to assign a rank to a list of items with 1 indicating his or her first choice, a 2 the second choice, etc. 441

Rating scale A type of question in which a respondent is asked to assign a qualitative assessment to the elements in a list. The qualitative assessment categories are specified in advance. 441

Ratio scale. An interval scale that has a true zero point. 146

Raw data Unorganized data that has not been processed, arranged or ordered. 2

Regression coefficient The value in a regression equation for b (i.e. beta or the slope of the regression line). 360

Regression equation In regression, a formula computed from a data set, that shows the linear relationship between the dependent and independent variable. 360

Rejection region The set of values for a test statistic that cause a rejection of the null hypothesis. 218

Reliability The degree that a measurement gives consistent and dependable responses. 150

Repeated measurement design A variation of the ANOVA that allows an experimenter to compare treatments when all subjects receive each treatment. 282

Residual In regression, the distance between the observed Y for an observation and the predicted Y. 365

Risk averse Behavior in which a person avoids an investment even though the expected gain is positive because of their fear of suffering a loss. 79

Sample A subset of a population that is used to provide the researcher with information about the population. 166

Sample size The number of people or items selected for in a sample. It is possible to calculate the sample size before doing a study after defining the amount of error that can be tolerated and determining the degree of certainty that that level will not be exceeded (Mathematical symbol = N). 13, 166, 184, 187

Sample standard deviation The standard deviation for items in a sample. The mathematical symbol is S and the calculation uses N-1 as the denominator. (In contrast see Population standard deviation). 16

Sample variance The variance for the items in a sample. The mathematical symbol is S^2 and the calculation uses N-1 as the denominator. (In contrast see Population variance). 16

Sampling error The inaccuracy of sample statistics relative to the population statistics they are meant to represent. 169

Sampling plan The process used to select items for a sample. 166

Scales A measurement characteristic based on how the units of measurement are assigned. The simplest system just counts items (nominal) whereas more advanced scales assign uniform numeric values to the scale (interval). 146

Scatter graph A graph in which the values for two variables associated with a series of items are displayed. For each item a plotted point is produced using the X axis scale for one variable and the Y axis scale for the second variable. (Synonym = XY graph) 36

Seasonal component The pattern in a time series which is repeated and the repetitions are based on the seasons in a year. 404

Sensitivity The capability of a measurement to detect important differences between treatments. 157

Simple addition rule The probability calculation for an OR probability in which the events are mutually exclusive. 61

Simple linear regression Regression analysis in which only one independent variable is included in the regression equation and the relationship between the independent and dependent variables is linear (i.e. a straight line). 357

Simple multiplication rule The probability calculation for an AND probability in which the events are independent. 65

Simple random sampling Sampling plan which uses all elements in a frame and each element has an equal chance of being selected. 170

Skew A distribution that is not symmetrical — either the left or right side has a long tail. 9, 117

Slope The term used in regression for the amount that Y changes for a 1 unit change in X (Synonyms = beta and regression coefficient) (Mathematical symbol = b) 360

Smoothing Techniques used to reduce the irregular pattern of a time series and help an analyst identify long term trends in the series. 399

Solomon four group design A design in which there are four groups of subjects and two groups receive the treatment and two do not. One of the treatment groups and one of the non-treatment groups is pre-tested, but the other groups are not pre-tested. Post treatment observations are made on all groups. 343

Spearman rank correlation coefficient A non-parametric procedure using ordinal measurements which shows the strength and direction of the relationship between two variables. (Mathematical symbol = r_s) 299

Stability A form of reliability which examines the consistency of a measurement over time. 150

Standard deviation A measure of the degree to which items in a data set are dispersed. The greater the dispersion, the greater the standard deviation. It is also the square root of the variance. (See Sample standard deviation and Population standard deviation). 15

Standard normal curve The bell shaped symmetrical curve that uses Z values as the X axis and has a mean of 0 and a standard deviation of 1. 122

Standard normal distribution The distribution that is derived from the normal curve and uses Z values as the X axis. 121

Standard error The standard deviation for a set of means. (Mathematical symbol = SE) 178

Standard error of the estimate A measure of how well a regression model fits the sample data by identifying the standard deviation for the residuals. 366

Standard score A transformed value based on how many standard deviation units an observed value is from the mean. (Synonyms = Z value, Z score and Standardized normal value) 123

Statistics There is more than one meaning for the term - it can mean a number that provides information about some characteristics in a data set. It may also refer to the mean, standard deviation, etc. when derived from a sample or it may represent a field of study.   1, 175

Statistical index A set of variables which are combined and used to measure a complex concept. 345

Statistical inference The use of a sample results to draw conclusion about a population. Two methods of statistical inference are estimation and hypothesis testing. 203

Statistical regression threat A threat to internal validity because subjects may be selected for a study due to an extreme value or characteristic and the relative changes during the study can be mistaken as a treatment effect. 157

Statistical significance A result that leads to the rejection of the null hypothesis because the corresponding p-value is less than or equal to the pre-set Alpha Risk. 283

Stratified sampling A sampling plan in which all elements in a frame are divided into groups or strata. Within each group/strata a random sample is then taken. 171

Subject The person who participates in a research study. 156

Subject selection threat A threat to internal validity because subjects selected for a study may not be representative of the population of interest and, in a multiple group study, the groups may not be equivalent resulting in the possibility of incorrect conclusions. 155

Systematic sampling Sampling plan in which every nth item is selected from the frame where n would be a number greater than or equal to 2. 170

t test A parametric test for comparing the difference between the means of two groups. (See one sample t test, paired t test and two sample t test) 236

Test for independence A name for the chi square test which tests to see if there is a relationship between the independent and dependent variables. 315

Testing threat A threat to internal validity because the performance of the subjects improves as they get more experience with the testing device and, any change in response during a study for a variable subject to this effect, can be mistaken as a treatment effect. 156

Thermometer scale A type of question in which a respondent is asked to place a mark on a line with only the extreme ends of the scale defined/identified. The respondent may mark any level at or between the extremes. 440

Time series A set of observations of a quantitative variable which are taken over a period of time. 393

Total sum of squares A statistic calculated in an ANOVA that reflects the variation for all the observations for all treatment groups. Is also the sum of the between groups and within groups sum of squares. 270

Treatment The procedures or agents that are administered to subjects to determine the type of response produced in the dependent variable. (Synonyms = Independent variable, Experimental condition). 55

Trend component The pattern in a time series which shows the long term direction of the series. 404

Tuskeegee study An example of an unethical study because the rights of the subjects in this investigation of the natural course of syphilis were abridged. 445

Two tail test A test that doesn't specify the direction of group differences. Thus the Alpha Risk is divided between both tails of the test statistic. The researcher can determine if one treatment is better than or worse than the other treatment in one study. 222

Two way analysis of variance An ANOVA test in which there are two main factors -the treatment factor and a second factor. (Synonym = Randomized block design), 278

Two way table A presentation of the interrelationship between two variables in a table (Synonyms = Bivariate table and Cross-tabs table). 34

Type 1 error Concluding that the treatments are different when in fact they are the same (i.e. accepting the alternative hypothesis when it is false.) 215

Type 2 error Concluding that the treatments are the same when in fact they are different (i.e. accepting the null hypothesis when it is false). 215

Unit free A statistic for which the measurement unit has been eliminated. Typically the measurement unit is present in the numerator and the denominator of an equation and when division occurs the measurement unit becomes 1. 346

Univariate table A presentation of just one variable in a table. (Synonym = One way table) 34

US radiation experiments An example of an unethical study because the subjects were given radiation, sometimes repeated doses as well as very high doses, without their permission. 446

Validity The degree that a measurement actually assesses the concept or phenomenon of interest. 152

Variable An observable attribute that when measured/viewed can have more than one value/assessment. 3

Variance A measure of the degree to which items in a data set are dispersed. The greater the dispersion, the greater the variance. It is also the square of the standard deviation. Mathematically it is the mean of a data set in which the difference between each observation and the mean has been squared. (See Sample variance and Population variance) 15

Weighted index An index in which the elements are assigned weights to reflect their relative importance. 349

Wilcoxon rank sum test A non-parametric test for ordinal measurement which compares the ranks of different treatment groups. 315

Within groups sum of squares A statistic calculated in an ANOVA that reflects the variation in the individual observations within each treatment group. 272

X axis The horizontal line in a graph. (Synonym = Abscissa) 36

Y axis The vertical line in a graph (Synonym = Ordinate) 36

Z score A transformed value based on how many standard deviation units an observed point is from the mean. (Synonyms = Z value, Standard score and Standardized normal value). 121

Z test A test that relies on calculation of a Z value. A known or population value for the standard deviation is required. 177

Z value A transformed value based on how many standard deviation units an observed point is from the mean. (Synonyms = Z score, Standard score and standardized normal value) 121